FROM ETERNITY TO HERE

SEAN CARROLL is a theoretical physicist at Caltech and the author of *The Particle at the End of the Universe: How the Hunt for the Higgs Boson Leads Us to the Edge of a New World*. After receiving his doctorate from Harvard University, he pursued research on particle physics, cosmology, and gravitation. He is one of the founders of the group blog *Cosmic Variance*. He lives in Los Angeles with his wife, writer Jennifer Ouellette.

"Carroll keeps it real, getting at the complex guts of cutting-edge cosmology in discussions that will challenge fans of Hawking's *A Brief History of Time*."   —*The Washington Post*

"[Carroll's] writing is accessible and peppered with cultural references—quotes from *Dumb and Dumber* and *Slaughterhouse-Five*. But don't be fooled . . . Carroll isn't afraid to wade into topics that have befuddled even brand-name physicists."   —*Wired*

"An engrossing, well-crafted introduction to the universe and the foundations of modern physics."   —*Nature*

"Tackles the daunting subject [of time] and has a lot of fun along the way."
—DailyKos.com

"Entertaining, enlightening, and headache-free explanations . . . a masterful overview . . . Unifying cosmology, thermodynamics, and information science into a refreshingly accessible whole."   —*Seed*

"Carroll explains time's fascinating subtleties in a lucid and entertaining manner."
—*New Scientist*

"With equal parts patience, enthusiasm, and humor, Carroll presents a mini-course on the natural history of cosmology itself: what has been pieced together by evidence, then unraveled by new frames of reference, and the reasons we're edging closer to answers even though scientific revolutions often seem to take us back to square one."   —*The Onion AV Club*

"One of the most lucid popular overviews of modern theoretical cosmology."
—*Natural History*

"Sean Carroll's *From Eternity to Here* provides a wonderfully accessible account of some of the most profound mysteries of modern physics. While you may not agree with all his conclusions, you will find the discussion fascinating, and taken to much deeper levels than is normal in a work of popular science."
—Roger Penrose, author of *The Road to Reality* and *The Emperor's New Mind*

"The meaning of time at a fundamental level remains one of the most elusive mysteries physicists encounter today. There isn't a single agreed-upon answer to some of the more perplexing questions to which Sean Carroll exposes the reader, but there is an abundance of interesting excursions written in entertaining and engaging language that will keep fascinated those interested in profound questions in modern physics."
—Lisa Randall, Harvard University, author of *Warped Passages*

FROM ETERNITY TO HERE

The Quest for the Ultimate Theory of Time

SEAN CARROLL

DUTTON

DUTTON
An imprint of Penguin Random House LLC
375 Hudson Street
New York, New York 10014

Previously published as a Dutton hardcover and Plume paperback.

First Dutton paperback printing 2016

Photograph on page 37 by Martin Röll, licensed under the Creative Commons Attribution ShareAlike 2.0 License, from Wikimedia Commons. Photograph on page 47 courtesy of the Huntington Library. Image on page 53 by the NASA/WMAP Science Team. Photograph on page 67 courtesy of Corbis Images. Image on page 119 courtesy of Getty Images. Figures on pages 147, 153, 177, 213, 270, 379, and 382 by Sean Carroll. Photograph on page 204 courtesy of the Smithsonian Institution. Photograph on page 259 courtesy of Professor Stephen Hawking. Photograph on page 267 courtesy of Professor Jacob Bekenstein. Photograph on page 295 by Jerry Bauer, from Wikimedia Commons. Photograph on page 315 courtesy of the Massachusetts Institute of Technology. All other images courtesy of Jason Torchinsky.

The Library of Congress has catalogued the hardcover edition as follows:

Carroll, Sean M., 1966–
  From eternity to here : the quest for the ultimate theory of time / Sean Carroll.
    p. cm.
  Includes bibliographical references and index.
  ISBN 978-0-525-95133-9 (hc.)
  ISBN 978-0-452-29654-1 (pbk.)
  1. Space and time. I. Title.
  QC173.59.S65C37 2009
  530.11—dc22   2009023828

Printed in the United States of America
10 9 8 7 6 5 4 3

To Jennifer
For all time

CONTENTS

PROLOGUE

Does anybody really know what time it is?

—Chicago, "Does Anybody Really Know What Time It Is?"

This book is about the nature of time, the beginning of the universe, and the underlying structure of physical reality. We're not thinking small here. The questions we're tackling are ancient and honorable ones: Where did time and space come from? Is the universe we see all there is, or are there other "universes" beyond what we can observe? How is the future different from the past?

According to researchers at the *Oxford English Dictionary*, *time* is the most used noun in the English language. We live through time, keep track of it obsessively, and race against it every day—yet, surprisingly, few people would be able to give a simple explanation of what time actually *is*.

In the age of the Internet, we might turn to Wikipedia for guidance. As of this writing, the entry on "Time" begins as follows:

> Time is a component of a measuring system used to sequence events, to compare the durations of events and the intervals between them, and to quantify the motions of objects. Time has been a major subject of religion, philosophy, and science, but defining time in a non-controversial manner applicable to all fields of study has consistently eluded the greatest scholars.[1]

Oh, it's on. By the end of this book, we will have defined *time* very precisely, in ways applicable to all fields. Less clear, unfortunately, will be *why* time has the properties that it does—although we'll examine some intriguing ideas.

Cosmology, the study of the whole universe, has made extraordinary strides over the past hundred years. Fourteen billion years ago, our universe (or at least the part of it we can observe) was in an unimaginably hot, dense state that we call "the Big Bang." Ever since, it has been expanding and cooling, and it looks like that's going to continue for the foreseeable future, and possibly forever.

A century ago, we didn't know any of that—scientists understood basically nothing about the structure of the universe beyond the Milky Way galaxy. Now we have taken the measure of the observable universe and are able to describe in detail its size and shape, as well as its constituents and the outline of its history. But there are important questions we cannot answer, especially concerning the early moments of the Big Bang. As we will see, those questions play a crucial role in our understanding of time—not just in the far-flung reaches of the cosmos, but in our laboratories on Earth and even in our everyday lives.

TIME SINCE THE BIG BANG

It's clear that the universe evolves as time passes—the early universe was hot and dense; the current universe is cold and dilute. But I am going to be drawing a much deeper connection. The most mysterious thing about time is that it has a direction: the past is different from the future. That's the *arrow of time*—unlike directions in space, all of which are created pretty much equal, the universe indisputably has a preferred orientation in time. A major theme of this book is that the arrow of time exists because the universe evolves in a certain way.

The reason why time has a direction is because the universe is full of irreversible processes—things that happen in one direction of time, but never the other. You can turn an egg into an omelet, as the classic example goes, but you can't turn an omelet into an egg. Milk disperses into coffee; fuels undergo combustion and turn into exhaust; people are born, grow older, and die. Everywhere in Nature we find sequences of events where one kind of event always happens before, and another kind after; together, these define the arrow of time.

Remarkably, a single concept underlies our understanding of irreversible processes: something called *entropy*, which measures the "disorderliness" of an object or conglomeration of objects. Entropy has a stubborn tendency to increase, or at least stay constant, as time passes—that's the famous Second Law of Thermodynamics.[2] And the reason why entropy wants to increase is deceptively simple: There are more ways to be disorderly than to be orderly, so (all else being equal) an orderly arrangement will naturally tend toward increasing disorder. It's not that hard to scramble the egg molecules into the form of an omelet, but delicately putting them back into the arrangement of an egg is beyond our capabilities.

The traditional story that physicists tell themselves usually stops there. But there is one absolutely crucial ingredient that hasn't received enough attention: If everything in the universe evolves toward increasing disorder, it must have started out in an exquisitely ordered arrangement. This whole chain of logic, purporting to explain why you can't turn an omelet into an egg, apparently rests on a deep assumption about the very beginning of the universe: It was in a state of very low entropy, very high order.

The arrow of time connects the early universe to something we experience literally every moment of our lives. It's not just breaking eggs, or other irreversible processes like mixing milk into coffee or how an untended room tends to get messier over time. The arrow of time is the reason why time seems to flow around us, or why (if you prefer) we seem to move through time. It's why we remember the past, but not the future. It's why we evolve and metabolize and eventually die. It's why we believe in cause and effect, and is crucial to our notions of free will.

And it's all because of the Big Bang.

WHAT WE SEE ISN'T ALL THERE IS

The mystery of the arrow of time comes down to this: Why were conditions in the early universe set up in a very particular way, in a configuration of low entropy that enabled all of the interesting and irreversible processes to come? That's the question this book sets out to address. Unfortunately, no one yet knows the right answer. But we've reached a point in the development of modern science where we have the tools to tackle the question in a serious way.

Scientists and prescientific thinkers have always tried to understand time. In ancient Greece, the pre-Socratic philosophers Heraclitus and Parmenides staked out different positions on the nature of time: Heraclitus stressed the primacy of change, while Parmenides denied the reality of change altogether. The nineteenth century was the heroic era of statistical mechanics—deriving the behavior of macroscopic objects from their microscopic constituents—in which figures like Ludwig Boltzmann, James Clerk Maxwell, and Josiah Willard Gibbs worked out the meaning of entropy and its role in irreversible processes. But they didn't know about Einstein's general relativity, or about quantum mechanics, and certainly not about modern cosmology. For the first time in the history of science, we at least have a chance of putting together a sensible theory of time and the evolution of the universe.

I'm going to suggest the following way out: The Big Bang was *not* the beginning of the universe. Cosmologists sometimes say that the Big Bang represents a true boundary to space and time, before which nothing existed indeed, time itself did not exist, so the concept of "before" isn't strictly applicable. But we don't know enough about the ultimate laws of physics to make a statement like that with confidence. Increasingly, scientists are taking seriously the possibility that the Big Bang is not really a beginning—it's just a phase through which the universe goes, or at least our part of the universe. If that's true, the question of our low-entropy beginnings takes on a different cast: not "Why did the universe start out with such a low entropy?" but rather "Why did our part of the universe pass through a period of such low entropy?"

That might not sound like an easier question, but it's a different one, and it opens up a new set of possible answers. Perhaps the universe we see is only part of a much larger multiverse, which doesn't start in a low-entropy configuration at all. I'll argue that the most sensible model for the multiverse is one in which entropy increases because entropy can *always* increase—there is no state of maximum entropy. As a bonus, the multiverse can be completely symmetric in time: From some moment in the middle where entropy is high, it evolves in the past and future to states where the entropy is even higher. The universe we see is a tiny sliver of an enormously larger ensemble, and our particular journey from a dense Big Bang to an everlasting emptiness is all part of the wider multiverse's quest to increase its entropy.

That's one possibility, anyway. I'm putting it out there as an example of the kind of scenarios cosmologists need to be contemplating, if they want to take seriously the problems raised by the arrow of time. But whether or not this particular idea is on the right track, the problems themselves are fascinating and real. Through most of this book, we'll be examining the problems of time from a variety of angles— time travel, information, quantum mechanics, the nature of eternity. When we aren't sure of the final answer, it behooves us to ask the question in as many ways as possible.

THERE WILL ALWAYS BE SKEPTICS

Not everyone agrees that cosmology should play a prominent role in our understanding of the arrow of time. I once gave a colloquium on the subject to a large audience at a major physics department. One of the older professors in the department didn't find my talk very convincing and made sure that everyone in the room knew of his unhappiness. The next day he sent an e-mail around to the department faculty, which he was considerate enough to copy to me:

> Finally, the magnitude of the entropy of the universe as a function of
> time is a very interesting problem for cosmology, but to suggest that
> a law of physics depends on it is sheer nonsense. Carroll's statement

that the second law owes its existence to cosmology is one of the dum-
mest [sic] remarks I heard in any of our physics colloquia, apart from
[redacted]'s earlier remarks about consciousness in quantum mechanics. I am astounded that physicists in the audience always listen politely
to such nonsense. Afterwards, I had dinner with some graduate students who readily understood my objections, but Carroll remained
adamant.

I hope he reads this book. Many dramatic-sounding statements are contained
herein, but I'm going to be as careful as possible to distinguish among three different types: (1) remarkable features of modern physics that sound astonishing but
are nevertheless universally accepted as true; (2) sweeping claims that are not necessarily accepted by many working physicists but that should be, as there is no
question they are correct; and (3) speculative ideas beyond the comfort zone of
contemporary scientific state of the art. We certainly won't shy away from spec-
ulation, but it will always be clearly labeled. When all is said and done, you'll be
equipped to judge for yourself which parts of the story make sense.

The subject of time involves a large number of ideas, from the everyday to the
mind-blowing. We'll be looking at thermodynamics, quantum mechanics, special and general relativity, information theory, cosmology, particle physics, and
quantum gravity. Part One of the book can be thought of as a lightning tour of
the terrain—entropy and the arrow of time, the evolution of the universe, and different conceptions of the idea of "time" itself. Then we will get a bit more systematic; in Part Two we will think deeply about spacetime and relativity, including the
possibility of travel backward in time. In Part Three we will think deeply about
entropy, exploring its role in multiple contexts, from the evolution of life to the
mysteries of quantum mechanics.

In Part Four we will put it all together to confront head-on the mysteries that
entropy presents to the modern cosmologist: What should the universe look like,
and how does that compare to what it actually does look like? I'll argue that the
universe doesn't look anything like it "should," after being careful about what that
is supposed to mean—at least, not if the universe we see is all there is. If our universe began at the Big Bang, it is burdened with a finely tuned boundary condition for which we have no good explanation. But if the observed universe is part of
a bigger ensemble—the multiverse—then we might be able to explain why a tiny
part of that ensemble witnesses such a dramatic change in entropy from one end
of time to the other.

All of which is unapologetically speculative but worth taking seriously. The stakes are big—time, space, the universe—and the mistakes we are likely to make along the way will doubtless be pretty big as well. It's sometimes helpful to let our imaginations roam, even if our ultimate goal is to come back down to Earth and explain what's going on in the kitchen.

PART ONE

TIME, EXPERIENCE, AND THE UNIVERSE

1

THE PAST IS PRESENT MEMORY

What is time? If no one asks me, I know. If I wish to explain it to one that asketh, I know not.

—St. Augustine, *Confessions*

The next time you find yourself in a bar, or on an airplane, or standing in line at the Department of Motor Vehicles, you can pass the time by asking the strangers around you how they would define the word *time*. That's what I started doing, anyway, as part of my research for this book. You'll probably hear interesting answers: "Time is what moves us along through life," "Time is what separates the past from the future," "Time is part of the universe," and more along those lines. My favorite was "Time is how we know when things happen."

All of these concepts capture some part of the truth. We might struggle to put the meaning of "time" into words, but like St. Augustine we nevertheless manage to deal with time pretty effectively in our everyday lives. Most people know how to read a clock, how to estimate the time it will take to drive to work or make a cup of coffee, and how to manage to meet their friends for dinner at roughly the appointed hour. Even if we can't easily articulate what exactly it is we mean by "time," its basic workings make sense at an intuitive level.

Like a Supreme Court justice confronted with obscenity, we know time when we see it, and for most purposes that's good enough. But certain aspects of time remain deeply mysterious. Do we really know what the word means?

WHAT WE MEAN BY *TIME*

The world does not present us with abstract concepts wrapped up with pretty bows, which we then must work to understand and reconcile with other concepts. Rather, the world presents us with phenomena, things that we observe and make note of, from which we must then work to derive concepts that help us understand how those phenomena relate to the rest of our experience. For subtle concepts such as entropy, this is pretty clear. You don't walk down the street and bump into some entropy; you have to observe a variety of phenomena in nature and discern a pattern that is best thought of in terms of a new concept you label "entropy." Armed with this helpful new concept, you observe even more phenomena, and you are inspired to refine and improve upon your original notion of what entropy really is.

For an idea as primitive and indispensable as "time," the fact that we invent the concept rather than having it handed to us by the universe is less obvious—time is something we literally don't know how to live without. Nevertheless, part of the task of science (and philosophy) is to take our intuitive notion of a basic concept such as "time" and turn it into something rigorous. What we find along the way is that we haven't been using this word in a single unambiguous fashion; it has a few different meanings, each of which merits its own careful elucidation.

Time comes in three different aspects, all of which are going to be important to us.

1. **Time labels moments in the universe.**
   Time is a coordinate; it helps us locate things.
2. **Time measures the duration elapsed between events.**
   Time is what clocks measure.
3. **Time is a medium through which we move.**
   Time is the agent of change. We move through it, or—equivalently— time flows past us, from the past, through the present, toward the future.

At first glance, these all sound somewhat similar. Time labels moments, it measures duration, and it moves from past to future—sure, nothing controversial about any of that. But as we dig more deeply, we'll see how these ideas don't *need* to be related to one another—they represent logically independent concepts that happen to be tightly intertwined in our actual world. Why that is so? The answer matters more than scientists have tended to think.

1. Time labels moments in the universe

John Archibald Wheeler, an influential American physicist who coined the term *black hole*, was once asked how he would define "time." After thinking for a while, he came up with this: "Time is Nature's way of keeping everything from happening at once."

There is a lot of truth there, and more than a little wisdom. When we ordinarily think about the world, not as scientists or philosophers but as people getting through life, we tend to identify "the world" as a collection of *things*, located in various *places*. Physicists combine all of the places together and label the whole collection "space," and they have different ways of thinking about the kinds of things that exist in space—atoms, elementary particles, quantum fields, depending on the context. But the underlying idea is the same. You're sitting in a room, there are various pieces of furniture, some books, perhaps food or other people, certainly some air molecules—the collection of all those things, everywhere from nearby to the far reaches of intergalactic space, is "the world."

And the world changes. We find objects in some particular arrangement, and we also find them in some other arrangement. (It's very hard to craft a sensible sentence along those lines without referring to the concept of time.) But we don't see the different configurations "simultaneously," or "at once." We see one configuration—here you are on the sofa, and the cat is in your lap—and then we see another configuration—the cat has jumped off your lap, annoyed at the lack of attention while you are engrossed in your book. So the world appears to us again and again, in various configurations, but these configurations are somehow distinct. Happily, we can label the various configurations to keep straight which is which—Miss Kitty is walking away "now"; she was on your lap "then." That label is time.

So the world exists, and what is more, the world *happens*, again and again. In that sense, the world is like the different frames of a film reel—a film whose camera view includes the entire universe. (There are also, as far as we can tell, an infinite number of frames, infinitesimally separated.) But of course, a film is much more than a pile of individual frames. Those frames better be in the right order, which is crucial for making sense of the movie. Time is the same way. We can say much more than "that happened," and "that also happened," and "that happened, too." We can say that this happened *before* that happened, and the other thing is going to happen *after*. Time isn't just a label on each instance of the world; it provides a sequence that puts the different instances in order.

A real film, of course, doesn't include the entire universe within its field of view. Because of that, movie editing typically involves "cuts"—abrupt jumps from one scene or camera angle to another. Imagine a movie in which every single transition between two frames was a cut to a completely different scene. When shown through a projector, it would be incomprehensible—on the screen it would look like random static. Presumably there is some French avant-garde film that has already used this technique.

The real universe is not an avant-garde film. We experience a degree of continuity through time—if the cat is on your lap now, there might be some danger that she will stalk off, but there is little worry that she will simply dematerialize into nothingness one moment later. This continuity is not absolute, at the microscopic level; particles can appear and disappear, or at least transform under the right conditions into different kinds of particles. But there is not a wholesale rearrangement of reality from moment to moment.

This phenomenon of persistence allows us to think about "the world" in a different way. Instead of a collection of things distributed through space that keep changing into different configurations, we can think of the entire *history* of the world, or any particular thing in it, in one fell swoop. Rather than thinking of Miss

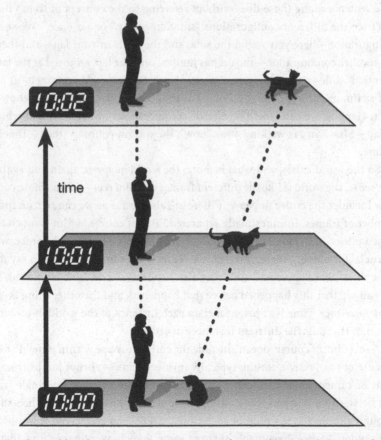

Figure 1: The world, ordered into different moments of time. Objects (including people and cats) persist from moment to moment, defining world lines that stretch through time.

Kitty as a particular arrangement of cells and fluids, we can think of her entire life stretching through time, from birth to death. The history of an object (a cat, a planet, an electron) through time defines its *world line*—the trajectory the object takes through space as time passes.[3] The world line of an object is just the complete set of positions the object has in the world, labeled by the particular time it was in each position.

Finding ourselves

Thinking of the entire history of the universe all at once, rather than thinking of the universe as a set of things that are constantly moving around, is the first step toward thinking of time as "kind of like space," which we will examine further in the chapters to come. We use both time and space to help us pinpoint things that happen in the universe. When you want to meet someone for coffee, or see a certain showing of a movie, or show up for work along with everyone else, you need to specify a time: "Let's meet at the coffee shop at 6:00 P.M. this Thursday."

If you want to meet someone, of course, it's not sufficient just to specify a time; you also need to specify a place. (Which coffee shop are we talking about here?) Physicists say that space is "three-dimensional." What that means is that we require three numbers to uniquely pick out a particular location. If the location is near the Earth, a physicist might give the latitude, longitude, and height above ground. If the location is somewhere far away, astronomically speaking, we might give its direction in the sky (two numbers, analogous to latitude and longitude), plus the

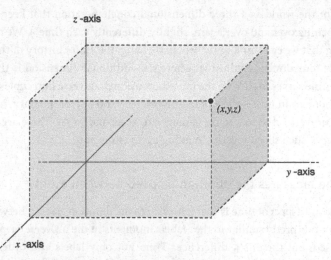

Figure 2: Coordinates attached to each point in space.

distance from Earth. It doesn't matter how we choose to specify those three numbers; the crucial point is that you will always need exactly three. Those three numbers are the *coordinates* of that location in space. We can think of a little label attached to each point, telling us precisely what the coordinates of that point are.

In everyday life, we can often shortcut the need to specify all three coordinates of space. If you say "the coffee shop at Eighth and Main Street," you're implicitly giving two coordinates—"Eighth" and "Main Street"—and you're assuming that we all agree the coffee shop is likely to be at ground level, rather than in the air or underground. That's a convenience granted to us by the fact that much of the space we use to locate things in our daily lives is effectively two-dimensional, confined near the surface of the Earth. But in principle, all three coordinates are needed to specify a point in space.

Each point in space occurs once at each moment of time. If we specify a certain location in space at one definite moment in time, physicists call that an *event*. (This is not meant to imply that it's an especially exciting event; any random point in empty space at any particular moment of time would qualify, so long as it's uniquely specified.) What we call the "universe" is just the set of all events—every point in space, at every moment of time. So we need four numbers—three coordinates of space, and one of time—to uniquely pick out an event. That's why we say that the universe is four-dimensional. This is such a useful concept that we will often treat the whole collection, every point in space at every moment of time, as a single entity called *spacetime*.

This is a big conceptual leap, so it's worth pausing to take it in. It's natural to think of the world as a three-dimensional conglomeration that keeps changing ("happening over and over again, slightly differently each time"). We're now suggesting that we can think of the whole shebang, the entire history of the world, as a single four-dimensional thing, where the additional dimension is time. In this sense, time serves to slice up the four-dimensional universe into copies of space at each moment in time—the whole universe at 10:00 A.M. on January 20, 2010; the whole universe at 10:01 A.M. on January 20, 2010; and so on. There are an infinite number of such slices, together making up the universe.

2. Time measures the duration elapsed between events

The second aspect of time is that it measures the duration elapsed between events. That sounds pretty similar to the "labels moments in the universe" aspect already discussed, but there is a difference. Time not only labels and orders different moments; it also measures the distance between them.

When taking up the mantle of philosopher or scientist and trying to make sense of a subtle concept, it's helpful to look at things operationally—how do we actually use this idea in our experience? When we use time, we refer to the measurements that we get by reading clocks. If you watch a TV show that is supposed to last one hour, the reading on your clock at the end of the show will be one hour later than what it read when the show began. That's what it *means* to say that one hour elapsed during the broadcast of that show: Your clock read an hour later when it ended than when it began.

But what makes a good clock? The primary criterion is that it should be consistent—it wouldn't do any good to have a clock that ticked really fast sometimes and really slowly at others. Fast or slow compared to what? The answer is: other clocks. As a matter of empirical fact (rather than logical necessity), there are some objects in the universe that are consistently periodic—they do the same thing over and over again, and when we put them next to one another we find them repeating in predictable patterns.

Think of planets in the Solar System. The Earth orbits around the Sun, returning to the same position relative to the distant stars once every year. By itself, that's not so meaningful—it's just the definition of a "year." But Mars, as it turns out, returns to the same position once every 1.88 years. That kind of statement is extremely meaningful—without recourse to the concept of a "year," we can say that Earth moves around the Sun 1.88 times every time Mars orbits just once.[4] Likewise, Venus moves around the Sun 1.63 times every time Earth orbits just once.

The key to measuring time is *synchronized repetition*—a wide variety of processes occur over and over again, and the number of times that one process repeats itself while another process returns to its original state is reliably predictable. The Earth spins on its axis, and it's going to do so 365.25 times every time the Earth moves around the Sun. The tiny crystal in a quartz watch vibrates 2,831,155,200 times every time the Earth spins on its axis. (That's 32,768 vibrations per second, 3,600 seconds in an hour, 24 hours in a day.[5]) The reason why quartz watches are reliable is that quartz crystal has extremely regular vibrations; even as the temperature or pressure changes, the crystal will vibrate the same number of times for every one rotation of the Earth.

So when we say that something is a good clock, we mean that it repeats itself in a predictable way relative to other good clocks. It is a fact about the universe that such clocks exist, and thank goodness. In particular, at the microscopic level where all that matters are the rules of quantum mechanics and the properties (masses, electric charges) of individual elementary particles, we find atoms and molecules that vibrate with absolutely predictable frequencies, forming a widespread array of excellent clocks marching in cheerful synchrony. A universe without good clocks—in which no processes repeated themselves a predictable number of times relative to other repeating processes—would be a scary universe indeed.[6]

Still, good clocks are not easy to come by. Traditional methods of timekeeping often referred to celestial objects—the positions of the Sun or stars in the sky—because things down here on Earth tend to be messy and unpredictable. In 1581, a young Galileo Galilei reportedly made a breakthrough discovery while he sat bored during a church service in Pisa. The chandelier overhead would swing gently back and forth, but it seemed to move more quickly when it was swinging widely (after a gust of wind, for example) and more slowly when wasn't moving as far. Intrigued, Galileo decided to measure how much time it took for each swing, using the only approximately periodic event to which he had ready access: the beating of his own pulse. He found something interesting: The number of heartbeats between swings of the chandelier was roughly the same, regardless of whether the swings were wide or narrow. The size of the oscillations—how far the pendulum swung back and forth—didn't affect the frequency of those oscillations. That's not unique to chandeliers in Pisan churches; it's a robust property of the kind of pendulum physicists call a "simple harmonic oscillator." And that's why pendulums form the centerpiece of grandfather clocks and other timekeeping devices: Their oscillations are extremely reliable. The craft of clock making involves the search for ever-more-reliable forms of oscillations, from vibrations in quartz to atomic resonances.

Our interest here is not really in the intricacies of clock construction, but in the meaning of time. We live in a world that contains all sorts of periodic processes, which repeat a predictable number of times in comparison to certain other periodic processes. And that's how we measure duration: by the number of repetitions of such a process. When we say that our TV program lasts one hour, we mean that the quartz crystal in our watch will oscillate 117,964,800 times between the start and end of the show (32,768 oscillations per second, 3,600 seconds in an hour).

Notice that, by being careful about defining time, we seem to have eradicated the concept entirely. That's just what any decent definition should do—you don't want to define something in terms of itself. The passage of time can be completely recast in terms of certain things happening together, in synchrony. "The program lasts one hour" is equivalent to "there will be 117,964,800 oscillations of the quartz crystal in my watch between the beginning and end of the program" (give or take a few commercials). If you really wanted to, you could reinvent the entire superstructure of physics in a way that completely eliminated the concept of "time," by replacing it with elaborate specifications of how certain things happen

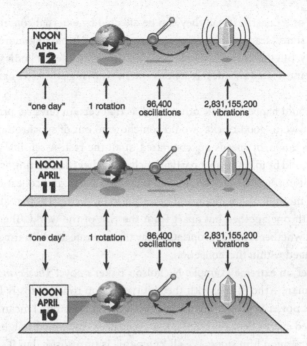

Figure 3: Good clocks exhibit synchronized repetition. Every time one day passes, the Earth rotates once about its axis, a pendulum with a period of 1 second oscillates 86,400 times, and a quartz watch crystal vibrates 2,831,155,200 times.

in coincidence with certain other things.[7] But why would we want to? Thinking in terms of time is convenient, and more than that, it reflects a simple underlying order in the way the universe works.

Slowing, stopping, bending time

Armed with this finely honed understanding of what we mean by the passage of time, at least one big question can be answered: What would happen if time were to slow down throughout the universe? The answer is: That's not a sensible question to ask. Slow down relative to what? If time is what clocks measure, and every clock were to "slow down" by the same amount, it would have absolutely no effect at all. Telling time is about synchronized repetition, and as long as the rate of one oscillation is the same relative to some other oscillation, all is well.

As human beings we *feel* the passage of time. That's because there are periodic processes occurring within our own metabolism—breaths, heartbeats, electrical pulses, digestion, rhythms of the central nervous system. We are a complicated, interconnected collection of clocks. Our internal rhythms are not as reliable as a pendulum or a quartz crystal; they can be affected by external conditions or our emotional states, leading to the impression that time is passing more quickly or more slowly. But the truly reliable clocks ticking away inside our bodies—vibrating molecules, individual chemical reactions—aren't moving any faster or slower than usual.[8]

What could happen, on the other hand, is that certain physical processes that we thought were "good clocks" would somehow go out of synchronization—one clock slows down, or speeds up, compared to all the rest. A sensible response in that case would be to blame that particular clock, rather than casting aspersions on time itself. But if we stretch a bit, we can imagine a particular collection of clocks (including molecular vibrations and other periodic processes) that all change in concert with one another, but apart from the rest of the world. Then we would have to ask whether it was appropriate to say that the rate at which time passes had really changed within that collection.

Consider an extreme example. Nicholson Baker's novel *The Fermata* tells the story of a man, Arno Strine, with the ability to "stop time." (Mostly he uses this miraculous power to go around undressing women.) It wouldn't mean anything if time stopped everywhere; the point is that Arno keeps moving through time, while everything around him stops. We all know this is unrealistic, but it's instructive to reflect upon the way in which it flouts the laws of physics. What this approach to stopping time entails is that every kind of motion and rhythm in Arno's body continues as usual, while every kind of motion and rhythm in the outside world freezes absolutely still. Of course we have to imagine that time continues for all of the air and fluids within Arno, otherwise he would instantly die. But if the air in the rest of the room has truly stopped experiencing time, each molecule must remain suspended precisely in its location; consequently, Arno would be unable to move, trapped in a prison of rigidly stationary air molecules. Okay, let's be generous and assume that time would proceed normally for any air molecules that came sufficiently close to Arno's skin. (The book alludes to something of the sort.) But everything outside, by assumption, is not changing in any way. In particular, no sound or light would be able to travel to him from the outside world; Arno would be completely deaf and blind. It turns out not to be a promising environment for a Peeping Tom.[9]

What if, despite all the physical and narrative obstacles, something like this really could happen? Even if we can't stop time around us, presumably we can imagine speeding up the motion of some local clocks. If we truly measure time by synchronized repetition, and we arranged an ensemble of clocks that were all running fast compared to the outside world while they remained in synchrony with one another, wouldn't that be something like "time running faster" within that arrangement?

It depends. We've wandered far afield from what might actually happen in the real world, so let's establish some rules. We're fortunate enough to live in a universe that features very reliable clocks. Without such clocks, we can't use time to measure the duration between events. In the world of *The Fermata*, we could say that time slowed down for the universe outside Arno Strine—or, equivalently and perhaps more usefully, that time for him sped up, while the rest of the world went on as usual. But just as well, we could say that "time" was completely unaffected, and what changed were the laws of particle physics (masses, charges on different particles) within Arno's sphere of influence. Concepts like "time" are not handed to us unambiguously by the outside world but are invented by human beings trying to make sense of the universe. If the universe were very different, we might have to make sense of it in a different way.

Meanwhile, there is a very real way for one collection of clocks to measure time differently than another: have them move along different paths through spacetime. That's completely compatible with our claim that "good clocks" should measure time in the same way, because we can't readily compare clocks unless they're next to one another in space. The total amount of time elapsed on two different trajectories can be different without leading to any inconsistencies. But it does lead to something important—the theory of relativity.

Twisty paths through spacetime

Through the miracle of synchronized repetition, time doesn't simply put different moments in the history of the universe into order; it also tells us "how far apart" they are (in time). We can say more than "1776 happened before 2010"; we can say "1776 happened 234 years before 2010."

I should emphasize a crucial distinction between "dividing the universe into different moments" and "measuring the elapsed time between events," a distinction that will become enormously important when we get to relativity. Let's imagine you are an ambitious temporal[10] engineer, and you're not satisfied to just have your wristwatch keep accurate time; you want to be able to know what time it is at every other event in spacetime as well. You might be tempted to wonder: Couldn't we (hypothetically) construct a time coordinate all throughout the universe, just by building an infinite number of clocks, synchronizing them to the same time, and scattering them throughout space? Then, wherever we went in spacetime, there would be a clock sitting at each point telling us what time it was, once and for all.

The real world, as we will see, doesn't let us construct an absolute universal time coordinate. For a long time people thought it did, under no less an authority than that of Sir Isaac Newton. In Newton's view of the universe, there was one particular right way to slice up the universe into slices of "space at a particular moment of time." And we could indeed, at least in a thought-experiment kind of way, send clocks all throughout the universe to set up a time coordinate that would uniquely specify when a certain event was taking place.

But in 1905, along comes Einstein with his special theory of relativity.[11] The central conceptual breakthrough of special relativity is that our two aspects of time, "time labels different moments" and "time is what clocks measure," are *not* equivalent, or even interchangeable. In particular, the scheme of setting up a time coordinate by sending clocks throughout the universe *would not work*: two clocks, leaving the same event and arriving at the same event but taking different paths to get there, will generally experience different durations along the journey, slipping out of synchronization. That's not because we haven't been careful enough to pick "good clocks," as defined above. It's because *the duration elapsed along two trajectories connecting two events in spacetime need not be the same.*

This idea isn't surprising, once we start thinking that "time is kind of like space." Consider an analogous statement, but for space instead of time: The distance traveled along two paths connecting two points in space need not be the same. Doesn't sound so surprising at all, does it? Of course we can connect two points in space by paths with different lengths; one could be straight and one could be curved, and we would always find that the distance along the curved path was greater. But the difference in *coordinates* between the same two points is always the same, regardless of how we get from one point to another. That's because, to drive home the obvious, the distance you travel is not the same as your change in coordinates. Consider a running back in football who zips back and forth while evading tacklers, and ends up advancing from the 30-yard line to the 80-yard line. (It should really be "the opponent's 20-yard line," but the point is clearer this way.)

Figure 4: Yard lines serve as coordinates on a football field. A running back who advances the ball from the 30-yard line to the 80-yard line has changed coordinates by 50 yards, even though the distance traveled may have been much greater.

The change in coordinates is 50 yards, no matter how long or short was the total distance he ran.

The centerpiece of special relativity is the realization that *time is like that*. Our second definition, time is duration as measured by clocks, is analogous to the total length of a path through space; the clock itself is analogous to an odometer or some other instrument that measures the total distance traveled. This definition is simply not the same as the concept of a coordinate labeling different slices of spacetime (analogous to the yard lines on a football field). And this is not some kind of technical problem that we can "fix" by building better clocks or making better choices about how we travel through spacetime; it's a feature of how the universe works, and we need to learn to live with it.

As fascinating and profound as it is that time works in many ways similar to space, it will come as no surprise that there are crucial differences as well. Two of them are central elements of the theory of relativity. First, while there are three dimensions of space, there is only one of time; that brute fact has important consequences for physics, as you might guess. And second, while a straight line between two points in space describes the shortest distance, a straight trajectory between two events in spacetime describes the *longest* elapsed duration.

But the most obvious, blatant, unmistakable difference between time and space is that time has a direction, and space doesn't. Time points from the past toward the future, while (out there in space, far away from local disturbances like the Earth) all directions of space are created equal. We can invert directions in space without doing damage to how physics works, but all sorts of real processes can happen in one direction of time but not the other. It's to this crucial difference that we now turn.

3. Time is a medium through which we move

The sociology experiment suggested at the beginning of this chapter, in which you ask strangers how they would define "time," also serves as a useful tool for distinguishing physicists from non-physicists. Nine times out of ten, a physicist will say something related to one of the first two notions above—time is a coordinate, or time is a measure of duration. Equally often, a non-physicist will say something related to the third aspect we mentioned—time is something that flows from past to future. Time whooshes along, from "back then" to "now" and on toward "later."

Or, conversely, someone might say that we move through time, as if time were a substance through which we could pass. In the Afterword to his classic *Zen and the Art of Motorcycle Maintenance*, Robert Pirsig relates a particular twist on this metaphor. The ancient Greeks, according to Pirsig, "saw the future as something that came upon them from behind their backs, with the past receding away before their eyes."[12] When you think about it, that seems a bit more honest than the conventional view that we march toward the future and away from the past. We know something about the past, from experience, while the future is more conjectural.

Common to these perspectives is the idea that time is a *thing*, and it's a thing that can *change*—flow around us, or pass by as we move through it. But conceptualizing time as some sort of substance with its own dynamics, perhaps even the ability to change at different rates depending on circumstances, raises one crucially important question.

What in the world is that supposed to *mean*?

Consider something that actually does flow, such as a river. We can think about the river from a passive or an active perspective: Either we are standing still as the water rushes by, or perhaps we are on a boat moving along with the river as the banks on either side move relative to us.

The river flows, no doubt about that. And what that means is that the location of some particular drop of river water *changes with time*—here it is at some moment, there it is just a bit later. And we can talk sensibly about the *rate* at which the river flows, which is just the velocity of the water—in other words, the distance that the water travels in a given amount of time. We could measure it in miles per hour, or meters per second, or whatever units of "distance traveled per interval of time" you prefer. The velocity may very well change from place to place or moment to moment—sometimes the river flows faster; sometimes it flows more slowly. When we are talking about the real flow of actual rivers, all this language makes perfect sense.

But when we examine carefully the notion that time itself somehow "flows," we hit a snag. The flow of the river was a change with time—but what is it supposed to mean to say that time changes with time? A literal flow is a change of location over time—but time doesn't have a "location." So what is it supposed to be changing with respect to?

Think of it this way: If time does flow, how would we describe its speed? It would have to be something like "x hours per hour"—an interval of time per unit time. And I can tell you what x is going to be—it's 1, all the time. The speed of time is 1 hour per hour, no matter what else might be going on in the universe.

The lesson to draw from all this is that it's not quite right to think of time as something that flows. It's a seductive metaphor, but not one that holds up under closer scrutiny. To extract ourselves from that way of thinking, it's helpful to stop picturing ourselves as positioned within the universe, with time flowing around us. Instead, let's think of the universe—all of the four-dimensional spacetime around us—as somehow a distinct entity, as if we were observing it from an external perspective. Only then can we appreciate time for what it truly is, rather than privileging our position right here in the middle of it.

The view from nowhen

We can't literally stand outside the universe. The universe is not some object that sits embedded in a larger space (as far as we know); it's the collection of everything that exists, space and time included. So we're not wondering what the universe would really look like from the point of view of someone outside it; no such being could possibly exist. Rather, we're trying to grasp the entirety of space and time as a single entity. Philosopher Huw Price calls this "the view from nowhen," a perspective separate from any particular moment in time.[13] We are all overly familiar with time, having dealt with it every day of our lives. But we can't help but situate ourselves within time, and it's useful to contemplate all of space and time in a single picture.

And what do we see, when looking down from nowhen? We don't see anything changing with time, because we are outside of time ourselves. Instead, we see all of history at once—past, present, and future. It's like thinking of space and time as a book, which we could in principle open to any passage, or even cut apart and spread out all the pages before us, rather than as a movie, where we are forced to watch events in sequence at specific times. We could also call this the Tralfamadorian perspective, after the aliens in Kurt Vonnegut's *Slaughterhouse-Five*. According to protagonist Billy Pilgrim,

> The Tralfamadorians can look at all the different moments just the way we can look at a stretch of the Rocky Mountains, for instance. They can see how permanent all the moments are, and they can look at any moment that interests them. It is just an illusion we have here on earth that one moment follows another like beads on a string, and that once a moment is gone it is gone forever.[14]

How do we reconstruct our conventional understanding of flowing time from this lofty timeless Tralfamadorian perch? What we see are correlated events, arranged in a sequence. There is a clock reading 6:45, and a person standing in their kitchen with a glass of water in one hand and an ice cube in the other. In another scene, the clock reads 6:46 and the person is again holding the glass of water, now with the ice cube inside. In yet another one, the clock reads 6:50 and the person is holding a slightly colder glass of water, now with the ice cube somewhat melted.

In the philosophical literature, this is sometimes called the "block time" or "block universe" perspective, thinking of all space and time as a single existing block of spacetime. For our present purposes, the important point is that we *can* think about time in this way. Rather than carrying a picture in the back of our minds in which time is a substance that flows around us or through which we move, we can think of an ordered sequence of correlated events, together constituting the entire universe. Time is then something we reconstruct from the correlations in these events. "This ice cube melted over the course of ten minutes" is equivalent to "the clock reads ten minutes later when the ice cube has melted than it does when the ice cube is put into the glass." We're not committing ourselves to some dramatic conceptual stance to the effect that it's *wrong* to think of ourselves as embedded within time; it just turns out to be more *useful*, when we get around to asking why time and the universe are the way they are, to be able to step outside and view the whole ball of wax from the perspective of nowhen.

Opinions differ, of course. The struggle to understand time is a puzzle of long standing, and what is "real" and what is "useful" have been very much up for debate. One of the most influential thinkers on the nature of time was St. Augustine, the fifth-century North African theologian and Father of the Church. Augustine is perhaps best known for developing the doctrine of original sin, but he was interdisciplinary enough to occasionally turn his hand to metaphysical issues. In Book XI of his *Confessions*, he discusses the nature of time.

> What is by now evident and clear is that neither future nor past exists, and it is inexact language to speak of three times—past, present, and future. Perhaps it would be exact to say: there are three times, a present of things past, a present of things present, a present of things to come. In the soul there are these three aspects of time, and I do not see them anywhere else. The present considering the past is memory, the present considering the present is immediate awareness, the present considering the future is expectation.[15]

Augustine doesn't like this block-universe business. He is what is known as a "presentist," someone who thinks that only the present moment is real—the past and future are things that we here in the present simply try to reconstruct, given the data and knowledge available to us. The viewpoint we've been describing, on the other hand, is (sensibly enough) known as "eternalism," which holds that past, present, and future are all equally real.[16]

Concerning the debate between eternalism and presentism, a typical physicist would say: "Who cares?" Perhaps surprisingly, physicists are not overly concerned with adjudicating which particular concepts are "real" or not. They care very much about how the real world works, but to them it's a matter of constructing comprehensive theoretical models and comparing them with empirical data. It's not the individual concepts characteristic of each model ("past," "future," "time") that matter; it's the structure as a whole. Indeed, it often turns out to be the case that one specific model can be described in two completely different ways, using an entirely different set of concepts. [17]

So, as scientists, our goal is to construct a model of reality that successfully accounts for all of these different notions of time—time is measured by clocks, time is a coordinate on spacetime, and our subjective feeling that time flows. The first two are actually very well understood in terms of Einstein's theory of relativity, as we will cover in Part Two of the book. But the third remains a bit mysterious. The reason why I am belaboring the notion of standing outside time to behold the entire universe as a single entity is because we need to distinguish the notion of time in and of itself from the perception of time as experienced from our parochial view within the present moment. The challenge before us is to reconcile these two perspectives.

2
THE HEAVY HAND OF ENTROPY

Eating is unattractive too . . . Various items get gulped into my mouth, and after skillful massage with tongue and teeth I transfer them to the plate for additional sculpture with knife and fork and spoon. That bit's quite therapeutic at least, unless you're having soup or something, which can be a real sentence. Next you face the laborious business of cooling, of reassembly, of storage, before the return of these foodstuffs to the Superette, where, admittedly, I am promptly and generously reimbursed for my pains. Then you tool down the aisles, with trolley or basket, returning each can and packet to its rightful place.

—Martin Amis, *Time's Arrow*[18]

Forget about spaceships, rocket guns, clashes with extraterrestrial civilizations. If you want to tell a story that powerfully evokes the feeling of being in an alien environment, you have to reverse the direction of time.

You could, of course, simply take an ordinary story and tell it backward, from the conclusion to the beginning. This is a literary device known as "reverse chronology" and appears at least as early as Virgil's *Aeneid*. But to really jar readers out of their temporal complacency, you want to have some of your characters experience time backward. The reason it's jarring, of course, is that all of us nonfictional characters experience time in the same way; that's due to the consistent increase of entropy throughout the universe, which defines the arrow of time.

THROUGH THE LOOKING GLASS

F. Scott Fitzgerald's short story "The Curious Case of Benjamin Button"—more recently made into a film starring Brad Pitt—features a protagonist who is born as an old man and gradually grows younger as time passes. The nurses of the hospital at which Benjamin is born are, understandably, somewhat at a loss.

Wrapped in a voluminous white blanket, and partly crammed into one of the cribs, there sat an old man apparently about seventy years of age. His sparse hair was almost white, and from his chin dripped a long smoke-coloured beard, which waved absurdly back and forth, fanned by the breeze coming in at the window. He looked up at Mr. Button with dim, faded eyes in which lurked a puzzled question.

"Am I mad?" thundered Mr. Button, his terror resolving into rage. "Is this some ghastly hospital joke?"

"It doesn't seem like a joke to us," replied the nurse severely. "And I don't know whether you're mad or not—but that is most certainly your child."

The cool perspiration redoubled on Mr. Button's forehead. He closed his eyes, and then, opening them, looked again. There was no mistake—he was gazing at a man of threescore and ten—a *baby* of threescore and ten, a baby whose feet hung over the sides of the crib in which it was reposing.[19]

No mention is made in the story of what poor Mrs. Button must have been feeling around this time. (In the movie version, at least the newborn Benjamin is baby-sized, albeit old and wrinkled.)

Because it is so bizarre, having time run backward for some characters in a story is often played for comic effect. In Lewis Carroll's *Through the Looking-Glass*, Alice is astonished upon first meeting the White Queen, who lives in both directions of time. The Queen is shouting and shaking her finger in pain:

"What IS the matter?" [Alice] said, as soon as there was a chance of making herself heard. "Have you pricked your finger?"

"I haven't pricked it YET," the Queen said, "but I soon shall—oh, oh, oh!"

"When do you expect to do it?" Alice asked, feeling very much inclined to laugh.

"When I fasten my shawl again," the poor Queen groaned out: "the brooch will come undone directly. Oh, oh!" As she said the words the brooch flew open, and the Queen clutched wildly at it, and tried to clasp it again.

"Take care!" cried Alice. "You're holding it all crooked!" And she caught at the brooch; but it was too late: the pin had slipped, and the Queen had pricked her finger.[20]

Carroll (no relation[21]) is playing with a deep feature of the nature of time—the fact that causes precede effects. The scene makes us smile, while serving as a reminder of how central the arrow of time is to the way we experience the world.

Time can be reversed in the service of tragedy, as well as comedy. Martin Amis's novel *Time's Arrow* is a classic of the reversing-time genre, even accounting for the fact that it's a pretty small genre.[22] Its narrator is a disembodied consciousness who lives inside another person, Odilo Unverdorben. The host lives life in the ordinary sense, forward in time, but the homunculus narrator experiences everything backward—his first memory is Unverdorben's death. He has no control over Unverdorben's actions, nor access to his memories, but passively travels through life in reverse order. At first Unverdorben appears to us as a doctor, which strikes the narrator as quite a morbid occupation—patients shuffle into the emergency room, where staff suck medicines out of their bodies and rip off their bandages, sending them out into the night bleeding and screaming. But near the end of the book, we learn that Unverdorben was an assistant at Auschwitz, where he created life where none had been before—turning chemicals and electricity and corpses into living persons. Only now, thinks the narrator, does the world finally make sense.

THE ARROW OF TIME

There is a good reason why reversing the relative direction of time is an effective tool of the imagination: In the actual, non-imaginary world, it never happens. Time has a direction, and it has the *same* direction for everybody. None of us has met a character like the White Queen, who remembers what we think of as "the future" rather than (or in addition to) "the past."

What does it mean to say that time has a direction, an arrow pointing from the past to the future? Think about watching a movie played in reverse. Generally, it's pretty clear if we are seeing something running the "wrong way" in time. A classic example is a diver and a pool. If the diver dives, and then there is a big splash, followed by waves bouncing around in the water, all is normal. But if we see a pool that starts with waves, which collect into a big splash, in the process lifting a diver up onto the board and becoming perfectly calm, we know something is up: The movie is being played backward.

Certain events in the real world always happen in the same order. It's dive, splash, waves; never waves, splash, spit out a diver. Take milk and mix it into a cup of black coffee; never take coffee with milk and separate the two liquids. Sequences of this sort are called *irreversible processes*. We are free to imagine that kind of sequence playing out in reverse, but if we actually see it happen, we suspect cinematic trickery rather than a faithful reproduction of reality.

Irreversible processes are at the heart of the arrow of time. Events happen in some sequences, and not in others. Furthermore, this ordering is perfectly consistent, as far as we know, throughout the observable universe. Someday we might find a planet in a distant solar system that contains intelligent life, but nobody suspects that we will find a planet on which the aliens regularly separate (the indigenous equivalents of) milk and coffee with a few casual swirls of a spoon. Why isn't that surprising? It's a big universe out there; things might very well happen in all sorts of sequences. But they don't. For certain kinds of processes—roughly speaking, complicated actions with lots of individual moving parts—there seems to be an allowed order that is somehow built into the very fabric of the world.

Tom Stoppard's play *Arcadia* uses the arrow of time as a central organizing metaphor. Here's how Thomasina, a young prodigy who was well ahead of her time, explains the concept to her tutor:

THOMASINA: When you stir your rice pudding, Septimus, the spoonful of jam spreads itself round making red trails like the picture of a meteor in my astronomical atlas. But if you need stir backward, the jam will not come together again. Indeed, the pudding does not notice and continues to turn pink just as before. Do you think this odd?

SEPTIMUS: No.

THOMASINA: Well, I do. You cannot stir things apart.

SEPTIMUS: No more you can, time must needs run backward, and since it will not, we must stir our way onward mixing as we go, disorder out of disorder into disorder until pink is complete, unchanging and unchangeable, and we are done with it for ever. This is known as free will or self-determination.[23]

The arrow of time, then, is a brute fact about our universe. Arguably *the* brute fact about our universe; the fact that things happen in one order and not in the reverse order is deeply ingrained in how we live in the world. Why is it like that? Why do we live in a universe where X is often followed by Y, but Y is never followed by X?

The answer lies in the concept of "entropy" that I mentioned above. Like energy or temperature, entropy tells us something about the particular state of a physical system; specifically, it measures how disorderly the system is. A collection of papers stacked neatly on top of one another has a low entropy; the same collection, scattered haphazardly on a desktop, has a high entropy. The entropy of a cup

of coffee along with a separate teaspoon of milk is low, because there is a particular orderly segregation of the molecules into "milk" and "coffee," while the entropy
of the two mixed together is comparatively large. All of the irreversible processes
that reflect time's arrow—we can turn eggs into omelets but not omelets into eggs,
perfume disperses through a room but never collects back into the bottle, ice cubes
in water melt but glasses of warm water don't spontaneously form ice cubes—
share a common feature: Entropy *increases* throughout, as the system progresses
from order to disorder. Whenever we disturb the universe, we tend to increase its
entropy.

A big part of our task in this book will be to explain how the single idea of entropy
ties together such a disparate set of phenomena, and then to dig more deeply into
what exactly this stuff called "entropy" really is, and why it tends to increase. The
final task—still a profound open question in contemporary physics—is to ask why
the entropy was so low in the past, so that it could be increasing ever since.

FUTURE AND PAST VS. UP AND DOWN

But first, we need to contemplate a prior question: Should we really be surprised
that certain things happen in one direction of time, but not in the other? Who ever
said that everything should be reversible, anyway?

Think of time as a label on events as they happen. That's one of the ways in
which time is like space—they both help us locate things in the universe. But from
that point of view, there is also a crucial difference between time and space—
directions in space are created equal, while directions in time (namely, "the past"
and "the future") are very different. Here on Earth, directions in space are easily distinguished—a compass tells us whether we are moving north, south, east,
or west, and nobody is in any danger of confusing up with down. But that's not
a reflection of deep underlying laws of nature—it's just because we live on a giant
planet, with respect to which we can define different directions. If you were floating in a space suit far away from any planets, all directions in space would truly be
indistinguishable—there would be no preferred notion of "up" or "down."

The technical way to say this is that there is a *symmetry* in the laws of nature—
every direction in space is as good as every other. It's easy enough to "reverse the
direction of space"—take a photograph and print it backward, or for that matter
just look in a mirror. For the most part, the view in a mirror appears pretty unremarkable. The obvious counterexample is writing, for which it's easy to tell that we
are looking at a reversed image; that's because writing, like the Earth, does pick out
a preferred direction (you're reading this book from left to right). But the images

of most scenes not full of human creations look equally "natural" to us whether we see them directly or we see them through a mirror.

Contrast that with time. The equivalent of "looking at an image through a mirror" (reversing the direction of space) is simply "playing a movie backward" (reversing the direction of time). And in that case, it's easy to tell when time has been inverted—the irreversible processes that define the arrow of time are suddenly occurring in the wrong order. What is the origin of this profound difference between space and time?

While it's true that the presence of the Earth beneath our feet picks out an "arrow of space" by distinguishing up from down, it's pretty clear that this is a local, parochial phenomenon, rather than a reflection of the underlying laws of nature. We can easily imagine ourselves out in space where there is no preferred direction. But the underlying laws of nature do not pick out a preferred direction of time, any more than they pick out a preferred direction in space. If we confine our attention to very simple systems with just a few moving parts, whose motion reflects the basic laws of physics rather than our messy local conditions, there is no arrow of time—we can't tell when a movie is being run backward. Think about Galileo's chandelier, rocking peacefully back and forth. If someone showed you a movie of the chandelier, you wouldn't be able to tell whether it was being shown forward or backward—its motion is sufficiently simple that it works equally well in either direction of time.

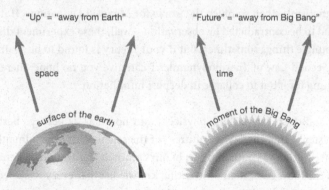

Figure 5: The Earth defines a preferred direction in space, while the Big Bang defines a preferred direction in time.

The arrow of time, therefore, is not a feature of the underlying laws of physics, at least as far as we know. Rather, like the up/down orientation space picked out by the Earth, the preferred direction of time is also a consequence of features of our environment. In the case of time, it's not that we live in the spatial vicinity of an influential object; it's that we live in the temporal vicinity of an influential event: the birth of the universe. The beginning of our observable universe, the hot dense state known as the Big Bang, had a very low entropy. The influence of that event orients us in time, just as the presence of the Earth orients us in space.

NATURE'S MOST RELIABLE LAW

The principle underlying irreversible processes is summed up in the Second Law of Thermodynamics:

> The entropy of an isolated system either remains constant or increases with time.

(The First Law states that energy is conserved.[24]) The Second Law is arguably the most dependable law in all of physics. If you were asked to predict what currently accepted principles of physics would still be considered inviolate a thousand years from now, the Second Law would be a good bet. Sir Arthur Eddington, a leading astrophysicist of the early twentieth century, put it emphatically:

> If someone points out to you that your pet theory of the universe is in disagreement with Maxwell's equations [the laws of electricity and magnetism]—then so much the worse for Maxwell's equations. If it is found to be contradicted by observation—well, these experimentalists do bungle things sometimes. But if your theory is found to be against the Second Law of Thermodynamics I can give you no hope; there is nothing for it but to collapse in deepest humiliation.[25]

C. P. Snow—British intellectual, physicist, and novelist—is perhaps best known for his insistence that the "Two Cultures" of the sciences and the humanities had grown apart and should both be a part of our common civilization. When he came to suggest the most basic item of scientific knowledge that every educated person should understand, he chose the Second Law:

> A good many times I have been present at gatherings of people who, by the standards of the traditional culture, are thought highly educated and who have with considerable gusto been expressing their incredulity at the illiteracy of scientists. Once or twice I have been provoked

and have asked the company how many of them could describe the Second Law of Thermodynamics, the law of entropy. The response was cold: it was also negative. Yet I was asking something which is about the scientific equivalent of: "Have you read a work of Shakespeare's?"[26]

I'm sure Baron Snow was quite the hit at Cambridge cocktail parties. (To be fair, he did later admit that even physicists didn't really understand the Second Law.)

Our modern definition of entropy was proposed by Austrian physicist Ludwig Boltzmann in 1877. But the concept of entropy, and its use in the Second Law of Thermodynamics, dates back to German physicist Rudolf Clausius in 1865. And the Second Law itself goes back even earlier—to French military engineer Nicolas Léonard Sadi Carnot in 1824. How in the world did Clausius use entropy in the Second Law without knowing its definition, and how did Carnot manage to formulate the Second Law without even using the concept of entropy at all?

The nineteenth century was the heroic age of thermodynamics—the study of heat and its properties. The pioneers of thermodynamics studied the interplay between temperature, pressure, volume, and energy. Their interest was by no means abstract—this was the dawn of the industrial age, and much of their work was motivated by the desire to build better steam engines.

Today physicists understand that heat is a form of energy and that the temperature of an object is simply a measure of the average kinetic energy (energy of motion) of the atoms in the object. But in 1800, scientists didn't believe in atoms, and they didn't understand energy very well. Carnot, whose pride was wounded by the fact that the English were ahead of the French in steam engine technology, set himself the task of understanding how efficient such an engine could possibly be—how much useful work could you do by burning a certain amount of fuel? He showed that there is a fundamental limit to such extraction. By taking an intellectual leap from real machines to idealized "heat engines," Carnot demonstrated there was a best possible engine, which got the most work out of a given amount of fuel operating at a given temperature. The trick, unsurprisingly, was to minimize the production of waste heat. We might think of heat as useful in warming our houses during the winter, but it doesn't help in doing what physicists think of as "work"—getting something like a piston or a flywheel to move from place to place. What Carnot realized was that even the most efficient engine possible is not perfect; some energy is lost along the way. In other words, the operation of a steam engine is an irreversible process.

So Carnot appreciated that engines did something that could not be undone. It was Clausius, in 1850, who understood that this reflected a law of nature. He formulated his law as "heat does not spontaneously flow from cold bodies to warm ones." Fill a balloon with hot water and immerse it in cold water. Everyone knows that the temperatures will tend to average out: The water in the balloon will cool down as the surrounding liquid warms up. The opposite never happens. Physical systems evolve toward a state of *equilibrium*—a quiescent configuration that is as uniform as possible, with equal temperatures in all components. From this insight, Clausius was able to re-derive Carnot's results concerning steam engines.

So what does Clausius' law (heat never flows spontaneously from colder bodies to hotter ones) have to do with the Second Law (entropy never spontaneously decreases)? The answer is, they are the same law. In 1865 Clausius managed to reformulate his original maxim in terms of a new quantity, which he called the "entropy." Take an object that is gradually cooling down—emitting heat into its surroundings. As this process happens, consider at every moment the amount of heat being lost, and divide it by the temperature of the object. The entropy is then the accumulated amount of this quantity (the heat lost divided by the temperature) over the course of the entire process. Clausius showed that the tendency of heat to flow from hot objects to cold ones was precisely equivalent to the claim that the entropy of a closed system would only ever go up, never go down. An equilibrium configuration is simply one in which the entropy has reached its maximum value, and has nowhere else to go; all the objects in contact are at the same temperature.

If that seems a bit abstract, there is a simple way of summing up this view of entropy: It measures the *uselessness* of a certain amount of energy.[27] There is energy in a gallon of gasoline, and it's useful—we can put it to work. The process of burning that gasoline to run an engine doesn't change the total amount of energy; as long as we keep careful track of what happens, energy is always conserved.[28] But along the way, that energy becomes increasingly useless. It turns into heat and noise, as well as the motion of the vehicle powered by that engine, but even that motion eventually slows down due to friction. And as energy transforms from useful to useless, its entropy increases all the while.

The Second Law doesn't imply that the entropy of a system can never decrease. We could invent a machine that separated out the milk from a cup of coffee, for example. The trick, though, is that we can only decrease the entropy of one thing by creating more entropy elsewhere. We human beings, and the machines that we might use to rearrange the milk and coffee, and the food and fuel each consume—all of these also have entropy, which will inevitably increase along the way. Physicists draw a distinction between *open systems*—objects that interact significantly with the outside world, exchanging entropy and energy—and *closed systems*—objects that are essentially isolated from external influences. In an open system, like the coffee and milk we put into our machine, entropy can certainly decrease. But in a closed system—say, the total system of coffee plus milk plus machine plus human operators plus fuel and so on—the entropy will always increase, or at best stay constant.

THE RISE OF ATOMS

The great insights into thermodynamics of Carnot, Clausius, and their colleagues all took place within a "phenomenological" framework. They knew the big picture but not the underlying mechanisms. In particular, they didn't know about atoms, so they didn't think of temperature and energy and entropy as properties of some microscopic substrate; they thought of each of them as real things, in and of themselves. It was common in those days to think of energy in particular as a form of fluid, which could flow from one body to another. The energy-fluid even had a name: "caloric." And this level of understanding was perfectly adequate to formulating the laws of thermodynamics.

But over the course of the nineteenth century, physicists gradually became convinced that the many substances we find in the world can all be understood as different arrangements of a fixed number of elementary constituents, known as "atoms." (The physicists actually lagged behind the chemists in their acceptance of atomic theory.) It's an old idea, dating back to Democritus and other ancient Greeks, but it began to catch on in the nineteenth century for a simple reason: The existence of atoms could explain many observed properties of chemical reactions, which otherwise were simply asserted. Scientists like it when a single simple idea can explain a wide variety of observed phenomena.

These days it is elementary particles such as quarks and leptons that play the role of Democritus's atoms, but the idea is the same. What a modern scientist calls an "atom" is the smallest possible unit of matter that still counts as a distinct chemical element, such as carbon or nitrogen. But we now understand that such atoms are not indivisible; they consist of electrons orbiting the atomic nucleus, and the nucleus is made of protons and neutrons, which in turn are made of different combinations of quarks. The search for rules obeyed by these elementary building blocks of matter is often called "fundamental" physics, although "elementary" physics would be more accurate (and arguably less self-aggrandizing). Henceforth, I'll use *atoms* in the established nineteenth-century sense of chemical elements, not the ancient Greek sense of elementary particles.

The fundamental laws of physics have a fascinating feature: Despite the fact that they govern the behavior of all the matter in the universe, you don't need to know them to get through your everyday life. Indeed, you would be hard-pressed to discover them, merely on the basis of your immediate experiences. That's because very large collections of particles obey distinct, autonomous rules of behavior, which don't really depend on the smaller structures underneath. The underlying rules are referred to as "microscopic" or simply "fundamental," while the separate rules that apply only to large systems are referred to as "macroscopic" or "emergent." The behavior of temperature and heat and so forth can certainly be understood in terms of atoms: That's the subject known as "statistical mechanics." But it can equally well be understood without knowing anything whatsoever about atoms: That's the phenomenological approach we've been discussing, known as "thermodynamics." It is a common occurrence in physics that in complex, macroscopic systems, regular patterns emerge dynamically from underlying microscopic rules. Despite the way it is sometimes portrayed, there is no competition between fundamental physics and the study of emergent phenomena; both are fascinating and crucially important to our understanding of nature.

One of the first physicists to advocate atomic theory was a Scotsman, James Clerk Maxwell, who was also responsible for the final formulation of the modern theory of electricity and magnetism. Maxwell, along with Boltzmann in Austria (and following in the footsteps of numerous others), used the idea of atoms to explain the behavior of gases, according to what was known as "kinetic theory." Maxwell and Boltzmann were able to figure out that the atoms in a gas in a container, fixed at some temperature, should have a certain distribution of velocities— this many would be moving fast, that many would be moving slowly, and so on. These atoms would naturally keep banging against the walls of the container, exerting a tiny force each time they did so. And the accumulated impact of those tiny forces has a name: It is simply the pressure of the gas. In this way, kinetic theory explained features of gases in terms of simpler rules.

ENTROPY AND DISORDER

But the great triumph of kinetic theory was its use by Boltzmann in formulating a microscopic understanding of entropy. Boltzmann realized that when we look at some macroscopic system, we certainly don't keep track of the exact properties of every single atom. If we have a glass of water in front of us, and someone sneaks in and (say) switches some of the water molecules around without changing the overall temperature and density and so on, we would never notice. There are many different arrangements of particular atoms that are *indistinguishable* from our macroscopic perspective. And then he noticed that low-entropy objects are more delicate with respect to such rearrangements. If you have an egg, and start exchanging bits of the yolk with bits of the egg white, pretty soon you will notice. The situations that we characterize as "low-entropy" seem to be easily disturbed by rearranging the atoms within them, while "high-entropy" ones are more robust.

Figure 6: Ludwig Boltzmann's grave in the Zentralfriedhof, Vienna. The inscribed equation, $S = k \log W$, is his formula for entropy in terms of the number of ways you can rearrange microscopic components of a system without changing its macroscopic appearance. (See Chapter Eight for details.)

So Boltzmann took the concept of entropy, which had been defined by Clausius and others as a measure of the uselessness of energy, and redefined it in terms of atoms:

> Entropy is a measure of the number of particular microscopic arrangements of atoms that appear indistinguishable from a macroscopic perspective.[29]

It would be difficult to overemphasize the importance of this insight. Before Boltzmann, entropy was a phenomenological thermodynamic concept, which followed its own rules (such as the Second Law). After Boltzmann, the behavior of entropy could be *derived* from deeper underlying principles. In particular, it suddenly makes perfect sense why entropy tends to increase:

> In an isolated system entropy tends to increase, because there are more ways to be high entropy than to be low entropy.

At least, that formulation sounds like it makes perfect sense. In fact, it sneaks in a crucial assumption: that we start with a system that has a low entropy. If we start with a system that has a high entropy, we'll be in equilibrium—nothing will happen at all. That word *start* sneaks in an asymmetry in time, by privileging earlier times over later ones. And this line of reasoning takes us all the way back to the low entropy of the Big Bang. For whatever reason, of the many ways we could arrange the constituents of the universe, at early times they were in a very special, low-entropy configuration.

This caveat aside, there is no question that Boltzmann's formulation of the concept of entropy represented a great leap forward in our understanding of the arrow of time. This increase in understanding, however, came at a cost. Before Boltzmann, the Second Law was absolute—an ironclad law of nature. But the definition of entropy in terms of atoms comes with a stark implication: entropy doesn't necessarily increase, even in a closed system; it is simply *likely* to increase. (Overwhelmingly likely, as we shall see, but still.) Given a box of gas evenly distributed in a high-entropy state, if we wait long enough, the random motion of the atoms will eventually lead them all to be on one side of the box, just for a moment—a "statistical fluctuation." When you run the numbers, it turns out that the time you would have to wait before expecting to see such a fluctuation is much larger than the age of the universe. It's not something we have to worry about, as a practical matter. But it's there.

Some people didn't like that. They wanted the Second Law of Thermodynamics, of all things, to be utterly inviolate, not just something that holds true most of the time. Boltzmann's suggestion met with a great deal of controversy, but these days it is universally accepted.

ENTROPY AND LIFE

This is all fascinating stuff, at least to physicists. But the ramifications of these ideas go far beyond steam engines and cups of coffee. The arrow of time manifests itself in many different ways—our bodies change as we get older, we remember the past but not the future, effects always follow causes. It turns out that *all* of these phenomena can be traced back to the Second Law. Entropy, quite literally, makes life possible.

The major source of energy for life on Earth is light from the Sun. As Clausius taught us, heat naturally flows from a hot object (the Sun) to a cooler object (the Earth). But if that were the end of the story, before too long the two objects would come into equilibrium with each other—they would attain the same temperature.

In fact, that is just what would happen if the Sun filled our entire sky, rather than describing a disk about half a degree across. The result would be an unhappy world indeed. It would be completely inhospitable to the existence of life—not simply because the temperature was high, but because it would be *static*. Nothing would ever change in such an equilibrium world.

In the real universe, the reason why our planet doesn't heat up until it reaches the temperature of the Sun is because the Earth loses heat by radiating it out into space. And the only reason it can do that, Clausius would proudly note, is because space is much colder than Earth.[30] It is because the Sun is a hot spot in a mostly cold sky that the Earth doesn't just heat up, but rather can absorb the Sun's energy, process it, and radiate it into space. Along the way, of course, entropy increases; a fixed amount of energy in the form of solar radiation has a much lower entropy than the same amount of energy in the form of the Earth's radiation into space.

This process, in turn, explains why the biosphere of the Earth is not a static place.[31] We receive energy from the Sun, but it doesn't just heat us up until we reach equilibrium; it's very low entropy radiation, so we can make use of it and then release it as high-entropy radiation. All of which is possible only because the universe as a whole, and the Solar System in particular, have a relatively low entropy at the present time (and an even lower entropy in the past). If the universe were anywhere near thermal equilibrium, nothing would ever happen.

Nothing good lasts forever. Our universe is a lively place because there is plenty of room for entropy to increase before we hit equilibrium and everything grinds to a halt. It's not a foregone conclusion—entropy might be able to simply grow forever. Alternatively, entropy may reach a maximum value and stop. This scenario is known as the "heat death" of the universe and was contemplated as long ago as the 1850s, amidst all the exciting theoretical developments in thermodynamics. William Thomson, Lord Kelvin, was a British physicist and engineer who played an important role in laying the first transatlantic telegraph cable. But in his more reflective moments, he mused on the future of the universe:

> The result would inevitably be a state of universal rest and death, if the universe were finite and left to obey existing laws. But it is impossible to conceive a limit to the extent of matter in the universe; and therefore science points rather to an endless progress, through an endless space, of action involving the transformation of potential energy into palpable motion and hence into heat, than to a single finite mechanism, running down like a clock, and stopping for ever.[32]

Here, Lord Kelvin has put his finger quite presciently on the major issue in these kinds of discussions, which we will revisit at length in this book: Is the capacity of the universe to increase in entropy finite or infinite? If it is finite, then the universe will eventually wind down to a heat death, once all useful energy has been converted to high-entropy useless forms of energy. But if the entropy can increase without bound, we are at least allowed to contemplate the possibility that the universe continues to grow and evolve forever, in one way or another.

In a famous short story entitled simply "Entropy," Thomas Pynchon had his characters apply the lessons of thermodynamics to their social milieu.

> "Nevertheless," continued Callisto, "he found in entropy, or the measure of disorganization of a closed system, an adequate metaphor to apply to certain phenomena in his own world. He saw, for example, the younger generation responding to Madison Avenue with the same spleen his own had once reserved for Wall Street: and in American 'consumerism' discovered a similar tendency from the least to the most probable, from differentiation to sameness, from ordered individuality to a kind of chaos. He found himself, in short, restating Gibbs' prediction in social terms, and envisioned a heat-death for his culture in which ideas, like heat-energy, would no longer be transferred, since each point in it would ultimately have the same quantity of energy; and intellectual motion would, accordingly, cease."[33]

To this day, scientists haven't yet determined to anyone's satisfaction whether the universe will continue to evolve forever, or whether it will eventually settle into a placid state of equilibrium.

WHY CAN'T WE REMEMBER THE FUTURE?

So the arrow of time isn't just about simple mechanical processes; it's a necessary property of the existence of life itself. But it's also responsible for a deep feature of what it means to be a conscious person: the fact that we remember the past but not the future. According to the fundamental laws of physics, the past and future are treated on an equal footing, but when it comes to how we perceive the world, they couldn't be more different. We carry in our heads representations of the past in the form of memories. Concerning the future, we can make predictions, but those predictions have nowhere near the reliability of our memories of the past.

Ultimately, the reason why we can form a reliable memory of the past is because the entropy was lower then. In a complicated system like the universe, there are many ways for the underlying constituents to arrange themselves into the form of "you, with a certain memory of the past, plus the rest of the universe." If that's all you know—that you exist right now, with a memory of going to the beach that summer between sixth and seventh grade—you simply don't have enough information to reliably conclude that you really did go to the beach that summer. It turns out to be overwhelmingly more likely that your memory is just a random fluctuation, like the air in a room spontaneously congregating over on one side. To make sense of your memories, you need to assume as well that the universe was ordered in a certain way—that the entropy was lower in the past.

Imagine that you are walking down the street, and on the sidewalk you notice a broken egg that appears as though it hasn't been sitting outside for very long. Our presumption of a low-entropy past allows us to say with an extremely high degree of certainty that not long ago there must have been an unbroken egg, which someone dropped. Since, as far as the future is concerned, we have no reason to suspect that entropy will decrease, there's not much we can say about the future of the egg—too many possibilities are open. Maybe it will stay there and grow moldy, maybe someone will clean it up, maybe a dog will come by and eat it. (It's unlikely that it will spontaneously reassemble itself into an unbroken egg, but strictly speaking that's among the possibilities.) That egg on the sidewalk is like a memory in your brain—it's a record of a prior event, but only if we assume a low-entropy boundary condition in the past.

We also distinguish past from future through the relationship between cause and effect. Namely, the causes come first (earlier in time), and then come the effects. That's why the White Queen seems so preposterous to us—how could she be yelping in pain *before* pricking her finger? Again, entropy is to blame. Think of the diver splashing into the pool—the splash always comes after the dive. According to the microscopic laws of physics, however, it is possible to arrange all of the molecules in the water (and the air around the pool, through which the sound of the splash travels) to precisely "unsplash" and eject the diver from the pool. To do this would require an unimaginably delicate choice of the position and velocity of every single one of those atoms—if you pick a random splashy configuration, there is almost no chance that the microscopic forces at work will correctly conspire to spit out the diver.

In other words, part of the distinction we draw between "effects" and "causes" is that "effects" generally involve an increase in entropy. If two billiard balls collide and go their separate ways, the entropy remains constant, and neither ball deserves to be singled out as the cause of the interaction. But if you hit the cue ball into a stationary collection of racked balls on the break (provoking a noticeable increase in entropy), you and I would say "the cue ball caused the break"—even though the laws of physics treat all of the balls perfectly equally.

THE ART OF THE POSSIBLE

In the last chapter we contrasted the block time view—the entire four-dimensional history of the world, past, present, and future, is equally real—with the presentist view—only the current moment is truly real. There is yet another perspective, sometimes called *possibilism*: The current moment exists, and the *past* exists, but the future does not (yet) exist.

The idea that the past exists in a way the future does not accords well with our informal notion of how time works. The past has already happened, while the future is still up for grabs in some sense—we can sketch out alternative possibilities, but we don't know which one is real. More particularly, when it comes to the past we have recourse to memories and records of what happened. Our records may have varying degrees of reliability, but they fix the actuality of the past in a way that isn't available when we contemplate the future.

Think of it this way: A loved one says, "I think we should change our vacation plans for next year. Instead of going to Cancún, let's be adventurous and go to Rio." You may or may not go along with the plan, but the strategy should you choose to implement it isn't that hard to work out: You change plane reservations, book a new hotel, and so forth. But if your loved one says, "I think we should change our vacation plans for last year. Instead of having gone to Paris, let's have been adventurous and have gone to Istanbul," your strategy would be very different—you'd think about taking your loved one to the doctor, not rearranging your past travel plans. The past is gone, it's in the books, there's no way we can set about changing it. So it makes perfect sense to us to treat the past and future on completely different footings. Philosophers speak of the distinction between Being—existence in the world—and Becoming—a dynamical process of change, bringing reality into existence.

That distinction between the fixedness of the past and the malleability of the future is nowhere to be found in the known laws of physics. The deep-down microscopic rules of nature run equally well forward or backward in time from any given situation. If you know the exact state of the universe, and all of the laws of physics, the future as well as the past is rigidly determined beyond John Calvin's wildest dreams of predestination.

The way to reconcile these beliefs—the past is once-and-for-all fixed, while the future can be changed, but the fundamental laws of physics are reversible—

ultimately comes down to entropy. If we knew the precise state of every particle in the universe, we could deduce the future as well as the past. But we don't; we know something about the universe's macroscopic characteristics, plus a few details here and there. With that information, we can predict certain broad-scale phenomena (the Sun will rise tomorrow), but our knowledge is compatible with a wide spectrum of specific future occurrences. When it comes to the past, however, we have at our disposal both our knowledge of the current macroscopic state of the universe, *plus* the fact that the early universe began in a low-entropy state. That one extra bit of information, known simply as the "Past Hypothesis," gives us enormous leverage when it comes to reconstructing the past from the present.

The punch line is that our notion of *free will*, the ability to change the future by making choices in a way that is not available to us as far as the past is concerned, is only possible because the past has a low entropy and the future has a high entropy. The future seems open to us, while the past seems closed, even though the laws of physics treat them on an equal footing.

Because we live in a universe with a pronounced arrow of time, we treat the past and future not just as different from a practical perspective, but as deeply and fundamentally different things. The past is in the books, but the future can be influenced by our actions. Of more direct importance for cosmology, we tend to conflate "explaining the history of the universe" with "explaining the state of the early universe"—leaving the state of the late universe to work itself out. Our unequal treatment of past and future is a form of *temporal chauvinism*, which can be hard to eradicate from our mind-set. But that chauvinism, like so many others, has no ultimate justification in the laws of nature. When thinking about important features of the universe, whether deciding what is "real" or why the early universe had a low entropy, it is a mistake to prejudice our explanations by placing the past and future on unequal footings. The explanations we seek should ultimately be timeless.

The major lesson of this overview of entropy and the arrow of time should be clear: The existence of the arrow of time is both a profound feature of the physical universe and a pervasive ingredient of our everyday lives. It's a bit embarrassing, frankly, that with all of the progress made by modern physics and cosmology, we still don't have a final answer for why the universe exhibits such a profound asymmetry in time. I'm embarrassed, at any rate, but every crisis is an opportunity, and by thinking about entropy we might learn something important about the universe.

3

THE BEGINNING AND END OF TIME

What has the universe got to do with it? You're here in Brooklyn! Brooklyn is not expanding!

—Alvy Singer's mom, *Annie Hall*

Imagine that you are wandering around in the textbook section of your local university bookstore. Approaching the physics books, you decide to leaf through some volumes on thermodynamics and statistical mechanics, wondering what they have to say about entropy and the arrow of time. To your surprise (having been indoctrinated by the book you're currently reading, or at least the first two chapters and the jacket copy), there is nothing there about cosmology. Nothing about the Big Bang, nothing about how the ultimate explanation for the arrow of time is to be found in the low-entropy boundary condition at the beginning of our observable universe.

There is no real contradiction, nor is there a nefarious conspiracy on the part of textbook writers to keep the central role of cosmology hidden from students of statistical mechanics. For the most part, people interested in statistical mechanics care about experimental situations in laboratories or kitchens here on Earth. In an experiment, we can control the conditions before us; in particular, we can arrange systems so that the entropy is much lower than it could be, and watch what happens. You don't need to know anything about cosmology and the wider universe to understand how that works.

But our aims are more grandiose. The arrow of time is much more than a feature of some particular laboratory experiments; it's a feature of the entire world around us. Conventional statistical mechanics can account for why it's easy to turn an egg into an omelet but hard to turn an omelet into an egg. What it can't account for is why, when we open our refrigerator, we are able to find an egg in the first place. Why are we surrounded by exquisitely ordered objects such as eggs and pianos and science books, rather than by featureless chaos?

Part of the answer is straightforward: The objects that populate our everyday experience are not closed systems. Of course an egg is not a randomly chosen configuration of atoms; it's a carefully constructed system, the assembly of which required a certain set of resources and available energy, not to mention a chicken. But we could ask the same question about the Solar System, or about the Milky Way galaxy. In each case, we have systems that are for all practical purposes isolated, but nevertheless have a much lower entropy than they could.

The answer, as we know, is that the Solar System hasn't always been a closed system; it evolved out of a protostellar cloud that had an even lower entropy. And that cloud came from the earlier galaxy, which had an even lower entropy. And the galaxy was formed out of the primordial plasma, which had an even lower entropy. And that plasma originated in the very early universe, which had an even lower entropy still.

And the early universe came out of the Big Bang. The truth is, we don't know much about why the early universe was in the configuration it was; that's one of the questions motivating us in this book. The ultimate explanation for the arrow of time as it manifests itself in our kitchens and laboratories and memories relies crucially on the very low entropy of the early universe.

You won't usually find any discussion of this story in conventional textbooks on statistical mechanics. They assume that we are interested in systems that start with relatively low entropy, and take it from there. But we want more—why did our universe have such a small entropy at one end of time, thereby setting the stage for the subsequent arrow of time? It makes sense to start by considering what we do know about how the universe has evolved from its beginning up to today.

THE VISIBLE UNIVERSE

Our universe is expanding, filled with galaxies gradually moving apart from one another. We experience only a small part of the universe directly, and in trying to comprehend the bigger picture it's tempting to reach for analogies. The universe, we are told, is like the surface of a balloon, on which small dots have been drawn to represent individual galaxies. Or the universe is like a loaf of raisin bread rising in the oven, with each galaxy represented by one of the raisins.

These analogies are terrible. And not only because it seems demeaning to have something as majestic as a galaxy be represented by a tiny, wrinkled raisin. The real problem is that any such analogy brings along with it associations that do not apply to the actual universe. A balloon, for example, has an inside and an outside, as well as a larger space into which it is expanding; the universe has none of those things. Raisin bread has an edge, and is situated inside an oven, and smells yummy; there are no corresponding concepts in the case of the universe.

So let's take another tack. To understand the universe around us, let's consider the real thing. Imagine standing outside on a clear, cloudless night, far away from the lights of the city. What do we see when we look into the sky? For the purposes of this thought experiment, we can grant ourselves perfect vision, infinitely sensitive to all the different forms of electromagnetic radiation.

We see stars, of course. To the unaided eye they appear as points of light, but we have long since figured out that each star is a massive ball of plasma, glowing through the energy of internal nuclear reactions, and that our Sun is a star in its own right. One problem is that we don't have a sense of depth—it's hard to tell how far away any of those stars are. But astronomers have invented clever ways to determine the distances to nearby stars, and the answers are impressively large. The closest star, Proxima Centauri, is about 40 trillion kilometers away; traveling at the speed of light, it would take about four years to get there.

Stars are not distributed uniformly in every direction. On our hypothetical clear night, we could not help but notice the Milky Way—a fuzzy band of white stretching across the sky, from one horizon to the other. What we're seeing is actually a collection of many closely packed stars; the ancient Greeks suspected as much, and Galileo verified that idea when he turned his telescope on the heavens. In fact, the Milky Way is a giant spiral galaxy—a collection of hundreds of billions of stars, arranged in the shape of a disk with a bulge in the center, with our Solar System located as one of the distant suburbs on one edge of the disk.

For a long time, astronomers thought that "the galaxy" and "the universe" were the same thing. One could easily imagine that the Milky Way constituted an isolated collection of stars in an otherwise empty void. But it was well known that, in addition to pointlike stars, the night sky featured fuzzy blobs known as "nebulae," which some argued were giant collections of stars in their own right. After fierce debates between astronomers in the early years of the twentieth century,[34] Edwin Hubble was eventually able to measure the distance to the nebula M33 (the thirty-third object in Charles Messier's catalog of fuzzy celestial objects not to be confused by when one was searching for comets), and found that it is much farther away than any star. M33, the Triangulum Galaxy, is in fact a collection of stars comparable in size to the Milky Way.

Upon further inspection, the universe turns out to be teeming with galaxies. Just as there are hundreds of billions of stars in the Milky Way, there are hundreds of billions of galaxies in the observable universe. Some galaxies (including ours) are members of groups or clusters, which in turn describe sheets and filaments of large-scale structure. On average, however, galaxies are uniformly distributed through space. In every direction we look, and at every different distance from us, the number of galaxies is roughly equal. The observable universe looks pretty much the same everywhere.

BIG AND GETTING BIGGER

Hubble was undoubtedly one of the greatest astronomers of history, but he was also in the right place at the right time. He bounced around a bit after graduating from college, spending time variously as a Rhodes scholar, high school teacher, lawyer, soldier in World War I, and for a while as a basketball coach. But ultimately he earned a Ph.D. in astronomy from the University of Chicago in 1917 and moved to California to take up a position at the Mount Wilson Observatory outside Los Angeles. He arrived to find the brand-new Hooker telescope, featuring a mirror 100 inches across, at the time the world's largest. It was at the 100-inch that Hubble made the observations of variable stars in other galaxies, establishing for the first time their great distance from the Milky Way.

Meanwhile other astronomers, led by Vesto Slipher, had been measuring the velocity of spiral nebulae using the Doppler effect.[35] If an object is moving with respect to you, any wave it emits (such as light or sound) will get compressed if it's moving toward you, and stretched if it's moving away. In the case of sound, we experience the Doppler effect as a raising of the pitch of objects that are coming toward us, and a lowering of the pitch as they move away. Similarly, we see the light from objects moving toward us shifted toward the blue (shorter wavelengths) than we would expect, and light from objects moving away is shifted toward the red (longer wavelengths). So an approaching object is blueshifted, while a receding object is redshifted.

Figure 7: Edwin Hubble, surveyor of the universe, smoking a pipe.

What Slipher found was that the vast majority of nebulae were redshifted. If these objects were moving randomly through the universe, we would expect about as many blueshifts as redshifts, so this pattern came as a surprise. If the nebulae were small clouds of gas and dust, we might have concluded that they had been forcibly ejected from our galaxy by some unknown mechanism. But Hubble's result, announced in

1925, scotched that possibility—what we were seeing was a collection of galaxies the size of our own, all running away from us as if they were afraid or something.

Hubble's next discovery made it all snap into place. In 1929 he and his collaborator Milton Humason compared the redshifts of galaxies to the distances he had measured, and found a striking correlation: The farther the galaxies were, the faster they were receding. This is now known as *Hubble's Law*: The apparent recession velocity of a galaxy is proportional to its distance from us, and the constant of proportionality is known as the *Hubble constant*.[36]

Hidden within this simple fact—the farther away things are, the faster they are receding—lies a deep consequence: We are not at the center of some giant cosmic migration. You might get the impression that we are somehow special, what with all of these galaxies moving way from us. But put yourself in the place of an alien astronomer within one of those other galaxies. If that astronomer looks back at us, of course they would see the Milky Way receding from them. But if they look in the opposite direction in the sky, they will also see galaxies moving away from them—because, from our perspective, those more distant galaxies are moving even faster. This is a very profound feature of the universe in which we live. There isn't any particular special place, or central point away from which everything is moving. All of the galaxies are moving away from all of the other galaxies, and each of them sees the same kind of behavior. It's almost as if the galaxies aren't moving at all, but rather that the galaxies are staying put and space itself is expanding in between them.

Which is, indeed, precisely what's going on, from the modern way of looking at things. These days we think of space not as some fixed and absolute stage through which matter moves, but as a dynamical and lively entity in its own right, according to Einstein's general theory of relativity. When we say space is expanding, we mean that more space is coming into existence in between galaxies. Galaxies themselves are not expanding, nor are you, nor are individual atoms; anything that is held together by some local forces will maintain its size, even in an expanding universe. (Maybe you are expanding, but you can't blame the universe.) A light wave, which is not bound together by any forces, will be stretched, leading to the cosmological redshift. And, of course, galaxies that are sufficiently far apart not to be bound by their mutual gravitational attraction will be moving away from one another.

This is a magnificent and provocative picture of the universe. Subsequent observations have confirmed the idea that, on the very largest scales, the universe is homogeneous: It's more or less the same everywhere. Clearly the universe is "lumpy" on smaller scales (here's a galaxy, there's a void of empty space next to it), but if you consider a sufficiently large volume of space, the number of galaxies and the amount of matter within it will be essentially the same, no matter which volume you pick. And the whole shebang is gradually getting bigger; in about 14 billion years, every distant galaxy we observe will be twice as far away as it is today.

We find ourselves in the midst of an overall smooth distribution of galaxies, the space between them expanding so that every galaxy is moving away from every other.[37] If the universe is expanding, what's it expanding into? Nothing. When we're talking about the universe, there's no need to invoke something for it to expand into—it's the universe—it doesn't have to be embedded in anything else; it might very well be all there is. We're not used to thinking like this, because the objects we experience in our everyday lives are all situated *within* space; but the universe *is* space, and there's no reason for there to be any such thing as "outside."

Likewise, there doesn't have to be an edge—the universe could just continue on infinitely far in space. Or, for that matter, it could be finite, by wrapping back on itself, like the surface of a sphere. There is a good reason to believe we will never know, on the basis of actual observations. Light has a finite speed (1 light-year per year, or 300,000 kilometers per second), and there is only a finite time since the Big Bang. As we look out into space, we are also looking backward in time. Since the Big Bang occurred approximately 14 billion years ago, there is an absolute limit to how far we can peer in the universe.[38] What we see is a relatively homogeneous collection of galaxies, about 100 billion of them all told, steadily expanding away from one another. But outside our observable patch, things could be very different.

THE BIG BANG

I've been casually throwing around the phrase *the Big Bang*. It's a bit of physics lingo that has long since entered the popular lexicon. But of all the confusing aspects of modern cosmology, probably none has been the subject of more misleading or simply untrue statements—including by professional cosmologists who really should know better—than "the Big Bang." Let's take a moment to separate what we know from what we don't.

The universe is smooth on large scales, and it's expanding; the space between galaxies is growing. Assuming that the number of atoms in the universe stays the same,[39] matter becomes increasingly dilute as time goes by. Meanwhile, photons get redshifted to longer wavelengths and lower energies, which means that the temperature of the universe decreases. The future of our universe is dilute, cold, and lonely.

Now let's run the movie backward. If the universe is expanding and cooling now, it was denser and hotter in the past. Generally speaking (apart from some niceties concerning dark energy, more about which later), the force of gravity acts to pull things together. So if we extrapolate the universe backward in time to a state that was denser than it is today, we would expect such an extrapolation to continue to be good; in other words, there's no reason to expect any sort of "bounce." The universe should simply have been more and more dense further and further in the past. We might imagine that there would be some moment, only a finite amount of time ago, when the universe was *infinitely* dense—a "singularity." It's that hypothetical singularity that we call "the Big Bang."[40]

Note that we are referring to the Big Bang as a *moment* in the history of the universe, not as a *place* in space. Just as there is no special point in the current universe that defines a center of the expansion, there is no special point corresponding to "where the Bang happened." General relativity says that the universe can be squeezed into zero size at the moment of the singularity, but be infinitely big at every moment after the singularity.

So what happened before the Big Bang? Here is where many discussions of modern cosmology run off the rails. You will often read something like the following: "Before the Big Bang, time and space did not exist. The universe did not come into being at some moment in time, because time itself came into being. Asking what happened before the Big Bang is like asking what lies north of the North Pole."

That all sounds very profound, and it might even be right. But it might not. The truth is, we just don't know. The rules of general relativity are unambiguous: Given certain kinds of stuff in the universe, there must have been a singularity in the past. But that's not really an internally consistent conclusion. The singularity itself would be a moment when the curvature of spacetime and the density of matter were infinite, and the rules of general relativity simply would not apply. The correct deduction is not that general relativity predicts a singularity, but that general relativity predicts that the universe evolves into a configuration where general relativity itself breaks down. The theory cannot be considered to be complete; something happens where general relativity predicts singularities, but we don't know what.

Possibly general relativity is not the correct theory of gravity, at least in the context of the extremely early universe. Most physicists suspect that a quantum theory of gravity, reconciling the framework of quantum mechanics with Einstein's ideas about curved spacetime, will ultimately be required to make sense of what happens at the very earliest times. So if someone asks you what really happened at the moment of the purported Big Bang, the only honest answer would be: "I don't know." Once we have a reliable theoretical framework in which we can ask questions about what happens in the extreme conditions characteristic of the early universe, we should be able to figure out the answer, but we don't yet have such a theory.

It might be that the universe didn't exist before the Big Bang, just as conventional general relativity seems to imply. Or it might very well be—as I tend to believe, for reasons that will become clear—that space and time did exist before the Big Bang; what we call the Bang is a kind of transition from one phase to another. Our quest to understand the arrow of time, anchored in the low entropy of the early universe, will ultimately put this issue front and center. I'll continue to use the phrase "the Big Bang" for "that moment in the history of the very early universe just before conventional cosmology becomes relevant," whatever that moment might actually be like in a more complete theory, and whether or not there is some kind of singularity or boundary to the universe.

HOT, SMOOTH BEGINNINGS

While we don't know what happened at the very beginning of the universe, there's a tremendous amount that we do know about what happened after that. The universe started out in an incredibly hot, dense state. Subsequently, space expanded and matter diluted and cooled, passing through a variety of transitions. A suite of observational evidence indicates that it's been about 14 billion years from the Big Bang to the present day. Even if we don't claim to know the details of what happened at the earliest moments, it all happened within a very short period of time; most of the history of the universe has occurred long since its mysterious beginnings, so it's okay to talk about how many years a given event occurred after the Big Bang. This broad-stroke picture is known as the "Big Bang *model*" and is well understood theoretically and supported by mountains of observational data, in contrast with the hypothetical "Big Bang *singularity*," which remains somewhat mysterious.

Our picture of the early universe is not based simply on theoretical extrapolation; we can use our theories to make testable predictions. For example, when the universe was about 1 minute old, it was a nuclear reactor, fusing protons and neutrons into helium and other light elements in a process known as "primordial nucleosynthesis." We can observe the abundance of such elements today and obtain spectacular agreement with the predictions of the Big Bang model.

We also observe cosmic microwave background radiation. The early universe was hot as well as dense, and hot things give off radiation. The theory behind night-vision goggles is that human beings (or other warm things) give off infrared radiation that can be detected by an appropriate sensor. The hotter something is, the more energetic (short wavelength, high frequency) is the radiation it emits. The early universe was extremely hot and gave off a lot of energetic radiation.

What is more, the early universe was opaque. It was sufficiently hot that electrons could not stay bound to atomic nuclei, but flew freely through space; photons frequently bounced off the free electrons, so that (had you been around) you wouldn't have been able to see your hand in front of your face. But eventually the temperature cooled to a point where electrons could get stuck to nuclei and stay there—a process called recombination, about 400,000 years after the Big Bang. Once that happened, the universe was transparent, so light could travel essentially unimpeded from that moment until today. Of course, it still gets redshifted by the cosmological expansion, so the hot radiation from the period of recombination has been stretched into microwaves, about 1 centimeter in wavelength, reaching a current temperature of 2.7 Kelvin (−270.4 degrees Celsius).

The story of the evolution of the universe according to the Big Bang model (as distinguished from the mysterious moment of the Big Bang itself) therefore makes a strong prediction: Our universe should be suffused with microwave radiation from all directions, a relic from an earlier time when the universe was hot and dense. This radiation was finally detected by Arno Penzias and Robert Wilson in 1965 at Bell Labs in Holmdel, New Jersey. And they weren't even looking for it—they were radio astronomers who became somewhat annoyed at this mysterious background radiation they couldn't get rid of. Their annoyance was somewhat mollified when they won the Nobel Prize in 1978.[41] It was the discovery of the microwave background that converted most of the remaining holdouts for the Steady State theory of cosmology (in which the temperature of the universe would be constant through time, and new matter is continually created) over to the Big Bang point of view.

TURNING UP THE CONTRAST KNOB ON THE UNIVERSE

The universe is a simple place. True, it contains complicated things like galaxies and sea otters and federal governments, but if we average out the local idiosyncrasies, on very large scales the universe looks pretty much the same everywhere. Nowhere is this more evident than in the cosmic microwave background. Every direction we look in the sky, we see microwave background radiation that looks exactly like that from an object glowing serenely at some fixed temperature—what physicists call "blackbody" radiation. However, the temperature is ever so slightly different from point to point on the sky; typically, the temperature in one direction differs from that in some other direction by about 1 part in 100,000. These fluctuations are called *anisotropies*—tiny departures from the otherwise perfectly smooth temperature of the background radiation in every direction.

Figure 8: Temperature anisotropies in the cosmic microwave background, as measured by NASA's Wilkinson Microwave Anisotropy Probe. Dark regions are slightly colder than average, light regions are slightly hotter than average. The differences have been dramatically enhanced for clarity.

These variations in temperature reflect slight differences in the density of matter from place to place in the early universe. Saying that the early universe was smooth is not just a simplifying assumption; it's a testable hypothesis that is strongly supported by the data. On very large scales, the universe is still smooth today. But the scales have to be pretty large—over 300 million light-years or so. On smaller scales, like the size of a galaxy or the Solar System or your kitchen, the universe is quite lumpy. It wasn't always like that; at early times, even small scales were very smooth. How did we get here from there?

The answer lies in gravity, which acts to turn up the contrast knob on the universe. In a region that has slightly more matter than average, there is a gravitational force that pulls things together; in regions that are slightly underdense, matter tends to flow outward to the denser regions. By this process—the evolution of structure in the universe—the tiny primordial fluctuations revealed in the microwave background anisotropies grow into the galaxies and structures we see today.

Imagine that we lived in a universe much like our current one, with the same kind of distribution of galaxies and clusters, but that was contracting rather than expanding. Would we expect that the galaxies would smooth out toward the future as the universe contracted, creating a homogeneous plasma such as we see in the past of our real (expanding) universe? Not at all. We would expect the contrast knob to continue to be turned up, even as the universe contracted—black holes and other massive objects would gather matter from the surrounding regions. Growth of structure is an irreversible process that naturally happens toward the future, whether the universe is expanding or contracting: It represents an increase in entropy. So the relative smoothness of the early universe, illustrated in the image of the cosmic microwave background, reflects the very low entropy of those early times.

THE UNIVERSE IS NOT STEADY

The Big Bang model seems like a fairly natural picture, once you believe in an approximately uniform universe that is expanding in time. Just wind the clock backward, and you get a hot, dense beginning. Indeed, the basic framework was put together in the late 1920s by Georges Lemaître, a Catholic priest from Belgium who had studied at Cambridge and Harvard before eventually earning his doctorate from MIT.[42] (Lemaître, who dubbed the beginning of the universe the "Primeval Atom," refrained from drawing any theological conclusions from his cosmological model, despite the obvious temptation.)

But there is a curious asymmetry in the Big Bang model, one that should come as no surprise to us by now: the difference between time and space. The idea that matter is smooth on large scales can be elevated into the "Cosmological Principle": There is no such thing as a special place in the universe. But it seems clear that there is a special *time* in the universe: the moment of the Big Bang.

Some mid-century cosmologists found this stark distinction between smoothness in space and variety in time to be a serious shortcoming of the Big Bang model, so they set about developing an alternative. In 1948, three leading astrophysicists— Hermann Bondi, Thomas Gold, and Fred Hoyle—suggested the Steady State model of the universe.[43] They based this model on the "Perfect Cosmological Principle"— there is no special place and no special time in the universe. In particular, they suggested that the universe wasn't any hotter or denser in the past than it is today.

The pioneers of the Steady State theory (unlike some of their later followers) were not crackpots. They understood that Hubble had discovered the expansion of the universe, and they respected the data. So how can the universe be expanding without diluting and cooling down? The answer they suggested was that matter was continually being created in between the galaxies, precisely balancing the dilution due to the expansion of the universe. (You don't need to make much: about one hydrogen atom per cubic meter every billion years. It's not like your living room will start filling up.) Creation of matter wouldn't happen all by itself; Hoyle invented a new kind of field, called the *C-field*, which he hoped would do the trick, but the idea never really caught on among physicists.

From our jaded modern perspective, the Steady State model seems like a lot of superstructure constructed on the basis of some fairly insubstantial philosophical presuppositions. But many great theories begin that way, before they are confronted with the harsh realities of data; Einstein certainly leaned on his own philosophical preferences during the construction of general relativity. But unlike relativity, when the data ultimately confronted the Steady State model, the result was not pretty.[44] The last thing you would expect from a model in which the temperature of the universe remains constant is a relic background radiation that indicates a hot beginning. After Penzias and Wilson discovered the microwave background, support for the Steady State theory crumbled, although there remains to this day a small cadre of true believers who invent ingenious ways of avoiding the most straightforward interpretation of the data.

Nevertheless, thinking about the Steady State model brings home the perplexing nature of time in the Big Bang model. In the Steady State cosmology, there was still unmistakably an arrow of time: Entropy increased, without limit, in the same direction, forever and ever. In a very legitimate sense, the problem of explaining the low-entropy initial conditions of the universe would be *infinitely bad* in a Steady State universe; whatever those conditions were, they were infinitely far in the past, and the entropy of any finite-sized system today would have been infinitesimally small. One could imagine that considerations of this form might have undermined the Steady State model from the start, if cosmologists had taken the need to explain the low entropy of the early universe seriously.

In the Big Bang picture, things don't seem quite as hopeless. We still don't know why the early universe had a low entropy, but at least we know when the early universe was: It was 14 billion years ago, and its entropy was small but not strictly zero. Unlike in the Steady State model, in the context of the Big Bang you can at least put your finger directly on where (really "when") the problem is located. Whether or not this is really an improvement can't be decided until we understand cosmology within a more comprehensive framework.

BUT IT *IS* ACCELERATING

We know a good deal about the evolution of the universe over the last 14 billion years. What's going to happen in the future?

Right now the universe is expanding, becoming increasingly colder and ever more dilute. For many years the big question in cosmology had been, "Will expansion continue forever, or will the universe eventually reach a maximum size and begin to contract toward a Big Crunch at the end of time?" Debating the relative merits of these alternatives was a favorite parlor game among cosmologists ever since the early days of general relativity. Einstein himself favored a universe that was finite in both space and time, so he liked the idea of an eventual re-collapse. Lemaître, in contrast, preferred the idea that the universe would continue to cool off and expand forever: ice, rather than fire.

Performing measurements that would decide the question empirically turned out to be more difficult. General relativity would seem to make a clear prediction: As the universe expands, the gravitational force between galaxies pulls all of them together, working to slow the expansion down. The question was simply whether there was enough matter in the universe to actually cause a collapse, or whether it would expand ever more gradually but for all eternity. For a long time it was a hard question to answer, as observations seemed to indicate that there was *almost* enough matter to reverse the expansion of the universe—but not quite enough.

The breakthrough occurred in 1998, from a completely different method. Rather than measuring the total amount of mass in the universe, and comparing with theory to determine whether there was enough to eventually reverse the universe's expansion, one could go out and directly measure the rate at which the expansion was slowing down. Easier said than done, of course. Basically what one had to do was what Hubble had done years before—measure both distances and apparent velocities of galaxies, and look at the relationship between them—but to enormously higher precision and at much greater distances. The technique eventually used was to search for Type Ia supernovae, exploding stars that not only have the virtue of being very bright (and therefore visible over cosmological distances), but also have almost the same brightness in every event (so that the apparent brightness can be used to deduce the distance to the supernova).[45]

The hard work was done by two teams: one led by Saul Perlmutter of Lawrence Berkeley National Laboratory, and one led by Brian Schmidt of Mount Stromlo Observatory in Australia. Perlmutter's group, which contained a number of particle physicists converted to the cause of cosmology, started earlier, and had championed the supernova technique in the face of considerable skepticism. Schmidt's group, which included a number of experts on supernova astronomy, started later but managed to catch up. The teams maintained a rivalry that was often friendly and occasionally less so, but they both made crucial contributions, and rightfully share the credit for the ultimate discovery.

As it happens, Brian Schmidt and I were office mates in graduate school at

Harvard in the early 1990s. I was the idealistic theorist, and he was the no-nonsense observer. In those days, when the technology of large-scale surveys in astronomy was just in its infancy, it was a commonplace belief that measuring cosmological parameters was a fool's errand, doomed to be plagued by enormous uncertainties that would prevent us from determining the size and shape of the universe with anything like the precision we desired. Brian and I made a bet concerning whether we would be able to accurately measure the total matter density of the universe within twenty years. I said we would; Brian was sure we wouldn't. We were poor graduate students at the time, but purchased a small bottle of vintage port, to be secreted away for two decades before we knew who had won. Happily for both of us, we learned the answer long before then; I won the bet, due in large part to the efforts of Brian himself. We split the bottle of port on the roof of Harvard's Quincy House in 2005.

And the answer is: The universe isn't decelerating at all; it's actually accelerating! If you were to measure the apparent recession velocity of a galaxy, and (hypothetically) came back a billion years later to measure it again, you would find that the velocity was now higher.[46] How can that be reconciled with the supposed prediction of general relativity that the universe should be slowing down? Like most such predictions of general relativity, there are hidden assumptions: in this case, that the primary source of energy in the universe consists of matter.

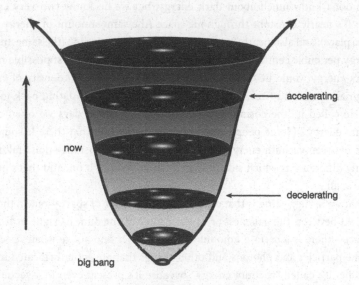

Figure 9: The accelerating universe.

To a cosmologist, *matter* is shorthand for "any collection of particles, each of which is moving much more slowly than the speed of light." (If particles are moving at or close to the speed of light, cosmologists refer to them as "radiation," whether or not they are actually electromagnetic radiation in the usual sense.) Einstein taught us long ago that particles have energy, even when they're not moving at all: $E = mc^2$ means that the energy of a perfectly stationary massive particle is given by its mass times the speed of light squared. For our present purposes, the crucial aspect of matter is that it dilutes away as the universe expands.[47] What general relativity actually predicts is that the expansion should be decelerating, as long as the energy is diluting away. If it's not—if the density of energy, the amount of energy in each cubic centimeter or cubic light-year of space, is approximately constant—then that energy provides a perpetual impulse to the expansion of space, and the universe will actually be accelerating.

It's possible, of course, that general relativity is not the correct theory of gravity on cosmological scales, and that possibility is one that physicists take very seriously. It seems more likely, however, that general relativity is correct, and the observations are telling us that most of the energy in the universe is not in the form of "matter" at all, but rather in the form of some stubbornly persistent stuff that sticks around even as space expands. We've dubbed that mysterious stuff "dark energy," and the nature of the dark energy is very much a favorite research topic for modern cosmologists, both theorists and observers.

We don't know much about dark energy, but we do know two very crucial things: It's nearly constant throughout space (the same amount of energy from place to place), and also nearly constant in density through time (the same amount of energy per cubic centimeter at different times). So the simplest possible model for dark energy would be one featuring an *absolutely constant* density of energy through all space and time. And in fact, that's an old idea, dating back to Einstein: He called it "the cosmological constant," and these days we often call it "vacuum energy." (Some people may try to convince you that there is some difference between vacuum energy and the cosmological constant—don't fall for it. The only difference is which side of the equation you put it on, and that's no difference at all.)

What we're suggesting is that every cubic centimeter of space—out in the desolate cold between the galaxies, or at the center of the Sun, or right in front of your face—there is a certain amount of energy, over and above whatever comes from the particles and photons and other things that are actually located in that little cube. It's called "vacuum energy" because it's present even in a vacuum, in a perfectly empty space—a minimum amount of energy inherent in the fabric of spacetime itself.[48] You can't feel it, you can't see it, you can't do anything with it, but it is there. And we know it is there because it exerts a crucial influence on the universe, imparting a gentle push that causes distant galaxies to accelerate away from us.

Unlike the gravity caused by ordinary matter, the effect of vacuum energy is to push things apart rather than pull them together. When Einstein first proposed the cosmological constant in 1917, his motivation was to explain a static universe, one that wasn't expanding or contracting. This wasn't a misguided philosophical stance—it was the best understanding according to the astronomy of the day; Hubble wouldn't discover the expansion of the universe until 1929. So Einstein imagined a universe in delicate balance between the pull of gravity among galaxies and the push of the cosmological constant. Once he learned of Hubble's discovery, he regretted ever introducing the cosmological constant—had he resisted the temptation, he might have predicted the expansion of the universe before it was discovered.

THE MYSTERY OF VACUUM ENERGY

In theoretical physics, it's not easy to un-invent a concept. The cosmological constant is the same as the idea of vacuum energy, the energy of empty space itself. The question is not "Is vacuum energy a valid concept?"—it's "What value should we expect the vacuum energy to have?"

Modern quantum mechanics implies that the vacuum is not a boring place; it's alive with *virtual particles*. A crucial consequence of quantum mechanics is Werner Heisenberg's uncertainty principle: It's impossible to pin down the observable features of any system into one unique state with perfect precision, and that includes the state of empty space. So if we were to look closely enough at empty space, we would see particles flashing into and out of existence, representing quantum fluctuations of the vacuum itself. These virtual particles are not especially mysterious or hypothetical—they are definitely there, and they have measurable effects in particle physics that have been observed many times over.

Virtual particles carry energy, and that energy contributes to the cosmological constant. We can add up the effects of all such particles to obtain an estimate for how large the cosmological constant should be. But it wouldn't be right to include the effects of particles with arbitrarily high energies. We don't believe that our conventional understanding of particle physics is adequate for very high-energy events—at some point, we have to take account of the effects of quantum gravity, the marriage of general relativity with quantum mechanics, which remains an incomplete theory at the moment.

So instead of appealing to the correct theory of quantum gravity, which we still don't have, we can simply examine the contributions to the vacuum energy of virtual particles at energies below where quantum gravity becomes important. That's the *Planck energy*, named after German physicist Max Planck, one of the pioneers of quantum theory, and it turns out to be about 2 billion joules (a conventional unit of energy).[49] We can add up the contributions to the vacuum energy from virtual particles with energies ranging from zero up to the Planck energy, and then cross our fingers and compare with what we actually observe.

The result is a complete fiasco. Our simple estimate of what the vacuum energy should be comes out to about 10^{105} joules per cubic centimeter. That's a lot of vacuum energy. What we actually observe is about 10^{-15} joules per cubic centimeter. So our estimate is larger than the experimental value by a factor of 10^{120}—a 1 followed by 120 zeroes. Not something we can attribute to experimental error. This has been called the biggest disagreement between theoretical expectation and experimental reality in all of science. For comparison, the total number of particles in the observable universe is only about 10^{88}; the number of grains of sand on all the Earth's beaches is only about 10^{20}.

The fact that the vacuum energy is so much smaller than it should be is a serious problem: the "cosmological constant problem." But there is also another problem: the "coincidence problem." Remember that vacuum energy maintains a constant density (amount of energy per cubic centimeter) as the universe expands, while the density of matter dilutes away. Today, they aren't all that different: Matter makes up about 25 percent of the energy of the universe, while vacuum energy makes up the other 75 percent. But they are changing appreciably with respect to each other, as the matter density dilutes away with the expansion and the vacuum energy does not. At the time of recombination, for example, the energy density in matter was a billion times larger than that in vacuum energy. So the fact that they are somewhat comparable today, uniquely in the history of the universe, seems like a remarkable coincidence indeed. Nobody knows why.

These are serious problems with our theoretical understanding of vacuum energy. But if we put aside our worries concerning why the vacuum energy is so small, and why it's comparable in density to the energy in matter, we are left with a phenomenological model that does a remarkable job of fitting the data. (Just like Carnot and Clausius didn't need to know about atoms to say useful things about entropy, we don't need to understand the origin of the vacuum energy to understand what it does to the expansion of the universe.) The first direct evidence for dark energy came from observations of supernovae in 1998, but since then a wide variety of methods have independently confirmed the basic picture. Either the universe is accelerating under the gentle influence of vacuum energy, or something even more dramatic and mysterious is going on.

THE DEEPEST FUTURE

As far as we can tell, the density of vacuum energy is unchanging as the universe expands. (It could be changing very slowly, and we just haven't been able to measure the changes yet—that's a major goal of modern observational cosmology.) We don't know enough about vacuum energy to say for sure what will happen to it indefinitely into the future, but the obvious first guess is that it will simply stay at its current value forever.

If that's true, and the vacuum energy is here to stay, it's straightforward to predict the very far future of our universe. The details get complicated in an interesting way, but the outline is relatively simple.[50] The universe will continue to expand, cool off, and become increasingly dilute. Distant galaxies will accelerate away from us, becoming more and more redshifted as they go. Eventually they will fade from view, as the time between photons that could possibly reach us becomes longer and longer. The entirety of the observable universe will just be our local group of gravitationally bound galaxies.

Galaxies don't last forever. The stars in them burn their nuclear fuel and die. Out of the remnant gas and dust more stars can form, but a point of diminishing returns is reached, after which all of the stars in the galaxy are dead. We are left with white dwarfs (stars that once burned, and ran out of fuel), brown dwarfs (stars that never burned in the first place), and neutron stars (stars that used to be white dwarfs but collapsed further under the pull of gravity). These objects may or may not be stable in their own right; our best current theoretical guess is that the protons and neutrons that make them up aren't perfectly stable themselves but will eventually decay into lighter particles. If that's true (and admittedly, we're not sure), the various forms of dead stars will eventually dissipate into a thin gas of particles that disperse into the void. It won't be quick; a reasonable estimate is 10^{40} years from now. For comparison, the current universe is about 10^{10} years old.

Besides stars, there are also black holes. Most large galaxies, including our own, have giant black holes at the center. In a galaxy the size of the Milky Way, with about 100 billion stars, the black hole might be a few million times as massive as the Sun—big compared to any individual star, but still small compared to the galaxy as a whole. But it will continue to grow, sweeping up whatever unfortunate stars happen to fall into it. Ultimately, however, all of the stars will have been used up. At that point, the black hole itself begins to evaporate into elementary particles.

That's the remarkable discovery of Stephen Hawking from 1976, which we'll discuss in detail in Chapter Twelve: "black holes ain't so black." Due once again to quantum fluctuations, a black hole can't help but gradually radiate out into the space around it, slowly losing energy in the process. If we wait long enough—and now we're talking 10^{100} years or so—even the supermassive black holes at the centers of galaxies will evaporate away to nothing.

Regardless of how the details play out, we are left with the same long-term picture. Other galaxies move away from us and disappear; our own galaxy will evolve through various stages, but the end result is a thin gruel of particles dissipating into the void. In the very far future, the universe becomes once again a very simple place: It will be completely empty, as empty as space can be. That's the diametric opposite of the hot, dense state in which the universe began; a vivid cosmological manifestation of the arrow of time.

THE ENTROPY OF THE UNIVERSE

An impressive number of brain-hours on the part of theoretical physicists have been devoted to the question of why the universe evolved in this particular fashion, rather than in some other way. It's certainly possible that this question simply has no answer; perhaps the universe is what it is, and the best we can do is to accept it. But we are hopeful, not without reason, that we can do more than accept it—we can explain it.

Given perfect knowledge of the laws of physics, the question "Why has the universe evolved in the fashion it has?" is equivalent to "Why were the initial conditions of the universe arranged in the way they were?" But that latter formulation is already sneaking in an implicit notion of time asymmetry, by privileging past conditions over future conditions. If our understanding of the fundamental, microscopic laws of nature is correct, we can specify the state of the universe at *any* time, and from there derive both the past and the future. It would be better to characterize our task as that of understanding what would count as a natural history of the universe as a whole.[51]

There is some irony in the fact that cosmologists have underappreciated the importance of the arrow of time, since it is arguably the single most blatant fact about the evolution of the universe. Boltzmann was able to argue (correctly) for the need for a low-entropy boundary condition in the past, without knowing anything about general relativity, quantum mechanics, or even the existence of other galaxies. Taking the problem of entropy seriously helps us look at cosmology in a new light, which might suggest some resolutions to long-standing puzzles.

But first, we need to be a little more clear about what exactly we mean about "the entropy of the universe." In Chapter Thirteen we will discuss the evolution of entropy in our observable universe in great detail, but the basic story goes as follows:

1. In the early universe, before structure forms, gravity has little effect on the entropy. The universe is similar to a box full of gas, and we can use the conventional formulas of thermodynamics to calculate its entropy. The total entropy within the space corresponding to our observable universe turns out to be about 10^{88} at early times.

2. By the time we reach our current stage of evolution, gravity has become very important. In this regime we don't have an ironclad formula, but we can make a good estimate of the total entropy just by adding up the contributions from black holes (which carry an enormous amount of entropy). A single supermassive black hole has an entropy of order 10^{90}, and there are approximately 10^{11} such black holes in the observable universe; our total entropy today is therefore something like 10^{101}.

3. But there is a long way to go. If we took all of the matter in the observable universe and collected it into a single black hole, it would have an entropy of 10^{120}. That can be thought of as the maximum possible entropy obtainable by rearranging the matter in the universe, and that's the direction in which we're evolving.[52]

Our challenge is to explain this history. In particular, why was the early entropy, 10^{88}, so much lower than the maximum possible entropy, 10^{120}? Note that the former number is much, much, much smaller than the latter; appearances to the contrary are due to the miracle of compact notation.

The good news is, at least the Big Bang model provides a context in which we can sensibly address this question. In Boltzmann's time, before we knew about general relativity or the expansion of the universe, the puzzle of entropy was even harder, simply because there was no such event as "the beginning of the universe" (or even "the beginning of the observable universe"). In contrast, we are able to pinpoint exactly when the entropy was small, and the particular form that low-entropy state took; that's a crucial step in trying to explain why it was like that.

It's possible, of course, that the fundamental laws of physics simply aren't reversible (although we'll give arguments against that later on). But if they are, the low entropy of our universe near the Big Bang leaves us with two basic possibilities:

1. The Big Bang was truly the beginning of the universe, the moment when time began. That may be because the true laws of physics allow spacetime to have a boundary, or because what we call "time" is just an approximation, and that approximation ceases to be valid near the Big Bang. In either case, the universe began in a low-entropy state, for reasons over and above the dynamical laws of nature—we need a new, independent principle to explain the initial state.

2. There is no such thing as an initial state, because time is eternal. In this case, we are imagining that the Big Bang isn't the beginning of the entire universe, although it's obviously an important event in the history of our local region. Somehow our observable patch of spacetime must fit into a bigger picture. And the way it fits must explain why the entropy was small at one end of time, without imposing any special conditions on the larger framework.

As to which of these is the correct description of the real world, the only answer is that we don't know. I will confess to a personal preference for Option 2, as I think it would be more elegant if the world were described as a nearly inevitable result of a set of dynamical laws, without needing an extra principle to explain why it appears precisely this way. Turning this vague scenario into an honest cosmological model will require that we actually take advantage of the mysterious vacuum energy that dominates our universe. Getting there from here requires a deeper understanding of curved spacetime and relativity, to which we now turn.

PART TWO

TIME IN EINSTEIN'S UNIVERSE

4

TIME IS PERSONAL

Time travels in divers paces with divers persons.

—William Shakespeare, *As You Like It*

When most people hear "scientist," they think "Einstein." Albert Einstein is an iconic figure; not many theoretical physicists attain a level of celebrity in which their likeness appears regularly on T-shirts. But it's an intimidating, distant celebrity. Unlike, say, Tiger Woods, the precise achievements Einstein is actually famous *for* remain somewhat mysterious to many people who would easily recognize his name.[53] His image as the rumpled, absentminded professor, with unruly hair and baggy sweaters, contributes to the impression of someone who embodied the life of the mind, disdainful of the mundane realities around him. And to the extent that the substance of his contributions is understood—equivalence of mass and energy, warping of space and time, a search for the ultimate theory— it seems to be the pinnacle of abstraction, far removed from everyday concerns.

The real Einstein is more interesting than the icon. For one thing, the rumpled look with the Don King hair attained in his later years bore little resemblance to the sharply dressed, well-groomed young man with the penetrating stare who was responsible for overturning physics more than once in the early decades of the twentieth century.[54] For another, the origins of the theory of relativity go beyond armchair

Figure 10: Albert Einstein in 1912. His "miraculous year" was 1905, while his work on general relativity came to fruition in 1915.

speculations about the nature of space and time; they can be traced to resolutely practical concerns of getting persons and cargo to the right place at the right time.

Special relativity, which explains how the speed of light can have the same value for all observers, was put together by a number of researchers over the early years of the twentieth century. (Its successor, general relativity, which interpreted gravity as an effect of the curvature of spacetime, was due almost exclusively to Einstein.) One of the major contributors to special relativity was the French mathematician and physicist Henri Poincaré. While Einstein was the one who took the final bold leap into asserting that the "time" as measured by any moving observer was as good as the "time" measured by any other, both he and Poincaré developed very similar formalisms in their research on relativity.[55]

Historian Peter Galison, in his book *Einstein's Clocks, Poincaré's Maps: Empires of Time*, makes the case that Einstein and Poincaré were as influenced by their earthbound day jobs as they were by esoteric considerations of the architecture of physics.[56] Einstein was working at the time as a patent clerk in Bern, Switzerland, where a major concern was the construction of accurate clocks. Railroads had begun to connect cities across Europe, and the problem of synchronizing time across great distances was of pressing commercial interest. The more senior Poincaré, meanwhile, was serving as president of France's Bureau of Longitude. The growth of sea traffic and trade led to a demand for more accurate methods of determining longitude while at sea, both for the navigation of individual ships and for the construction of more accurate maps.

And there you have it: maps and clocks. Space and time. In particular, an appreciation that what matters is not questions of the form "Where are you really?" or "What time is it actually?" but "Where are you with respect to other things?" and "What time does your clock measure?" The rigid, absolute space and time of Newtonian mechanics accords pretty well with our intuitive understanding of the world; relativity, in contrast, requires a certain leap into abstraction. Physicists at the turn of the century were able to replace the former with the latter only by understanding that we should not impose structures on the world because they suit our intuition, but that we should take seriously what can be measured by real devices.

Special relativity and general relativity form the basic framework for our modern understanding of space and time, and in this part of the book we're going to see what the implications of "spacetime" are for the concept of "time."[57] We'll be putting aside, to a large extent, worries about entropy and the Second Law and the arrow of time, and taking refuge in the clean, precise world of fundamentally reversible laws of physics. But the ramifications of relativity and spacetime will turn out to be crucial to our program of providing an explanation for the arrow of time.

LOST IN SPACE

Zen Buddhism teaches the concept of "beginner's mind": a state in which one is free of all preconceptions, ready to apprehend the world on its own terms. One could debate how realistic the ambition of attaining such a state might be, but the concept is certainly appropriate when it comes to thinking about relativity. So let's put aside what we think we know about how time works in the universe, and turn to some thought experiments (for which we know the answers from real experiments) to figure out what relativity has to say about time.

To that end, imagine we are isolated in a sealed spaceship, floating freely in space, far away from the influence of any stars or planets. We have all of the food and air and basic necessities we might wish, and some high school–level science equipment in the form of pulleys and scales and so forth. What we're not able to do is to look outside at things far away. As we go, we'll consider what we can learn from various sensors aboard or outside the ship.

But first, let's see what we can learn just inside the spaceship. We have access to the ship's controls; we can rotate the vessel around any axis we choose, and we can

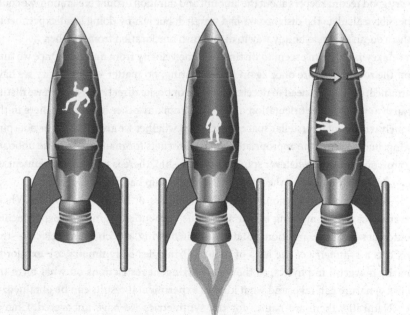

Figure 11: An isolated spaceship. From left to right: freely falling, accelerating, and spinning.

fire our engines to move in whatever direction we like. So we idle away the hours by alternating between moving the ship around in various ways, not really knowing or caring where we are going, and playing a bit with our experiments.

What do we learn? Most obviously, we can tell when we're accelerating the ship. When we're not accelerating, a fork from our dinner table would float freely in front of us, weightless; when we fire the rockets, it falls down, where "down" is defined as "away from the direction in which the ship is accelerating."[58] If we play a bit more, we might figure out that we can also tell when the ship is spinning around some axis. In that case, a piece of cutlery perfectly positioned on the rotational axis could remain there, freely floating; but anything at the periphery would be "pulled" to the hull of the ship and stay there.

So there are some things about the state of our ship we can determine observationally, just by doing simple experiments inside. But there are also things that we *can't* determine. For example, we don't know where we are. Say we do a bunch of experiments at one location in our unaccelerated, non-spinning ship. Then we fire the rockets for a bit, zip off somewhere else, kill the rockets so that we are once again unaccelerated and non-spinning, and do the same experiments again. If we have any skill at all as experimental physicists, we're going to get the same results. Had we been very good record keepers about the amount and duration of our acceleration, we could possibly calculate the distance we had traveled; but just by doing local experiments, there doesn't seem to be any way to distinguish one location from another.

Likewise, we can't seem to distinguish one velocity from another. Once we turn off the rockets, we are once again freely floating, no matter what velocity we have attained; there is no need to decelerate in the opposite direction. Nor can we distinguish any particular orientation of the ship from any other orientation, here in the lonely reaches of interstellar space. We can tell whether we are spinning or not spinning; but if we fire the appropriate guidance rockets (or manipulate some onboard gyroscopes) to stop whatever spin we gave the ship, there is no local experiment we can do that would reveal the angle by which the ship had rotated.

These simple conclusions reflect deep features of how reality works. Whenever we can do something to our apparatus without changing any experimental outcomes—shift its position, rotate it, set it moving at a constant velocity—this reflects a *symmetry* of the laws of nature. Principles of symmetry are extraordinarily powerful in physics, as they place stringent restrictions on what form the laws of nature can take, and what kind of experimental results can be obtained.

Naturally, there are names for the symmetries we have uncovered. Changing one's location in space is known as a "translation"; changing one's orientation in space is known as a "rotation"; and changing one's velocity through space is known as a "boost." In the context of special relativity, the collection of rotations and boosts are known as "Lorentz transformations," while the entire set including translations are known as "Poincaré transformations."

The basic idea behind these symmetries far predates special relativity. Galileo himself was the first to argue that the laws of nature should be invariant under what we now call translations, rotations, and boosts. Even without relativity, if Galileo and Newton had turned out to be right about mechanics, we would not be able to determine our position, orientation, or velocity if we were floating freely in an isolated spaceship. The difference between relativity and the Galilean perspective resides in what actually happens when we switch to the reference frame of a moving observer. The miracle of relativity, in fact, is that changes in velocity are seen to be close relatives of changes in spatial orientation; a boost is simply the spacetime version of a rotation.

Before getting there, let's pause to ask whether things could have been different. For example, we claimed that one's absolute position is unobservable, and one's absolute velocity is unobservable, but one's absolute acceleration can be measured.[59] Can we imagine a world, a set of laws of physics, in which absolute position is unobservable, but absolute velocity can be objectively measured?[60]

Sure we can. Just imagine moving through a stationary medium, such as air or water. If we lived in an infinitely big pool of water, our position would be irrelevant, but it would be straightforward to measure our velocity with respect to the water. And it wouldn't be crazy to think that there is such a medium pervading space.[61] After all, ever since the work of Maxwell on electromagnetism we have known that light is just a kind of wave. And if you have a wave, it's natural to think that there must be something doing the waving. For example, sound needs air to propagate; in space, no one can hear you scream. But light can travel through empty space, so (according to this logic, which will turn out not to be right) there must be some medium through which it is traveling.

So physicists in the late nineteenth century postulated that electromagnetic waves propagated through an invisible but all-important medium, which they called the "aether." And experimentalists set out to actually detect the stuff. But they didn't succeed—and that failure set the stage for special relativity.

THE KEY TO RELATIVITY

Imagine we're back out in space, but this time we've brought along some more sophisticated experimental apparatus. In particular, we have an impressive-looking contraption, complete with state-of-the-art laser technology, that measures the speed of light. While we are freely falling (no acceleration), to calibrate the thing we check that we get the same answer for the speed of light no matter how we orient our experiment. And indeed we do. Rotational invariance is a property of the propagation of light, just as we suspected.

But now we try to measure the speed of light while moving at different velocities. That is, first we do the experiment, and then we fire our rockets a bit and turn them off so that we've established some constant velocity with respect to our initial motion, and then we do the experiment again. Interestingly, no matter how much velocity we picked up, the speed of light that we measure is always the same. If there really were an aether medium through which light traveled just as sound travels through air, we should get different answers depending on our speed relative to the aether. But we don't. You might guess that the light had been given some sort of push by dint of the fact that it was created within your moving ship. To check that, we'll allow you to remove the curtains from the windows and let some light come in from the outside world. When you measure the velocity of the light that was emitted by some outside source, once again you find that it doesn't depend on the velocity of your own spaceship.

A real-world version of this experiment was performed in 1887 by Albert Michelson and Edward Morley. They didn't have a spaceship with a powerful rocket, so they used the next best thing: the motion of the Earth around the Sun. The Earth's orbital velocity is about 30 kilometers per second, so in the winter it has a net velocity of about 60 kilometers per second different from its velocity in the summer, when it's moving in the other direction. That's not much compared to the speed of light, which is about 300,000 kilometers per second, but Michelson designed an ingenious device known as an "interferometer" that was extremely sensitive to small changes in velocities along different directions. And the answer was: The speed of light seems to be the same, no matter how fast we are moving.

Advances in science are rarely straightforward, and the correct way to interpret the Michelson-Morley results was not obvious. Perhaps the aether is dragged along with the Earth, so that our relative velocity remains small. After some furious back-and-forth theorizing, physicists hit upon what we now regard to be the right answer: The speed of light is simply a universal invariant. Everyone measures light to be moving at the same speed, independent of the motion of the experimenter.[62] Indeed, the entire content of special relativity boils down to these two principles:

- No local experiment can distinguish between observers moving at constant velocities.
- The speed of light is the same to all observers.

When we use the phrase *the speed of light*, we are implicitly assuming that it's the speed of light through empty space that we're talking about. It's perfectly easy to make light move at some other speed, just by introducing a transparent medium— light moves more slowly through glass or water than it does through empty space, but that doesn't tell us anything profound about the laws of physics. Indeed, "light" is not all that important in this game. What's important is that there exists some unique preferred velocity through spacetime. It just so happens that light moves at that speed when it's traveling through empty space—but the existence of a speed limit is what matters, not that light is able to go that fast.

We should appreciate how astonishing all this is. Say you're in your spaceship, and a friend in a faraway spaceship is beaming a flashlight at you. You measure the velocity of the light from the flashlight, and the answer is 300,000 kilometers per second. Now you fire your rockets and accelerate toward your friend, until your relative velocity is 200,000 kilometers per second. You again measure the speed of the light coming from the flashlight, and the answer is: 300,000 kilometers per second. That seems crazy; anyone in their right mind should have expected it to be 500,000 kilometers per second. What's going on?

The answer, according to special relativity, is that it's not the speed of light that depends on your reference frame—it's your notion of a "kilometer" and a "second." If a meterstick passes by us at high velocity, it undergoes "length contraction"— it appears shorter than the meterstick that is sitting at rest in our reference frame. Likewise, if a clock moves by us at high velocity, it undergoes "time dilation"—it appears to be ticking more slowly than the clock that is sitting at rest. Together, these phenomena precisely compensate for any relative motion, so that everyone measures exactly the same speed of light.[63]

The invariance of the speed of light carries with it an important corollary: Nothing can move faster than light. The proof is simple enough; imagine being in a rocket that tries to race against the light being emitted by a flashlight. At first the rocket is stationary (say, in our reference frame), and the light is passing it at 300,000 kilometers per second. But then the rocket accelerates with all its might, attaining a tremendous velocity. When the crew in the rocket checks the light from the (now distant) flashlight, they see that it is passing them by at—300,000 kilometers per second. No matter what they do, how hard they accelerate or for how long, the light is always moving faster, and always moving faster by the same amount.[64] (From their point of view, that is. From the perspective of an external observer, they appear to be moving closer and closer to the speed of light, but they never reach it.)

However, while length contraction and time dilation are perfectly legitimate ways to think about special relativity, they can also get pretty confusing. When we think about the "length" of some physical object, we need to measure the distance between one end of it and the other, but implicitly we need to do so *at the same time*. (You can't make yourself taller by putting a mark on the wall by your feet, climbing a ladder, putting another mark by your head, and proclaiming the distance between the marks to be your height.) But the entire spirit of special relativity tells us to avoid making statements about separated events happening at the same time. So let's tackle the problem from a different angle: by taking "spacetime" seriously.

SPACETIME

Back to the spaceship with us. This time, however, instead of being limited to performing experiments inside the sealed ship, we have access to a small fleet of robot probes with their own rockets and navigation computers, which we can program to go on journeys and come back as we please. And each one of them is equipped with a very accurate atomic clock. We begin by carefully synchronizing these clocks with the one on our main shipboard computer, and verifying that they all agree and keep very precise time.

Then we send out some of our probes to zip away from us for a while and eventually come back. When they return, we notice something right away: The clocks on the probe ships no longer agree with the shipboard computer. Because this is a thought experiment, we can rest assured that the difference is not due to cosmic rays or faulty programming or tampering by mischievous aliens—the probes really did experience a different amount of time than we did.

Happily, there is an explanation for this unusual phenomenon. The time that clocks experience isn't some absolute feature of the universe, out there to be measured once and for all, like the yard lines on a football field. Instead, the time measured by a clock depends on the particular trajectory that the clock takes, much like the total distance covered by a runner depends on their path. If, instead of sending out robot probes equipped with clocks from a spaceship, we had sent out robots on wheels equipped with odometers from a base located on the ground, nobody would be surprised that different robots returned with different odometer readings. The lesson is that clocks are kind of like odometers, keeping track of some measure of distance traveled (through time or through space) along a particular path.

If clocks are kind of like odometers, then time is kind of like space. Remember that even before special relativity, if we believed in absolute space and time à la

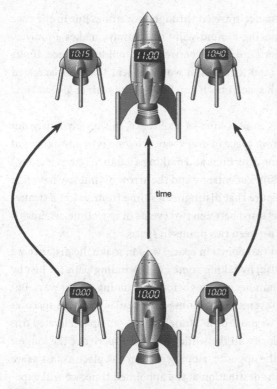

Figure 12: Time elapsed on trajectories that go out and come back is less than that elapsed according to clocks that stay behind.

Isaac Newton, there was nothing stopping us from combining them into one entity called "spacetime." It was still necessary to give four numbers (three to locate a position in space, and one time) to specify an event in the universe. But in a Newtonian world, space and time had completely separate identities. Given two distinct events, such as "leaving the house Monday morning" and "arriving at work later that same morning," we could separately (and uniquely, without fear of ambiguity) talk about the distance between them and the time elapsed between them. Special relativity says that this is not right. There are not two different things, "distance in space" measured by odometers and "duration in time" measured by clocks.

There is only one thing, the *interval in spacetime* between two events, which corresponds to an ordinary distance when it is mostly through space and to a duration measured by clocks when it is mostly through time.

What decides "mostly"? The speed of light. Velocity is measured in kilometers per second, or in some other units of distance per time; hence, having some special speed as part of the laws of nature provides a way to translate between space and time. When you travel more slowly than the speed of light, you are moving mostly through time; if you were to travel faster than light (which you aren't about to do), you would be moving mostly through space.

Let's try to flesh out some of the details. Examining the clocks on our probe ships closely, we realize that all of the traveling clocks are different in a similar way: They read *shorter* times than the one that was stationary. That is striking, as we were comforting ourselves with the idea that time is kind of like space, and the clocks were reflecting a distance traveled through spacetime. But in the case of good old ordinary space, moving around willy-nilly always makes a journey longer; a straight line is the shortest distance between two points in space. If our clocks are telling us the truth (and they are), it would appear that unaccelerated motion—a straight line through spacetime, if you like—is the path of *longest* time between two events.

Well, what did you expect? Time is kind of like space, but it's obviously not completely indistinguishable from space in every way. (No one is in any danger of getting confused by some driving directions and making a left turn into yesterday.) Putting aside for the moment issues of entropy and the arrow of time, we have just uncovered the fundamental feature that distinguishes time from space: Extraneous motion *decreases* the time elapsed between two events in spacetime, whereas it *increases* the distance traveled between two points in space.

If we want to move between two points in space, we can make the distance we actually travel as long as we wish, by taking some crazy winding path (or just by walking in circles an arbitrary number of times before continuing on our way). But consider traveling between two events in spacetime—particular points in space, at particular moments in time. If we move on a "straight line"—an unaccelerated trajectory, moving at constant velocity all the while—we will experience the longest duration possible. So if we do the opposite, zipping all over the place as fast as we can, but taking care to reach our destination at the appointed time, we will experience a shorter duration. If we zipped around at precisely the speed of light, we would never experience any duration at all, no matter how we traveled. We can't do exactly that, but we can come as close as we wish.[65]

That's the precise sense in which "time is kind of like space"—spacetime is a generalization of the concept of space, with time playing the role of one of the dimensions of spacetime, albeit one with a slightly different flavor than the spatial dimensions. None of this is familiar to us from our everyday experience, because we tend to move much more slowly than the speed of light. Moving much more slowly than light is like being a running back who only marched precisely up the football field, never swerving left or right. To a player like that, "distance traveled" would be identical to "number of yards gained," and there would be no ambiguity. That's what time is like in our everyday experience; because we and all of our friends move much more slowly than the speed of light, we naturally assume that time is a universal feature of the universe, rather than a measure of the spacetime interval along our particular trajectories.

STAYING INSIDE YOUR LIGHT CONE

One way of coming to terms with the workings of spacetime according to special relativity is to make a map: draw a picture of space and time, indicating where we are allowed to go. Let's warm up by drawing a picture of Newtonian spacetime. Because

Newtonian space and time are absolute, we can uniquely define "moments of constant time" on our map. We can take the four dimensions of space and time and slice them into a set of three-dimensional copies of space at constant time, as shown in Figure 13. (We're actually only able to show two-dimensional slices on the figure; use your imagination to interpret each slice as representing three-dimensional space.) Crucially, everyone agrees on the difference between space and time; we're not making any arbitrary choices.

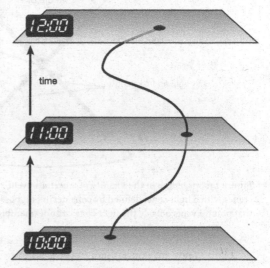

Every Newtonian object (a person, an atom, a rocket ship)

Figure 13: Newtonian space and time. The universe is sliced into moments of constant time, which unambiguously separate time into past and future. World lines of real objects can never double back across a moment of time more than once.

defines a world line—the path the object takes through spacetime. (Even if you sit perfectly still, you still move through spacetime; you're aging, aren't you?[66]) And those world lines obey a very strict rule: Once they pass through one moment of time, they can never double backward in time to pass through the same moment again. You can move as fast as you like—you can be here one instant, and a billion light-years away 1 second later—but you have to keep moving forward in time, your world line intersecting each moment precisely once.

Relativity is different. The Newtonian rule "you must move forward in time" is replaced by a new rule: You must move more slowly than the speed of light. (Unless you are a photon or another massless particle, in which case you always move exactly at the speed of light if you are in empty space.) And the structure we were able to impose on Newtonian spacetime, in the form of a unique slicing into moments of constant time, is replaced by another kind of structure: *light cones*.

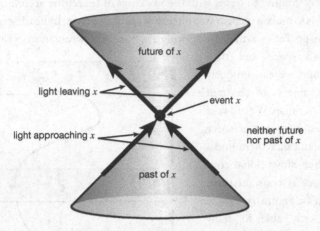

Figure 14: Spacetime in the vicinity of a certain event *x*. According to relativity, every event comes with a light cone, defined by considering all possible paths light could take to or from that point. Events outside the light cone cannot unambiguously be labeled "past" or "future."

Light cones are conceptually pretty simple. Take an event, a single point in spacetime, and imagine all of the different paths that light could take to or from that event; these define the light cone associated with that event. Hypothetical light rays emerging from the event define a future light cone, while those converging on the event define a past light cone, and when we mean both we just say "the light cone." The rule that you can't move faster than the speed of light is equivalent to saying that your world line must remain inside the light cone of every event through which it passes. World lines that do this, describing slower-than-light objects, are called "timelike"; if somehow you could move faster than light, your world line would be "spacelike," since it was covering more ground in space than in time. If you move exactly at the speed of light, your world line is imaginatively labeled "lightlike."

Starting from a single event in Newtonian spacetime, we were able to define a surface of constant time that spread uniquely throughout the universe, splitting the set of all events into the past and the future (plus "simultaneous" events precisely on the surface). In relativity we can't do that. Instead, the light cone associated with an event divides spacetime into the past of that event (events inside the past light cone), the future of that event (inside the future light cone), the light cone itself, and a bunch of points outside the light cone that are neither in the past nor in the future.

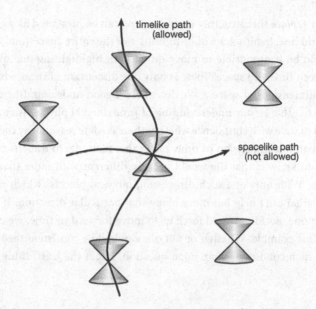

timelike path
(allowed)

spacelike path
(not allowed)

Figure 15: Light cones replace the moments of constant time from Newtonian spacetime. World lines of massive particles must come to an event from inside the past light cone, and leave inside the future light cone—a timelike path. Spacelike paths move faster than light and are therefore not allowed.

It's that last bit that really gets people. In our reflexively Newtonian way of thinking about the world, we insist that some faraway event happened either in the past, or in the future, or at the same time as some event on our own world line. In relativity, for spacelike separated events (outside one another's light cones), the answer is "none of the above." We could *choose* to draw some surfaces that sliced through spacetime and label them "surfaces of constant time," if we really wanted to. That would be taking advantage of the definition of time as a coordinate on spacetime, as discussed in Chapter One. But the result reflects our personal choice, not a real feature of the universe. In relativity, the concept of "simultaneous faraway events" does not make sense.[67]

There is a very strong temptation, when drawing maps of spacetime such as shown in Figure 15, to draw a vertical axis labeled "time," and a horizontal axis (or two) labeled "space." The absence of those axes on our version of the spacetime diagram is completely intentional. The whole point of spacetime according to relativity is that it is not fundamentally divided up into "time" and "space." The light cones, demarcating the accessible past and future of each event, are not added on top of the straightforward Newtonian decomposition of spacetime into time and space; they *replace* that structure entirely. Time can be measured along each individual world line, but it's not a built-in feature of the entire spacetime.

It would be irresponsible to move on without highlighting one other difference between time and space: There is only one dimension of time, whereas there are three dimensions of space.[68] We don't have a good understanding of why this should be so. That is, our understanding of fundamental physics isn't sufficiently developed to state with confidence whether there is some reason why there couldn't be more than one dimension of time, or for that matter zero dimensions of time. What we do know is that life would be very different with more than one time dimension. With only one such dimension, physical objects (which move along timelike paths) can't help but move along that particular direction. If there were more than one, nothing would force us to move forward in time; we could move in circles, for example. Whether or not one can build a consistent theory of physics under such conditions is an open question, but at the least, things would be different.

EINSTEIN'S MOST FAMOUS EQUATION

Einstein's major 1905 paper in which he laid out the principles of special relativity, "On the Electrodynamics of Moving Bodies," took up thirty pages in *Annalen der Physik*, the leading German scientific journal of the time. Soon thereafter, he published a two-page paper entitled "Does the Inertia of a Body Depend upon Its Energy Content?"[69] The purpose of this paper was to point out a straightforward but interesting consequence of his longer work: The energy of an object at rest is proportional to its mass. (Mass and inertia are here being used interchangeably.) That's the idea behind what is surely the most famous equation in history,

$$E = mc^2.$$

Let's think about this equation carefully, as it is often misunderstood. The factor c^2 is of course the speed of light squared. Physicists learn to think, *Aha, relativity must be involved*, whenever they see the speed of light in an equation. The factor m is the mass of the object under consideration. In some places you might read about the "relativistic mass," which increases when an object is in motion. That's not really the most useful way of thinking about things; it's better to consider m as the once-and-for-all mass that an object has when it is at rest. Finally, E is not exactly "the energy"; in this equation, it specifically plays the role of the energy of an object at rest. If an object is moving, its energy will certainly be higher.

So Einstein's famous equation tells us that the energy of an object when it is at rest is equal to its mass times the speed of light squared. Note the importance of the innocuous phrase *an object*. Not everything in the world is an object! For example, we've already spoken of dark energy, which is responsible for the acceleration of the universe. Dark energy doesn't seem to be a collection of particles or other objects; it pervades spacetime smoothly. So as far as dark energy is concerned, $E = mc^2$ simply doesn't apply. Likewise, some objects (such as a photon) can never be at rest, since they are always moving at the speed of light. In those cases, again, the equation isn't applicable.

Everyone knows the practical implication of this equation: Even a small amount of mass is equivalent to a huge amount of energy. (The speed of light, in everyday units, is a really big number.) There are many forms of energy, and what special relativity is telling us is that mass is one form that energy can take. But the various forms can be converted back and forth into one another, which happens all the time. The domain of validity of $E = mc^2$ isn't limited to esoteric realms of nuclear physics or cosmology; it's applicable to every kind of object at rest, on Mars or in your living room. If we take a piece of paper and burn it, letting the photons produced escape along with their energy, the resulting ashes will have a slightly lower mass (no matter how careful we are to capture all of them) than the combination of the original paper plus the oxygen it used to burn. $E = mc^2$ isn't just about atomic bombs; it's a profound feature of the dynamics of energy all around us.

5

TIME IS FLEXIBLE

The reason why the universe is eternal is that it does not live for itself; it gives life to others as it transforms.

—Lao Tzu, *Tao Te Ching*

The original impetus behind special relativity was not a puzzling experimental result (although the Michelson-Morley experiment certainly was that); it was an apparent conflict between two preexisting theoretical frameworks.[70] On the one hand you had Newtonian mechanics, the gleaming edifice of physics on which all subsequent theories had been based. On the other hand you had James Clerk Maxwell's unification of electricity and magnetism, which came about in the middle of the nineteenth century and had explained an impressive variety of experimental phenomena. The problem was that these two marvelously successful theories didn't fit together. Newtonian mechanics implied that the relative velocity of two objects moving past each other was simply the sum of their two velocities; Maxwellian electromagnetism implied that the speed of light was an exception to this rule. Special relativity managed to bring the two theories together into a single whole, by providing a framework for mechanics in which the speed of light did play a special role, but which reduced to Newton's model when particles were moving slowly.

Like many dramatic changes of worldview, the triumph of special relativity came at a cost. In this case, the greatest single success of Newtonian physics—his theory of gravity, which accounted for the motions of the planets with exquisite precision—was left out of the happy reconciliation. Along with electromagnetism, gravity is the most obvious force in the universe, and Einstein was determined to fit it in to the language of relativity. You might expect that this would involve modifying a few equations here and there to make Newton's equations consistent with invariance under boosts, but attempts along those lines fell frustratingly short.

Eventually Einstein hit on a brilliant insight, essentially by employing the spaceship thought experiment we've been considering. (He thought of it first.) In describing our travels in this hypothetical sealed spaceship, I was careful to note that we are far away from any gravitational fields, so we wouldn't have to worry about falling into a star or having our robot probes deflected by the pull of a nearby planet. But what if we were near a prominent gravitational field? Imagine our ship was, for example, in orbit around the Earth. How would that affect the experiments we were doing inside the ship?

Einstein's answer was: They wouldn't affect them at all, as long as we confined our attention to relatively small regions of space and brief intervals of time. We can do whatever kinds of experiments we like—measuring the rates of chemical reactions, dropping balls and watching how they fall, observing weights on springs— and we would get exactly the same answer zipping around in low-Earth orbit as we would in the far reaches of interstellar space. Of course if we wait long enough we can tell we are in orbit; if we let a fork and a spoon freely float in front of our noses, with the fork just slightly closer to the Earth, the fork will then feel just a slightly larger gravitational pull, and therefore move just ever so slightly away from the spoon. But effects like that take time to accumulate; if we confine our attention to sufficiently small regions of space and time, there isn't any experiment we can imagine doing that could reveal the presence of the gravitational pull keeping us in orbit around the Earth.

Contrast the difficulty of detecting a gravitational field with, for example, the ease of detecting an electric field, which is a piece of cake. Just take your same fork and spoon, but now give the fork some positive charge, and the spoon some negative charge. In the presence of an electric field, the opposing charges would be pushed in opposite directions, so it's pretty easy to check whether there are any electric fields in the vicinity.

The difference with gravity is that there is no such thing as a "negative gravitational charge." Gravity is *universal*—everything responds to it in the same way. Consequently, it can't be detected in a small region of spacetime, only in the difference between its

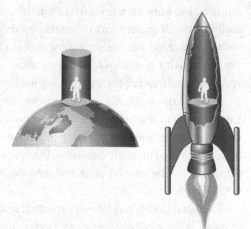

Figure 16: The gravitational field on a planet is locally indistinguishable from the acceleration of a rocket.

effects on objects at different events in spacetime. Einstein elevated this observation to the status of a law of nature, the *Principle of Equivalence*: No local experiment can detect the existence of a gravitational field.

I know what you're thinking: "I have no trouble detecting gravity at all. Here I am sitting in my chair, and it's gravity that's keeping me from floating up into the room." But how do you know it's gravity? Only by looking outside and checking that you're on the surface of the Earth. If you were in a spaceship that was accelerating, you would also be pushed down into your chair. Just as you can't tell the difference between freely falling in interstellar space and freely falling in low-Earth orbit, you also can't tell the difference between constant acceleration in a spaceship and sitting comfortably in a gravitational field. That's the "equivalence" in the Principle of Equivalence: The apparent effects of the force of gravity are equivalent to those of living in an accelerating reference frame. It's not the force of gravity that you feel when you are sitting in a chair; it's the force of the chair pushing up on your posterior. According to general relativity, free fall is the natural, unforced state of motion, and it's only the push from the surface of the Earth that deflects us from our appointed path.

CURVING STRAIGHT LINES

You or I, having come up with the bright idea of the Principle of Equivalence while musing over the nature of gravity, would have nodded sagely and moved on with our lives. But Einstein was smarter than that, and he appreciated what this insight really meant. If gravity isn't detectable by doing local experiments, then it's not really a "force" at all, in the same way that electricity or magnetism are forces. Because gravity is universal, it makes more sense to think of it as a feature of spacetime itself, rather than some force field stretching through spacetime.

In particular, realized Einstein, gravity can be thought of as a manifestation of the *curvature* of spacetime. We've talked quite a bit about spacetime as a generalization of space, and how the time elapsed along a trajectory is a measure of the distance traveled through spacetime. But space isn't necessarily rigid, flat, and rectilinear; it can be warped, stretched, and deformed. Einstein says that spacetime is the same way.

It's easiest to visualize two-dimensional space, modeled, for example, by a piece of paper. A flat piece of paper is not curved, and the reason we know that is that it obeys the principles of good old-fashioned Euclidean geometry. Two initially parallel lines, for example, never intersect, nor do they grow apart.

In contrast, consider the two-dimensional surface of a sphere. First we have to generalize the notion of a "straight line," which on a sphere isn't an obvious concept. In Euclidean geometry, as we were taught in high school, a straight line is the shortest distance between two points. So let's declare an analogous definition: A "straight line" on a curved geometry is the shortest curve connecting two points, which on a sphere would be a portion of a great circle. If we take two paths on a sphere that are initially parallel, and extend them along great circles, they will eventually intersect. That proves that the principles of Euclidean geometry are no longer at work, which is one way of seeing that the geometry of a sphere is curved.

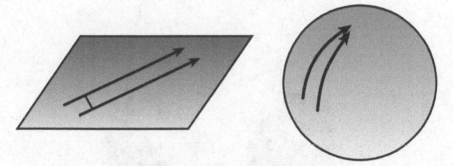

Figure 17: Flat geometry, with parallel lines extending forever; curved geometry, where initially parallel lines eventually intersect.

Einstein proposed that four-dimensional spacetime can be curved, just like the surface of a two-dimensional sphere. The curvature need not be uniform like a sphere, the same from point to point; it can vary in magnitude and in shape from place to place. And here is the kicker: When we see a planet being "deflected by the force of gravity," Einstein says it is really just traveling in a straight line. At least, as straight as a line can be in the curved spacetime through which the planet is moving. Following the insight that an unaccelerated trajectory yields the greatest possible time a clock could measure between two events, a straight line through spacetime is one that does its best to maximize the time on a clock, just like a straight line through space does its best to minimize the distance read by an odometer.

Let's bring this down to Earth, in a manner of speaking. Consider two clocks, synchronized with each other, sitting on the Earth's surface. While one clock stays put, we're going to throw the other clock high up into the air, and wait for it to come back down. (We can ignore the rotation of the Earth for the purposes of this utterly impractical thought experiment, so the clock returns to just where it started.) According to the viewpoint of general relativity, the clock that goes up and comes back down is not accelerating; it's in free fall, doing its best to move in a straight line through space-time. The clock on the ground, meanwhile, is accelerating—it's being prevented from freely falling because the Earth's surface keeps it up. Therefore, the clock we throw will experience more elapsed time than the Earthbound clock—compared to the accelerated clock on the ground, the freely falling one we throw appears to run more quickly.

Figure 18: Time as measured by a clock on the Earth's surface will be shorter than that measured by one thrown into the air, as the former clock is on an accelerated (non-free-falling) trajectory.

We don't often throw clocks into space and let them fall back to Earth. But there are clocks down here at the surface that regularly exchange signals with clocks on satellites. That, for example, is the basic mechanism behind the Global Positioning System (GPS) that helps modern cars give driving directions in real time. Your personal GPS receiver gets signals from a number of satellites orbiting the Earth, and determines its position by comparing the time between the different signals. That calculation would quickly go astray if the gravitational time dilation due to general relativity were not taken into account; the GPS satellites experience about 38 more microseconds per day in orbit than they would on the ground. Rather than teaching your receiver equations from general relativity, the solution actually adopted is to tune the satellite clocks so that they run a little bit more slowly than they should if they were to keep correct time down here on the surface.

EINSTEIN'S MOST IMPORTANT EQUATION

The saying goes that every equation cuts your book sales in half. I'm hoping that this page is buried sufficiently deeply in the book that nobody notices before purchasing it, because I cannot resist the temptation to display another equation: the Einstein field equation for general relativity.

$$R_{\mu\nu} - (1/2)Rg_{\mu\nu} = 8\pi G T_{\mu\nu}.$$

This is the equation that a physicist would think of if you said "Einstein's equation"; that $E = mc^2$ business is a minor thing, a special case of a broader principle. This one, in contrast, is a deep law of physics: It reveals how stuff in the universe causes spacetime to curve, and therefore causes gravity. Both sides of the equation are not simply numbers, but *tensors*—geometric objects that capture multiple things going on at once. (If you thought of them as 4x4 arrays of numbers, you would be pretty close to right.) The left-hand side of the equation characterizes the curvature of spacetime. The right-hand side characterizes all the various forms of stuff that make spacetime curve—energy, momentum, pressure, and so on. In one fell swoop, Einstein's equation reveals how any particular collection of particles and fields in the universe creates a certain kind of curvature in spacetime.

According to Isaac Newton, the source of gravity was mass; heavier objects gave rise to stronger gravitational fields. In Einstein's universe, things are more complicated. Mass gets replaced by energy, but there are also other properties that go into curving spacetime. Vacuum energy, for example, has not only energy, but also *tension*—a kind of negative pressure. A stretched string or rubber band has tension, pulling back rather than pushing out. It's the combined effect of the energy plus the tension that causes the universe to accelerate in the presence of vacuum energy.[71]

The interplay between energy and the curvature of spacetime has a dramatic consequence: In general relativity, energy is not conserved. Not every expert in the field would agree with that statement, not because there is any controversy over what the theory predicts, but because people disagree on how to define "energy" and "conserved." In a Newtonian absolute spacetime, there is a well-defined notion of the energy of individual objects, which we can add up to get the total energy of the universe, and that energy never changes (it's the same at every moment in time). But in general relativity, where spacetime is dynamical, energy can be injected into matter or absorbed from it by the motions of spacetime. For example, vacuum energy remains absolutely constant in density as the universe expands. So the energy per cubic centimeter is constant, while the number of cubic centimeters is increasing—the total energy goes up. In a universe dominated by radiation, in contrast, the total energy goes down, as each photon loses energy due to the cosmological redshift.

You might think we could escape the conclusion that energy is not conserved by including "the energy of the gravitational field," but that turns out to be much harder than you might expect—there simply is no well-defined local definition of the energy in the gravitational field. (That shouldn't be completely surprising, since the gravitational field can't even be detected locally.) It's easier just to bite the bullet and admit that energy is not conserved in general relativity, except in certain special circumstances.[72] But it's not as if chaos has been loosed on the world; given the curvature of spacetime, we can predict precisely how any particular source of energy will evolve.

HOLES IN SPACETIME

Black holes are probably the single most interesting dramatic prediction of general relativity. They are often portrayed as something relatively mundane: "Objects where the gravitational field is so strong that light itself cannot escape." The reality is more interesting.

Even in Newtonian gravity, there's nothing to stop us from contemplating an object so massive and dense that the escape velocity is greater than the speed of light, rendering the body "black." Indeed, the idea was occasionally contemplated, including by British geologist John Michell in 1783 and by Pierre-Simon Laplace in 1796.[73] At the time, it wasn't clear whether the idea quite made sense, as nobody knew whether light was even affected by gravity, and the speed of light didn't have the fundamental importance it attains in relativity. More important, though, there is a very big distinction hidden in the seemingly minor difference between "an escape velocity greater than light" and "light cannot escape." Escape velocity is the speed at which we would have to start an object moving upward in order for it to escape the gravitational field of a body *without any further acceleration*. If I throw a baseball up in the air in the hopes that it escapes into outer space, I have to throw it faster than escape velocity. But there is absolutely no reason why I couldn't put the same baseball on a rocket and gradually accelerate it into space without ever reaching escape velocity. In other words, it's not necessary to reach escape velocity in order to actually escape; given enough fuel, you can go as slowly as you like.

But a real black hole, as predicted by general relativity, is a lot more dramatic than that. It is a true region of no return—once you enter, there is no possibility of leaving, no matter what technological marvels you have at your disposal. That's because general relativity, unlike Newtonian gravity or special relativity, allows spacetime to curve. At every event in spacetime we find light cones that divide space into the past, future, and places we can't reach. But unlike in special relativity, the light cones are not fixed in a rigid alignment; they can tilt and stretch as spacetime curves under the influence of matter and energy. In the vicinity of a massive object, light cones tilt toward the object, in accordance with the tendency of things to be pulled in by the gravitational field. A black hole is a region of spacetime where the light cones have tilted so much that you would have to move faster than the speed of light to escape. Despite the similarity of language, that's an enormously stronger statement than "the escape velocity is larger than the speed of light." The boundary defining the black hole, separating places where you still have a chance to escape from places where you are doomed to plunge ever inward, is the *event horizon*.

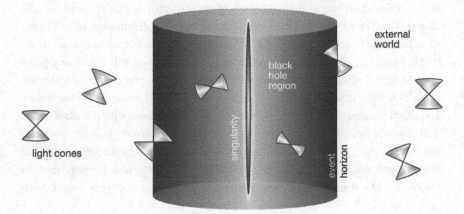

Figure 19: Light cones tilt in the vicinity of a black hole. The event horizon, demarcating the edge of the black hole, is the place where they tip over so far that nothing can escape without moving faster than light.

There may be any number of ways that black holes could form in the real world, but the standard scenario is the collapse of a sufficiently massive star. In the late 1960s, Roger Penrose and Stephen Hawking proved a remarkable feature of general relativity: If the gravitational field becomes sufficiently strong, a singularity *must* be formed.[74] You might think that's only sensible, since gravity becomes stronger and stronger and pulls matter into a single point. But in Newtonian gravity, for example, it's not true. You can get a singularity if you try hard enough, but the generic result of squeezing matter together is that it will reach some point of maximum density. But in general relativity, the density and spacetime curvature increase without limit, until they form a singularity of infinite curvature. Such a singularity lies inside every black hole.

It would be wrong to think of the singularity as residing at the "center" of the black hole. If we look carefully at the representation of spacetime near a black hole shown in Figure 19, we see that the future light cones inside the event horizon keep tipping toward the singularity. But that light cone *defines* what the observer at that event would call "the future." Like the Big Bang singularity in the past, the black hole singularity in the future is a moment of time, not a place in space. Once you are inside the event horizon, you have absolutely no choice but to continue on to the grim destiny of the singularity, because it lies ahead of you in time, not in some direction in space. You can no more avoid hitting the singularity than you can avoid hitting tomorrow.

When you actually do fall through the event horizon, you might not even notice. There is no barrier there, no sheet of energy that you pass through to indicate that you've entered a black hole. There is simply a diminution of your possible future life choices; the option of "returning to the outside universe" is no longer available, and "crashing into the singularity" is your only remaining prospect. In fact, if you knew how massive the black hole was, you could calculate precisely how long it will take (according to a clock carried along with you) before you reach the singularity and cease to exist; for a black hole with the mass of the Sun, it would be about one-millionth of a second. You might try to delay this nasty fate, for example, by firing rockets to keep yourself away from the singularity, but it would only be counterproductive. According to relativity, unaccelerated motion *maximizes* the time between two events. By struggling, you only hasten your doom.[75]

There is a definite moment on your infalling path when you cross the event horizon. If we imagine that you had been sending a constant stream of radio updates to your friends outside, they will never be able to receive anything sent after that time. They do not, however, see you wink out of existence; instead, they receive your signals at longer and longer intervals, increasingly redshifted to longer and longer wavelengths. Your final moment before crossing the horizon is (in principle) frozen in time from the point of view of an external observer, although it becomes increasingly dimmer and redder as time passes.

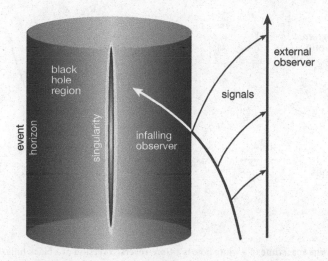

Figure 20: As an object approaches an event horizon, to a distant observer it appears to slow down and become increasingly redshifted. The moment on the object's world line when it crosses the horizon is the last moment it can be seen from the outside.

WHITE HOLES: BLACK HOLES RUN BACKWARD

If you think a bit about this black-hole story, you'll notice something intriguing: time asymmetry. We have been casually tossing around terminology that assumes a directionality to time; we say "once you pass the event horizon you can never leave," but not "once you leave the event horizon you can never return." That's not because we have been carelessly slipping into temporally asymmetric language; it's because the notion of a black hole is intrinsically time-asymmetric. The singularity is unambiguously in your future, not in your past.

The time asymmetry here isn't part of the underlying physical theory. General relativity is perfectly time-symmetric; for every specific spacetime that solves Einstein's equation, there is another solution that is identical except that the direction of time is reversed. A black hole is a particular solution to Einstein's equation, but there are equivalent solutions that run the other way: *white holes*.

The description of a white hole is precisely the same as that of a black hole, if we simply reverse the tenses of all words that refer to time. There is a singularity in the past, from which light cones emerge. The event horizon lies to the future of the singularity, and the external world lies to the future of that. The horizon represents a place past which, once you exit, you can never return to the white-hole region.

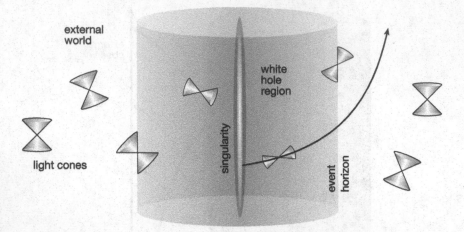

Figure 21: The spacetime of a white hole is a time-reversed version of a black hole.

So why do we hear about black holes in the universe all the time, and hardly ever hear about white holes? For one thing, notice that we can't "make" a white hole. Since we are in the external world, the singularity and event horizon associated with a white hole are necessarily in our *past*. So it's not a matter of wondering what we would do to create a white hole; if we're going to find one, it will already have been out there in the universe from the beginning.

But in fact, thinking slightly more carefully, we should be suspicious of that word *make*. Why, in a world governed by reversible laws of physics, do we think of ourselves as "making" things that persist into the future, but not things that extend into the past? It's the same reason why we believe in free will: A low-entropy boundary condition in the past dramatically fixes what possibly could have happened, while the absence of any corresponding future boundary condition leaves what can yet happen relatively open.

So when we ask, "Why does it seem relatively straightforward to make a black hole, while white holes are something that we would have to find already existing in the universe?" the answer should immediately suggest itself: because a black hole tends to have more entropy than the things from which you would make it. Actually calculating what the entropy is turns out to be a tricky business involving Hawking radiation, as we'll see in Chapter Twelve. But the key point is that black holes have a lot of entropy. Black holes turn out to provide the strongest connection we have between gravitation and entropy—the two crucial ingredients in an ultimate explanation of the arrow of time.

6

LOOPING THROUGH TIME

You see, my son, here time changes into space.

—Richard Wagner, *Parsifal*

Everyone knows what a time machine looks like: something like a steampunk sled with a red velvet chair, flashing lights, and a giant spinning wheel on the back. For those of a younger generation, a souped-up stainless-steel sports car is an acceptable substitute; our British readers might think of a 1950s-style London police box.[76] Details of operation vary from model to model, but when one actually travels in time, the machine ostentatiously dematerializes, presumably to be re-formed many millennia in the past or future.

That's not how it would really work. And not because time travel is impossible and the whole thing is just silly; whether or not time travel is possible is more of an open question than you might suspect. I've emphasized that time is kind of like space. It follows that, if you did stumble across a working time machine in the laboratory of some mad inventor, it would simply look like a "space machine"—an ordinary vehicle of some sort, designed to move you from one place to another. If you want to visualize a time machine, think of launching a rocket ship, not disappearing in a puff of smoke.

So what is actually entailed in traveling through time? There are two cases of possible interest: traveling to the future and traveling to the past. To the future is easy: Just keep sitting in your chair. Every hour, you will move one hour into the future. "But," you say, "that's boring. I want to move far into the future, really quickly, a lot faster than one hour per hour. I want to visit the twenty-fourth century before lunchtime." But we know it's impossible to move faster than one hour per hour, relative to a clock that travels along with you. You might be able to trick yourself, by going to sleep or entering suspended animation, but time will still be passing.

On the other hand, you could distort the total amount of time experienced along your world line as compared to the world lines of other people. In a Newtonian universe even that wouldn't be possible, as time is universal and every world line connecting the same two events experiences the same elapsed time. But in special relativity we can affect the elapsed time by moving through space. Unaccelerated motion gives us the longest time between events; so if you want to get to the future quickly (from your perspective), you just have to move on a highly non-straight path through spacetime. You could zip out on a rocket near the speed of light and then return; or, given sufficient fuel, you could just fly around in circles at ultra-high speed, never straying very far from your starting point in space. When you stopped and exited your spaceship, besides being dizzy you would have "traveled into the future"; more accurately, you would have experienced less time along your world line than was experienced by anyone who stayed behind. Traveling to the future is easy, and getting there more quickly is just a technology problem, not something that conflicts with the fundamental laws of physics.

But you might want to come back, and that's where the challenge lies. The problems with traveling through time arise when we contemplate traveling to the *past*.

CHEATING SPACETIME

Lessons learned from watching *Superman* movies notwithstanding, traveling backward in time is not a matter of reversing the rotation of the Earth. Spacetime itself has to cooperate. Unless, of course, you cheat, by moving faster than the speed of light.

In a Newtonian universe, traveling backward in time is simply out of the question. World lines extend through a spacetime that is uniquely divided into three-dimensional moments of equal time, and the one unbreakable rule is that they must never double back and return to the past. In special relativity, things aren't much better. Defining "moments of equal time" across the universe is highly arbitrary, but at every event we are faced with the restrictions enforced by light cones. If we are made of ordinary stuff, confined to move from each event forward into the interior of its light cone, there is no hope of traveling backward in time; in a spacetime diagram, we are doomed to march relentlessly upward.

Things would be a little bit more interesting if we were made of non-ordinary stuff. In particular, if we were made of *tachyons*—particles that always move faster than light. Sadly, we are not made of tachyons, and there are good reasons to believe that tachyons don't even exist. Unlike ordinary particles, tachyons are forced to always travel outside the light cone. In special relativity, whenever we move outside the light cone, from somebody's point of view we're also moving backward in time. More important, the light cones are the only structure defined on spacetime in relativity; there is no separate notion of "space at one moment of time." So if a particle can start at some event where you are momentarily located and move outside your light cone (faster than light), it can necessarily move into the past from your point of view. There's nothing to stop it.

Tachyons, therefore, can apparently do something scary and unpredictable: "start" from an event on the world line of some ordinary (slower-than-light) object, defined by some position in space and some moment in time, and travel on a path that takes them to a *previous* point on the same world line. If you had a flashlight that emitted tachyons, you could (in principle) construct an elaborate series of mirrors by which you could send signals in Morse code to yourself in the past. You could warn your earlier self not to eat the shrimp in that restaurant that one time, or to go on that date with the weirdo from the office, or to sink your life savings into Pets.com stock.

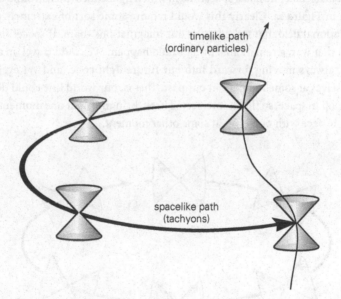

timelike path
(ordinary particles)

spacelike path
(tachyons)

Figure 22: If tachyons could exist, they could be emitted by ordinary objects and zip around to be absorbed in the past. At every event along its trajectory, the tachyon moves outside the light cone.

Clearly, the possibility of travel backward in time raises the possibility of paradoxes, which is unsettling. There is a cheap way out: Notice that tachyons don't seem to exist, and declare that they are simply incompatible with the laws of physics.[77]

That is both fruitful and accurate, at least as far as special relativity is concerned. When curved spacetime gets into the game, things become more interesting.

CIRCLES IN TIME

For those of us who are not made of tachyons, our trajectories through spacetime are limited by the speed of light. From whatever event defines our current location, we necessarily move "forward in time," toward some other event inside our light cone—in technical jargon, we move on a timelike path through spacetime. This is a local requirement, referring only to features of our neighborhood of the universe. But in general relativity, spacetime is curved. That means light cones in our neighborhood may be tilted with respect to those far away. It's that kind of tilting that leads to black holes.

But imagine that, instead of light cones tilting inward toward a singularity and creating a black hole, we had a spacetime in which light cones tilted around a circle, as shown in Figure 23. Clearly this would require some sort of extremely powerful gravitational field, but we're letting our imaginations roam. If spacetime were curved in that way, a remarkable thing would happen: We could travel on a timelike path, always moving forward into our future light cone, and yet eventually meet ourselves at some moment in our past. That is, our world line could describe a closed loop in space, so that it intersected itself, bringing us at one moment in our lives face-to-face with ourselves at some other moment.

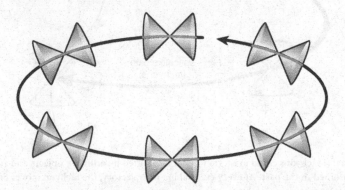

Figure 23: In curved spacetime, we can imagine light cones tilting around in a circle, creating closed timelike curves.

Such a world line—always moving forward in time from a local perspective, but managing to intersect itself in the past—is a *closed timelike curve*, or CTC.

That's what we really mean when we talk about a "time machine" in general relativity. To actually move along the closed timelike curve would involve ordinary travel through spacetime, on a spaceship or something even more mundane—perhaps even sitting "motionless" in your chair. It's the curvature of spacetime itself that brings you into contact with your own past. This is a central feature of general relativity that will become important later on when we return to the origin of the universe and the problem of entropy: Spacetime is not stuck there once and for all, but can change (and even come into or pop out of existence) in response to the effects of matter and energy.

It's not difficult to find spacetimes in general relativity that include closed timelike curves. All the way back in 1949, mathematician and logician Kurt Gödel found a solution to Einstein's equation that described a "spinning" universe, which contains closed timelike curves passing through every event. Gödel was friends with Einstein in his later years at the Institute for Advanced Study in Princeton, and the idea for his solution arose in part from conversations between the two men.[78] In 1963, New Zealand mathematician Roy Kerr found the exact solution for a rotating black hole; interestingly, the singularity takes the form of a rapidly spinning ring, the vicinity of which is covered by closed timelike curves.[79] And in 1974, Frank Tipler showed that an infinitely long, rotating cylinder of matter would create closed timelike curves around it, if the cylinder were sufficiently dense and rotating sufficiently rapidly.[80]

But you don't have to work nearly so hard to construct a spacetime with closed timelike curves. Consider good old garden-variety flat spacetime, familiar from special relativity. But now imagine that the timelike direction (as defined by some particular unaccelerated observer) is a *circle*, rather than stretching on forever. In such a universe, something that moved forward in time would find itself coming back to the same moment in the universe's history over and over. In the Harold Ramis movie *Groundhog Day*, Bill Murray's character keeps waking up every morning to experience the exact same situations he had experienced the day before. The circular-time universe we're imagining here is something like that, with two important exceptions: First, it would be truly identical every day, including the actions of the protagonist himself. And second, there would not be any escape. In particular, winning the love of Andie MacDowell would not save you.

The circular-time universe isn't just an amusing playground for filmmakers; it's an exact solution to Einstein's equation. We know that, by choosing some unaccelerated reference frame, we can "slice" four-dimensional flat spacetime into three-dimensional moments of equal time. Take two such slices: say, midnight February 2, and midnight February 3—two moments of time that extend

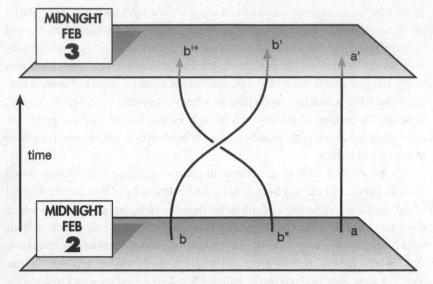

Figure 24: A circular-time universe, constructed by identifying two different moments in flat spacetime. Two closed timelike curves are shown: one that loops through only once before closing, from (a) to (a'), and another that loops twice, from (b) to (b') to (b'') to (b''').

throughout the universe (in this special case of flat spacetime, in this particular reference frame). Now take just the one-day's-worth of spacetime that is between those slices, and throw everything else away. Finally, *identify* the beginning time with the final time. That is, make a rule that says whenever a world line hits a particular point in space on February 3, it instantly reemerges at the same point in space back on February 2. At heart, it's nothing more than rolling up a piece of paper and taping together opposite sides to make a cylinder. At every event, even at midnight when we've identified different slices, everything looks perfectly smooth and spacetime is flat—time is a circle, and no point on a circle is any different than any other point. This spacetime is rife with closed timelike curves, as illustrated in Figure 24. It might not be a realistic universe, but it demonstrates that the rules of general relativity alone do not prohibit the existence of closed timelike curves.

THE GATE INTO YESTERDAY

There are two major reasons why most people who have given the matter a moment's thought would file the possibility of time travel under "Science Fiction," not "Serious Research." First, it's hard to see how to actually create a closed time-like curve, although we'll see that some people have ideas. But second, and more fundamentally, it's hard to see how the notion could make sense. Once we grant the possibility of traveling into our own past, it's just too easy to invent nonsensical or paradoxical situations.

To fix our ideas, consider the following simple example of a time machine: the gate into yesterday. ("Gate into tomorrow" would be equally accurate—just go the other way.) We imagine there is a magical gate, standing outside in a field. It's a perfectly ordinary gate in every way, with one major exception: When you walk through from what we'll call "the front," you emerge in the same field on the other side, but *one day earlier*—at least, from the point of view of the "background time" measured by outside observers who don't ever step through the gate. (Imagine fixed clocks standing in the field, never passing through the gate, synchronized in the rest frame of the field itself.) Correspondingly, when you walk through the back side of the gate, you emerge through the front one day later than when you left.

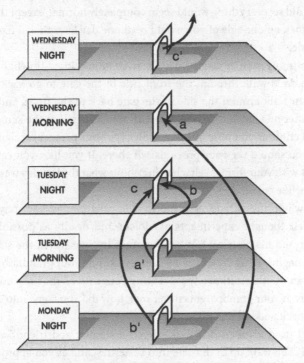

Figure 25: The gate into yesterday, showing one possible world line. A traveler walks through the front of the gate from the right (a) and appears out the back one day in the past (a'). The person spends half a day walking around the side of the gate to enter from the front again (b) and reappears one day earlier (b'). Then the person waits a day and enters the back side of the gate (c), emerging from the front one day in the future (c').

It sounds magical and wondrous, but all we've done is describe a particular sort of unusual spacetime; we've identified a set of points in space at unequal times. Nobody is disappearing in any puffs of smoke; from the point of view of any particular observer, their own world line marches uninterruptedly into the future, one second per second. When you look through the gate from the front, you don't see inky blackness, or swirling psychedelic colors; you see the field on the other side, just as you would if you looked through any other door. The only difference is, you see what it looked like *yesterday*. If you peer around the side of the gate, you see the field today, while peering through the gate gives you a view of one day before. Likewise, if you move around the other side and peer through the gate from the back, you see just the other side of the field, but you see what it will look like tomorrow. There is nothing to stop you from walking through the gate and immediately back again, any number of times you like, or for that matter from standing with one foot on either side of the threshold. You wouldn't feel any strange tingling sensations if you did so; everything would seem completely normal, except that an accurate fixed clock on one side of you would read one day later than a fixed clock on the other side.

The gate-into-yesterday spacetime clearly contains closed timelike curves. All you have to do is walk through the front side of the gate to go back in time by one day, then walk around the side of the gate back to the front, and wait there patiently. After one day has passed, you will find yourself at the same place and time in spacetime as you were one day earlier (by your personal reckoning)—and of course, you should see your previous self there. If you like, you can exchange pleasantries with your former self, chatting about what the last day was like. That's a closed timelike curve.

This is where the paradoxes come in. For whatever reason, physicists love to make their thought experiments as violent and deadly as possible; think of Schrödinger and his poor cat.[81] When it comes to time travel, the standard scenario is to imagine going back into the past and killing your grandfather before he met your grandmother, ultimately preventing your own birth. The paradox itself is then obvious: If your grandparents never met, how did you come into existence in order to go back and kill one of them?[82]

We don't need to be so dramatic. Here's a simpler and friendlier version of the paradox. You walk up to the gate into yesterday, and as you approach you see a version of yourself waiting for you there, looking about one day older than you presently are. Since you know about the closed timelike curves, you are not too surprised; obviously you lingered around after passing through the gate, looking forward to the opportunity to shake hands with a previous version of yourself. So the two versions of you exchange pleasantries, and then you leave your other self behind as you walk through the front of the gate into yesterday. But after passing through, out of sheer perverseness, you decide not to go along with the program. Rather than hanging around to meet up with your younger self, you wander off, catching a taxi to the airport and hopping on a flight to the Bahamas. You never do meet up with the version of yourself that went through the gate in the first place. But that version of yourself did meet with a future version of itself—indeed, you still carry the memory of the meeting. What is going on?

ONE SIMPLE RULE

There is a simple rule that resolves all possible time travel paradoxes.[83] Here it is:

• Paradoxes do not happen.

It doesn't get much simpler than that.

At the moment, scientists don't really know enough about the laws of physics to say whether they permit the existence of macroscopic closed timelike curves. If they don't, there's obviously no need to worry about paradoxes. The more interesting question is, do closed timelike curves *necessarily* lead to paradoxes? If they do, then they can't exist, simple as that.

But maybe they don't. We all agree that logical contradictions cannot occur. More specifically, in the classical (as opposed to quantum mechanical[84]) setup we are now considering, there is only one correct answer to the question "What happened at the vicinity of this particular event in spacetime?" In every part of spacetime, something happens—you walk through a gate, you are all by yourself, you meet someone else, you somehow never showed up, whatever it may be. And that something is whatever it is, and was whatever it was, and will be whatever it will be, now and forever. If, at a certain event, your grandfather and grandmother were getting it on, that's what happened at that event. There is nothing you can do to change it, because it happened. You can no more change events in your past in a spacetime with closed timelike curves than you can change events that already happened in an ordinary, no-closed-timelike-curves spacetime.[85]

It should be clear that consistent stories are *possible*, even in spacetimes with closed timelike curves. Figure 25 depicts the world line of one intrepid adventurer who jumps back in time twice, then gets bored and jumps forward once, before walking away. There's nothing paradoxical about that. And we can certainly imagine a non-paradoxical version of the scenario from the end of the previous section.

You approach the gate, where you see an older version of yourself waiting for you there; you exchange pleasantries, and then you leave your other self behind as you walk through the front of the gate into yesterday. But instead of obstinately wandering off, you wait around a day to meet up with the younger version of yourself, with whom you exchange pleasantries before going on your way. Everyone's version of every event would be completely consistent.

We can have much more dramatic stories that are nevertheless consistent. Imagine that we have been appointed Guardian of the Gate, and our job is to keep vigilant watch over who passes through. One day, as we are standing off to the side, we see a stranger emerge from the rear side of the gate. That's no surprise; it just means that the stranger will enter ("has entered"?—our language doesn't have the tenses to deal with time travel) the front side of the gate tomorrow. But as you keep vigilant watch, you see that the stranger who emerged simply loiters around for one day, and when precisely twenty-four hours have passed, walks calmly through the front of the gate. Nobody ever approached from elsewhere—the entering and exiting strangers formed a closed loop, and that twenty-four hours constituted the stranger's entire life span. That may strike you as weird or unlikely, but there is nothing paradoxical or logically inconsistent about it.[86]

The real question is, what happens if we try to cause trouble? That is, what if we choose not to go along with the plan? In the story where you meet a slightly older version of yourself just before you cross through the front of the gate and jump backward in time, the crucial point is that you seem to have a *choice* once you pass through. You can obediently fulfill your apparent destiny, or you can cause trouble by wandering off. If that's the choice you make, what is to stop you? That is where the paradoxes seem to get serious.

We know what the answer is: That can't happen. If you met up with an older version of yourself, we know with absolute metaphysical certainty that once you age into that older self, you will be there to meet with your younger self. Imagine that we remove messy human beings from the problem by just considering simple inanimate objects, like series of billiard balls passing through the gate. There may be more than one consistent set of things that could happen at the various events in spacetime—but one and only one set of things will actually occur.[87] Consistent stories happen; inconsistent ones do not.

ENTROPY AND TIME MACHINES

The issue that troubles us, when we get right down to it, isn't anything about the laws of physics; it's about free will. We have a strong feeling that we can't be predestined to do something we choose not to do; that becomes a difficult feeling to sustain, if we've already seen ourselves doing it.

There are times when our free will must be subjugated to the laws of physics. If we get thrown out of a window on the top floor of a skyscraper, we expect to hurtle to the ground, no matter how much we would rather fly away and land safely elsewhere. That kind of predestination we're willing to accept. But the much more detailed kind implied by closed timelike curves, where it seems that the working out of a consistent history through spacetime simply forbids us from making free choices that would otherwise be possible, is bothersome. Sure, we could be committed determinists and imagine that all of the atoms in our bodies and in the external world, following the unbending dictates of Newton's laws of motion, will conspire to force us to behave in precisely the required way in order to avoid paradoxes, but it seems somewhat at variance with the way we think about ourselves.[88]

The nub of the problem is that you can't have a consistent arrow of time in the presence of closed timelike curves. In general relativity, the statement "We remember the past and not the future" becomes "We remember what happened within our past light cone, but not within our future light cone." But on a closed timelike curve, there are spacetime events that are both in our past light cone and in our future light cone, since those overlap. So do we remember such events or not? We might be able to guarantee that events along a closed timelike curve are consistent with the microscopic laws of physics, but in general they cannot be compatible with an uninterrupted increase of entropy along the curve.

To emphasize this point, think about the hypothetical stranger who emerges from the gate, only to enter it from the other side one day later, so that their entire life story is a one-day loop repeated ad infinitum. Take a moment to contemplate the exquisite level of precision required to pull this off, if we were to think about the loop as "starting" at one point. The stranger would have to ensure that, one day later, every single atom in his body was in precisely the right place to join up smoothly with his past self. He would have to make sure, for example, that his clothes didn't accumulate a single extra speck of dust that wasn't there one day earlier in his life, that the contents of his digestive tract was precisely the same, and that his hair and toenails were precisely the same length. This seems incompatible with our experience of how entropy increases—to put it mildly—even if it's not strictly a violation of the Second Law (since the stranger is not a closed system). If we merely shook hands with our former selves, rather than joining up with them, the required precision doesn't seem quite so dramatic; but in either case the insistence that we be in the right place at the right time puts a very stringent constraint on our possible future actions.

Our concept of free will is intimately related to the idea that the past may be set in stone, but the future is up for grabs. Even if we believe that the laws of physics in principle determine the future evolution of some particular state of the universe with perfect fidelity, we don't know what that state is, and in the real world the increase of entropy is consistent with any number of possible futures. The kind of predestination seemingly implied by consistent evolution in the presence of closed timelike curves is precisely the same we would get into if there really were a low-entropy future boundary condition in the universe, just on a more local scale.

In other words: If closed timelike curves were to exist, consistent evolution in their presence would seem just as strange and unnatural to us as a movie played backward, or any other example of evolution that decreases entropy. It's not impossible; it's just highly unlikely. So either closed timelike curves can't exist, or big macroscopic things can't travel on truly closed paths through spacetime—or everything we think we know about thermodynamics is wrong.

PREDICTIONS AND WHIMSY

Life on a closed timelike curve seems depressingly predestined: If a system moves on a closed loop along such a curve, it is required to come back to precisely the state in which it started. But from the point of view of an observer standing outside, closed timelike curves also raise what is seemingly the opposite problem: What happens along such a curve cannot be uniquely predicted from the prior state of the universe. That is, we have the very strong constraint that evolution along a closed timelike curve must be consistent, but there can be a large number of consistent evolutions that are possible, and the laws of physics seem powerless to predict which one will actually come to pass.[89]

We've talked about the contrast between a presentist view of the universe, holding that only the current moment is real, and an eternalist or block-universe view, in which the entire history of the universe is equally real. There is an interesting philosophical debate over which is the more fruitful version of reality; to a physicist, however, they are pretty much indistinguishable. In the usual way of thinking, the laws of physics function as a computer: You give as input the present state, and the laws return as output what the state will be one instant later (or earlier, if we wish). By repeating this process multiple times, we can build up the entire predicted history of the universe from start to finish. In that sense, complete knowledge of the present implies complete knowledge of all of history.

Closed timelike curves make that program impossible, as a simple thought experiment reveals. Hearken back to the stranger who appeared out of the gate into yesterday, then jumped back in the other side a day later to form a closed loop. There would be no way to predict the existence of such a stranger from the state of the universe at an earlier time. Let's say that we start in a universe that, at some particular moment, has no closed timelike curves. The laws of physics purportedly allow us to predict what happens in the future of that moment. But if someone creates closed timelike curves, that ability vanishes. Once the closed timelike curves are established, mysterious strangers and other random objects can consistently appear and travel around them—or not. There is no way to predict what will happen, just from knowing the complete state of the universe at a previous time.

We can insist all we like, in other words, that what happens in the presence of closed timelike curves be *consistent*—there are no paradoxes. But that's not enough to make it *predictable*, with the future determined by the laws of physics and the state of the universe at one moment in time. Indeed, closed timelike curves can make it impossible to define "the universe at one moment in time." In our previous discussions of spacetime, it was crucially important that we were allowed to "slice" our four-dimensional universe into three-dimensional "moments of time," the complete set of which was labeled with different values of the time coordinate. But in the presence of closed timelike curves, we generally won't be able to slice spacetime that way.[90] Locally—in the near vicinity of any particular event—the division of spacetime into "past" and "future" as defined by light cones is perfectly normal. Globally, we can't divide the universe consistently into moments of time.

In the presence of closed timelike curves, therefore, we have to abandon the concept of "determinism"—the idea that the state of the universe at any one time determines the state at all other times. Do we value determinism so highly that this conflict means we should reject the possibility of closed timelike curves entirely? Not necessarily. We could imagine a different way in which the laws of physics could be formulated—not as a computer that calculates the next moment from the present moment, but as some set of conditions that are imposed on the history of the universe as a whole. It's not clear what such conditions might be, but we have no way of excluding the idea on the basis of pure thought.

All this vacillation might come off as unseemly, but it reflects an important lesson. Some of our understanding of time is based on logic and the known laws of physics, but some of it is based purely on convenience and reasonable-sounding assumptions. We *think* that the ability to uniquely determine the future from knowledge of our present state is important, but the real world might end up having other ideas. If closed timelike curves could exist, we would have a definitive answer to the debate between eternalism and presentism: The eternalist block universe would win hands down, for the straightforward reason that the universe can't be nicely divided into a series of "presents" if there are closed timelike curves lurking around.

The ultimate answer to the puzzles raised by closed timelike curves is probably that they simply don't (and can't) exist. But if that's true, it's because the laws of physics won't let you warp spacetime enough to create them, not because they let you kill your ancestors. So it's to the laws of physics we should turn.

FLATLAND

Closed timelike curves offer an interesting thought-experiment laboratory in which to explore the nature of time. But if we're going to take them seriously, we need to ask whether or not they could exist in the real world, at least according to the rules of general relativity.

I've already mentioned a handful of solutions to Einstein's equation that feature closed timelike curves—the circular-time universe, the Gödel universe, the inner region near the singularity of a rotating black hole, and an infinite spinning cylinder. But these all fall short of our idea of what it would mean to "build" a time machine—to create closed timelike curves where there weren't any already. In the case of the circular-time universe, the Gödel universe, and the rotating cylinder, the closed timelike curves are built into the universe from the start.[91] The real question is, can we make closed timelike curves in a local region of spacetime?

Glancing all the way back at Figure 23, it's easy to see why all of these solutions feature some kind of rotation—it's not enough to tilt light cones; we want them to tilt around in a circle. So if we were to sit down and guess how to make a closed timelike curve in spacetime, we might think to start something rotating—if not an infinite cylinder or a black hole, then perhaps a pretty long cylinder, or just a very massive star. We might be able to get even more juice by starting with two giant masses, and shooting them by each other at an enormous relative speed. And then, if we got lucky, the gravitational pull of those masses would distort the light cones around them enough to create a closed timelike curve.

That all sounds a bit loosey-goosey, and indeed we're faced with an immediate problem: General relativity is complicated. Not just conceptually, but technically; the equations governing the curvature of spacetime are enormously difficult to solve in any real-world situation. What we know about the exact predictions of the theory comes mostly from highly idealized cases with a great deal of symmetry, such as a static star or a completely smooth universe. Determining the spacetime curvature caused by two black holes passing by each other near the speed of light is beyond our current capabilities (although the state of the art is advancing rapidly).

In this spirit of dramatic simplification, we can ask, what would happen if two massive objects passed by each other at high relative velocity, but in a universe with only *three dimensions of spacetime*? That is, instead of the three dimensions of space and one dimension of time in our real four-dimensional spacetime, let's pretend that there are only two dimensions of space, to make three spacetime dimensions in total.

Throwing away a dimension of space in the interest of simplicity is a venerable move. Edwin A. Abbott, in his book *Flatland*, conceived of beings who lived in a two-dimensional space as a way of introducing the idea that there could be *more* than three dimensions, while at the same time taking the opportunity to satirize Victorian culture.[92] We will borrow Abbott's terminology, and refer to a universe with two spatial dimensions and one time dimension as "Flatland," even if it's not really flat—we care about cases where spacetime is curved, and light cones can tip, and timelike curves can be closed.

STUDYING TIME MACHINES IN FLATLAND (AND IN CAMBRIDGE)

Consider the situation portrayed in Figure 26, where two massive objects in Flatland are zooming past each other at high velocity. The marvelous feature of a three-dimensional universe is that Einstein's equation simplifies enormously, and what would have been an impossibly complicated problem in the real four-dimensional world can now be solved exactly. In 1991, astrophysicist Richard Gott rolled up his sleeves and calculated the spacetime curvature for this situation. Remarkably, he found that heavy objects moving by each other in Flatland *do* lead to closed timelike curves, if they are moving fast enough. For any particular value of the mass of the two bodies, Gott calculated a speed at which they would have to be moving in order to tilt the surrounding light cones sufficiently to open up the possibility of time travel.[93]

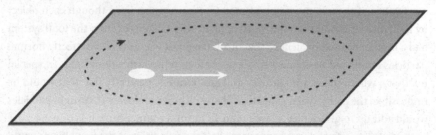

Figure 26: A Gott time machine in Flatland. If two objects pass by each other with sufficiently high relative velocity, the dashed loop will be a closed timelike curve. Note that the plane illustrated here is truly two-dimensional, not a projection of three-dimensional space.

This is an intriguing result, but it doesn't quite count as "building" a time machine. In Gott's spacetime, the objects start out far apart, pass by each other, and then zip back out to infinity again. Ultimately, the closed timelike curves were destined to occur; there is no point in the evolution where their formation could have been avoided. So the question still lingers—can we build a Gott time machine? For example, we could imagine starting with two massive objects in Flatland that were at rest with respect to each other, and hooking up rocket engines to each of them. (Keep telling yourself: "thought experiment.") Could we accelerate them fast enough to create closed timelike curves? That would really count as "building a time machine," albeit in somewhat unrealistic circumstances.

The answer is fascinating, and I was lucky enough to be in on the ground floor when it was worked out.[94] When Gott's paper appeared in 1991, I was a graduate student at Harvard, working mostly with my advisor, George Field. But like many Harvard students, I frequently took the Red Line subway down to MIT to take courses that weren't offered at my home institution. (Plenty of MIT students came the other way for similar reasons.) Among these were excellent courses on theoretical particle physics from Edward ("Eddie") Farhi, and on early-universe cosmology from Alan Guth. Eddie was a younger guy with a Bronx accent and a fairly no-nonsense attitude toward physics, at least for someone who wrote papers like "Is it Possible to Create a Universe in the Laboratory by Quantum Tunneling?"[95] Alan was an exceptionally clear-minded physicist who was world-famous as the inventor of the inflationary universe scenario. They were both also friendly and engaged human beings, guys with whom you'd be happy to socialize with, even without interesting physics to talk about.

So I was thrilled and honored when the two of them pulled me into a collaboration to tackle the question of whether it was possible to build a Gott time machine. Another team of theorists—Stanley Deser, Roman Jackiw, and Nobel laureate Gerard 't Hooft—were also working on the problem, and they had uncovered a curious feature of the two moving bodies in Gott's universe: Even though each object by itself moved slower than the speed of light, when taken together the total system had a momentum equivalent to that of a tachyon. It was as if two perfectly normal particles combined to create a single particle moving faster than light. In special relativity, where there is no gravity and spacetime is perfectly flat, that would be impossible; the combined momentum of any number of slower-than-light particles would add up to give a nice slower-than-light total momentum. It is only because of the peculiarities of curved spacetime that the velocities of the two objects could add together in that funny way. But to us, it wasn't quite the final word; who is to say that the peculiarities of curved spacetime didn't allow you to make tachyons?

We tackled the rocket-ship version of the problem: Could you start with slowly moving objects and accelerate them fast enough to make a time machine? When put that way, it's hard to see what could possibly go wrong—with a big enough rocket, what's to stop you from accelerating the heavy objects to whatever speed you like?

The answer is, there's not enough energy in the universe. We started by assuming an "open universe"—the plane in Flatland through which our particles were moving extended out to infinity. But it is a peculiar feature of gravity in Flatland that there is an absolute upper limit on the total amount of energy that you can fit in an open universe. Try to fit more, and the spacetime curvature becomes too much, so that the universe closes in on itself.[96] In four-dimensional spacetime, you can fit as much energy in the universe as you like; each bit of energy curves spacetime nearby, but the effect dilutes away as you go far from the source. In three-dimensional spacetime, by contrast, the effect of gravity doesn't dilute away; it just builds up. In an open three-dimensional universe, therefore, there is a maximum amount of energy you can possibly have—and it is not enough to make a Gott time machine if you don't have one to start with.

That's an interesting way for Nature to manage to avoid creating a time machine. We wrote two papers, one by the three of us that gave reasonable-sounding arguments for our result, and another with Ken Olum that proved it in greater generality. But along the way we noticed something interesting. There's an upper limit to how much energy you can have in an open Flatland universe, but what about a closed universe? If you try to stick too much energy into an open universe, the problem is that it closes in on itself. But turn that bug into a feature by considering closed universes, where space looks something like a sphere instead of like a plane.[97] Then there is precisely one value of the total amount of allowed energy—there is no wriggle room; the total curvature of space has to add up to be exactly that of a sphere—and that value is twice as large as the most you can fit in an open universe.

When we compared the total amount of energy in a closed

Figure 27: Particles moving in a closed Flatland universe, with the topology of a sphere. Think of ants crawling over the surface of a beach ball.

Flatland universe to the amount you would need to create a Gott time machine, we found there was enough. This was after we had already submitted our first paper and it had been accepted for publication in *Physical Review Letters*, the leading journal in the field. But journals allow you to insert small notes "added in proof" to your papers before they are published, so we tacked on a couple of sentences mentioning that we thought you *could* make a time machine in a closed Flatland universe, even if it were impossible in an open universe.

We goofed. (The single best thing about working with famous senior collaborators as a young scientist is that, when you goof, you can think to yourself, "Well if even *those* guys didn't catch this one, how dumb could it have been?") It did seem a little funny to us that Nature had been so incredibly clever in avoiding Gott time machines in open universes but didn't seem to have any problem with them in closed universes. But there was certainly enough energy to accelerate the objects to sufficient velocity, so again—what could possibly go wrong?

Very soon thereafter, Gerard 't Hooft figured out what could go wrong. A closed universe, unlike an open universe, has a finite total volume—really a "finite total area," since we have only two spatial dimensions, but you get the idea. What 't Hooft showed was that, if you set some particles moving in a closed Flatland universe in an attempt to make a Gott time machine, that volume starts to rapidly decrease. Basically, the universe starts to head toward a Big Crunch. Once that possibility occurs to you, it's easy to see how spacetime avoids making a time machine—it crunches to zero volume before the closed timelike curves are created. The equations don't lie, and Eddie and Alan and I acknowledged our mistake, submitting an erratum to *Physical Review Letters*. The progress of science marched on, seemingly little worse for the wear.

Between our result about open universes and 't Hooft's result about closed universes, it was clear that you couldn't make a Gott time machine in Flatland by starting from a situation where such a time machine wasn't already there. It may seem that much of the reasoning used to derive these results is applicable only to the unrealistic case of three-dimensional spacetime, and you would be right. But it was very clear that general relativity was trying to tell us something: It doesn't like closed timelike curves. You can try to make them, but something always seems to go wrong. We would certainly like to ask how far you could push that conclusion into the real world of four-dimensional spacetime.

WORMHOLES

In the spring of 1985, Carl Sagan was writing a novel—*Contact*, in which astrophysicist Ellie Arroway (later to be played by Jodie Foster in the movie version) makes first contact with an alien civilization.[98] Sagan was looking for a way to move Ellie quickly over interstellar distances, but he didn't want to take the science fiction writer's lazy way out and invoke warp drive to move her faster than light. So he did what any self-respecting author would do: He threw his heroine into a black hole, hoping that she would pop out unharmed twenty-six light-years away.

Not likely. Poor Ellie would have been "spaghettified"—stretched to pieces by the tidal forces near the singularity of the black hole, and not spit out anywhere at all. Sagan wasn't ignorant of black-hole physics; he was thinking about rotating black holes, where the light cones don't actually force you to smack into the singularity, at least according to the exact solution that had been found by Roy Kerr back in the sixties. But he recognized that he wasn't the world's expert, either, and he wanted to be careful about the science in his novel. Happily, he was friends with the person who was the world's expert: Kip Thorne, a theoretical physicist at Caltech who is one of the foremost authorities on general relativity.

Thorne was happy to read Sagan's manuscript, and noticed the problem: Modern research indicates that black holes in the real world aren't quite as well behaved as the pristine Kerr solution. An actual black hole that might have been created by physical processes in our universe, whether spinning or not, would chew up an intrepid astronaut and never spit her out. But there might be an alternative idea: a *wormhole.*

Unlike black holes, which almost certainly exist in the real world and for which we have a great deal of genuine observational evidence, wormholes are entirely conjectural playthings of theorists. The idea is more or less what it sounds like: Take advantage of the dynamical nature of spacetime in general relativity to imagine a "bridge" connecting two different regions of space.

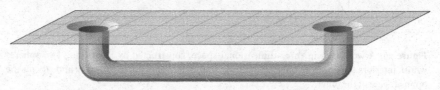

Figure 28: A wormhole connecting two distant parts of space. Although it can't be accurately represented in a picture, the physical distance through the wormhole could be much shorter than the ordinary distance between the wormhole mouths.

A typical representation of a wormhole is depicted in Figure 28. The plane represents three-dimensional space, and there is a sort of tunnel that provides a shortcut between two distant regions; the places where the wormhole connects with the external space are the "mouths" of the wormhole, and the tube connecting them is called the "throat." It doesn't *look* like a shortcut—in fact, from the picture, you might think it would take longer to pass through the wormhole than to simply travel from one mouth to the other through the rest of space. But that's just a limitation on our ability to draw interesting curved spaces by embedding them in our boring local region of three-dimensional space. We are certainly welcome to contemplate a geometry that is basically of the form shown in the previous figure, but in which the distance through the wormhole is anything we like—including much shorter than the distance through ordinary space.

In fact, there is a much more intuitive way of representing a wormhole. Just imagine ordinary three-dimensional space, and "cut out" two spherical regions of equal size. Then identify the surface of one sphere with the other. That is, proclaim that anything that enters one sphere immediately emerges out of the opposite side of the other. What we end up with is portrayed in Figure 29; each sphere is one of the mouths of a wormhole. This is a wormhole of precisely zero length; if you enter one sphere, you instantly emerge out of the other. (The word *instantly* in that sentence should set off alarm bells—instantly to *whom*?)

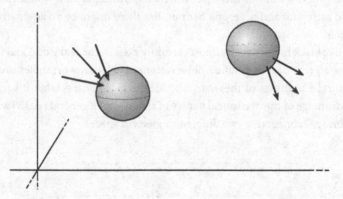

Figure 29: A wormhole in three-dimensional space, constructed by identifying two spheres whose interiors have been removed. Anything that enters one sphere instantly appears on the opposite side of the other.

The wormhole is reminiscent of our previous gate-into-yesterday example. If you look through one end of the wormhole, you don't see swirling colors or flashing lights; you see whatever is around the other end of the wormhole, just as if you were looking through some sort of periscope (or at a video monitor, where the camera is at the other end). The only difference is that you could just as easily put your hand through, or (if the wormhole were big enough), jump right through yourself.

This sort of wormhole is clearly a shortcut through spacetime, connecting two distant regions in no time at all. It performs exactly the trick that Sagan needed for his novel, and on Thorne's advice he rewrote the relevant section. (In the movie version, sadly, there were swirling colors and flashing lights.) But Sagan's question set off a chain of ideas that led to innovative scientific research, not just a more accurate story.

TIME MACHINE CONSTRUCTION MADE EASY

A wormhole is a shortcut through space; it allows you to get from one place to another much faster than you would if you took a direct route through the bulk of spacetime. You are never moving faster than light from your local point of view, but you get to your destination sooner than light would be able to if the wormhole weren't there. We know that faster-than-light travel can be used to go backward in time; travel through a wormhole isn't literally that, but certainly bears a family resemblance. Eventually Thorne, working with Michael Morris, Ulvi Yurtsever, and others, figured out how to manipulate a wormhole to create closed time-like curves. [99]

The secret is the following: When we toss around a statement like, "the wormhole connects two distant regions of space," we need to take seriously the fact that it really connects two sets of events in *spacetime*. Let's imagine that spacetime is perfectly flat, apart from the wormhole, and that we have defined a "background time" in some rest frame. When we identify two spheres to make a wormhole, we do so "at the same time" with respect to this particular background time coordinate. In some other frame, they wouldn't be at the same time.

Now let's make a powerful assumption: We can pick up and move each mouth of the wormhole independently of the other. There is a certain amount of hand-waving justification that goes into this assumption, but for the purposes of our thought experiment it's perfectly okay. Next, we let one mouth sit quietly on an unaccelerated trajectory, while we move the other one out and back at very high speed.

To see what happens, imagine that we attach a clock to each wormhole mouth. The clock on the stationary mouth keeps time along with the background time coordinate. But the clock on the out-and-back wormhole mouth experiences less time along its path, just like any other moving object in relativity. So when the two mouths are brought back next to each other, the clock that moved now seems to be behind the clock that stayed still.

Now consider exactly the same situation, but think of it from the point of view that you would get by *looking through the wormhole*. Remember, you don't see anything spooky when you look through a wormhole mouth; you just see whatever view is available to the other mouth. If we compare the two clocks as seen through the wormhole mouth, they don't move with respect to each other. That's because the length of the wormhole throat doesn't change (in our simplified example it's exactly zero), even when the mouth moves. Viewed through the wormhole, there are just two clocks that are sitting nearby each other, completely stationary. So they remain in synchrony, keeping perfect time as far as they are each concerned.

How can the two clocks continue to agree with each other, when we previously said that the clock that moved and came back would have experienced less elapsed time? Easy—the clocks appear to differ when we look at them as an external observer, but they appear to match when we look at them through the wormhole. This puzzling phenomenon has a simple explanation: Once the two wormhole mouths move on different paths through spacetime, the identification between them is no longer at the same time from the background point of view. The sphere

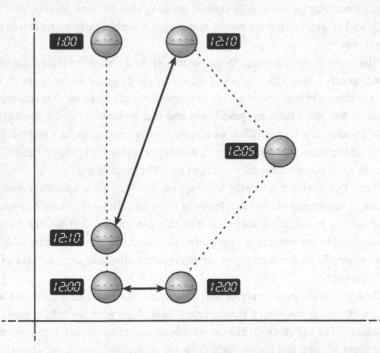

Figure 30: A wormhole time machine. Double-sided arrows represent identifications between the spherical wormhole mouths. The two mouths start nearby, identified at equal background times. One remains stationary, while the other moves away and returns near the speed of light, so that they become identified at very different background times.

representing one mouth is still identified with the sphere representing the other mouth, but now they are identified *at different times*. By passing through one, you move into the past, as far as the background time is concerned; by passing through in the opposite direction, you move into the future.

This kind of wormhole, therefore, is exactly like the gate into yesterday. By manipulating the ends of a wormhole with a short throat, we have connected two different regions of spacetime with very different times. Once we've done that, it's easy enough to travel through the wormhole in such a way as to describe a closed timelike curve, and all of the previous worries about paradoxes apply. This procedure, if it could be carried out in the real world, would unambiguously count as "building a time machine" by the standards of our earlier discussion.

PROTECTION AGAINST TIME MACHINES

The wormhole time machines make it sound somewhat plausible that closed timelike curves could exist in the real world. The problem seemingly becomes one of technological ability, rather than restrictions placed by the laws of physics; all we need is to find a wormhole, keep it open, move one of the mouths in the right way . . . Well, perhaps it's not completely plausible after all. As one might suspect, there turn out to be a number of good reasons to believe that wormholes don't provide a very practical route to building time machines.

First, wormholes don't grow on trees. In 1967, theoretical physicist Robert Geroch investigated the question of wormhole construction, and he showed that you actually could create a wormhole by twisting spacetime in the appropriate way—but only if, as an intermediate step in the process, you created a closed timelike curve. In other words, the first step to building a time machine by manipulating a wormhole is to build a time machine so that you can make a wormhole.[100] But even if you were lucky enough to stumble across a wormhole, you'd be faced with the problem of keeping it open. Indeed, this difficulty is recognized as the single biggest obstacle to the plausibility of the wormhole time machine idea.

The problem is that keeping a wormhole open requires negative energies. Gravity is attractive: The gravitational field caused by an ordinary positive-energy object works to pull things together. But look back at Figure 29 and see what the wormhole does to a collection of particles that pass through it—it "defocuses" them, taking particles that were initially coming together and now pushing them apart. That's the opposite of gravity's conventional behavior, and a sign that negative energies must be involved.

Do negative energies exist in Nature? Probably not, at least not in the ways necessary to sustain a macroscopic wormhole—but we can't say for sure. Some people have proposed ideas for using quantum mechanics to create pockets of negative energy, but they're not on a very firm footing. A big hurdle is that the question necessarily involves both gravity and quantum mechanics, and the intersection of those two theories is not very well understood.

As if that weren't enough to worry about, even if we found a wormhole and knew a way to keep it open, chances are that it would be unstable—the slightest disturbance would send it collapsing into a black hole. This is another question for which it's hard to find a clear-cut answer, but the basic idea is that any tiny ripple in energy can zoom around a closed timelike curve an arbitrarily large number of times. Our best current thinking is that this kind of repeat journey is inevitable, at least for some small fluctuations. So the wormhole doesn't just feel the mass of a single speck of dust passing through—it feels that effect over and over again, creating an enormous gravitational field, enough to ultimately destroy our would-be time machine.

Nature, it seems, tries very hard to stop us from building a time machine. The accumulated circumstantial evidence prompted Stephen Hawking to propose what he calls the "Chronology Protection Conjecture": The laws of physics (whatever they may be) prohibit the creation of closed timelike curves.[101] We have a lot of evidence that something along those lines is true, even if we fall short of a definitive proof.

Time machines fascinate us, in part because they seem to open the door to paradoxes and challenge our notions of free will. But it's likely that they don't exist, so the problems they present aren't the most pressing (unless you're a Hollywood screenwriter). The arrow of time, on the other hand, is indisputably a feature of the real world, and the problems it presents demand an explanation. The two phenomena are related; there can be a consistent arrow of time throughout the observable universe only because there are no closed timelike curves, and many of the disconcerting properties of closed timelike curves arise from their incompatibility with the arrow of time. The absence of time machines is necessary for a consistent arrow of time, but it's by no means sufficient to explain it. Having laid sufficient groundwork, it's time to confront the mystery of time's direction head-on.

PART THREE

ENTROPY AND TIME'S ARROW

7

RUNNING TIME BACKWARD

*This is what I mean when I say I would like to swim against the stream of
time: I would like to erase the consequences of certain events and restore
an initial condition.*

—Italo Calvino, *If on a Winter's Night a Traveler*

Pierre-Simon Laplace was a social climber at a time when social climbing was a
risky endeavor.[102] When the French Revolution broke out, Laplace had established
himself as one of the greatest mathematical minds in Europe, as he would frequently remind his colleagues at the Académie des Sciences. In 1793 the Reign of
Terror suppressed the Académie; Laplace proclaimed his Republican sympathies,
but he also moved out of Paris just to be safe. (Not without reason; his colleague
Antoine Lavoisier, the father of modern
chemistry, was sent to the guillotine in
1794.) He converted to Bonapartism when
Napoleon took power, and dedicated his
Théorie Analytique des Probabilités to the
emperor. Napoleon gave Laplace a position as minister of the interior, but he
didn't last very long—something about
being too abstract-minded. After the res-
toration of the Bourbons, Laplace became
a Royalist, and omitted the dedication
to Napoleon from future editions of his
book. He was named a marquis in 1817.

Social ambitions notwithstanding,
Laplace could be impolitic when it came
to his science. A famous anecdote concerns his meeting with Napoleon, after he

Figure 31: Pierre Simon Laplace, mathe
matician, physicist, swerving politician,
and unswerving determinist.

had asked the emperor to accept a copy of his *Méchanique Céleste*—a five-volume treatise on the motions of the planets. It seems unlikely that Napoleon read the whole thing (or any of it), but someone at court did let him know that the name of God was entirely absent. Napoleon took the opportunity to mischievously ask, "M. Laplace, they tell me you have written this large book on the system of the universe, and have never even mentioned its Creator." To which Laplace answered stubbornly, "I had no need of that hypothesis."[103]

One of the central tenets of Laplace's philosophy was determinism. It was Laplace who truly appreciated the implications of Newtonian mechanics for the relationship between the present and the future: Namely, if you understood everything about the present, the future would be absolutely determined. As he put it in the introduction to his essay on probability:

> We may regard the present state of the universe as the effect of its past and the cause of its future. An intellect which at a certain moment would know all forces that set nature in motion, and all positions of all items of which nature is composed, if this intellect were also vast enough to submit these data to analysis, it would embrace in a single formula the movements of the greatest bodies of the universe and those of the tiniest atom; for such an intellect nothing would be uncertain and the future just like the past would be present before its eyes.[104]

These days we would probably say that a sufficiently powerful computer could, given all there was to know about the present universe, predict the future (and retrodict the past) with perfect accuracy. Laplace didn't know about computers, so he imagined a vast intellect. His later biographers found this a bit dry, so they attached a label to this hypothetical intellect: *Laplace's Demon.*

Laplace never called it a demon, of course; presumably he had no need to hypothesize demons any more than gods. But the idea captures some of the menace lurking within the pristine mathematics of Newtonian physics. The future is not something that has yet to be determined; our fate is encoded in the details of the current universe. Every moment of the past and future is fixed by the present. It's just that we don't have the resources to perform the calculation.[105]

There is a deep-seated impulse within all of us to resist the implications of Laplace's Demon. We don't want to believe that the future is determined, even if someone out there did have access to the complete state of the universe. Tom Stoppard's *Arcadia* once again expresses this anxiety in vivid terms.

VALENTINE: Yes. There was someone, forget his name, 1820s, who pointed out that from Newton's laws you could predict everything to come—I mean, you'd need a computer as big as the universe but the formula would exist.

CHLOË: But it doesn't work, does it?

VALENTINE: No. It turns out the maths is different.

CHLOË: No, it's all because of sex.

VALENTINE: Really?

CHLOË: That's what I think. The universe is deterministic all right, just like Newton said, I mean it's trying to be, but the only thing going wrong is people fancying other people who aren't supposed to be in that part of the plan.

VALENTINE: Ah. The attraction Newton left out. All the way back to the apple in the garden. Yes. (Pause.) Yes, I think you're the first person to think of this.[106]

We won't be exploring whether sexual attraction helps us wriggle free of the iron grip of determinism. Our concern is with why the past seems so demonstrably different from the future. But that wouldn't be nearly the puzzle it appears to be if it weren't for the fact that the underlying laws of physics seem perfectly reversible; as far as Laplace's Demon is concerned, there's no difference between reconstructing the past and predicting the future.

Reversing time turns out to be a surprisingly subtle concept for something that would appear at first glance to be relatively straightforward. (Just run the movie backward, right?) Blithely reversing the direction of time is *not* a symmetry of the laws of nature—we have to dress up what we really mean by "reversing time" in order to correctly pinpoint the underlying symmetry. So we'll approach the topic somewhat circuitously, through simplified toy models. Ultimately I'll argue that the important concept isn't "time reversal" at all, but the similar-sounding notion of "reversibility"—our ability to reconstruct the past from the present, as Laplace's Demon is purportedly able to do, even if it's more complicated than simply reversing time. And the key concept that ensures reversibility is *conservation of information*—if the information needed to specify the state of the world is preserved as time passes, we will always be able to run the clock backward and recover any previous state. That's where the real puzzle concerning the arrow of time will arise.

CHECKERBOARD WORLD

Let's play a game. It's called "checkerboard world," and the rules are extremely simple. You are shown an array of squares—the checkerboard—with some filled in white, and some filled in gray. In computer-speak, each square is a "bit"—we could label the white squares with the number "0," and the gray squares with "1." The checkerboard stretches infinitely far in every direction, but we get to see only some finite part of it at a time.

The point of the game is to *guess the pattern*. Given the array of squares before you, your job is to discern patterns or rules in the arrangements of whites and grays. Your guesses are then judged by revealing more checkerboard than was originally shown, and comparing the predictions implied by your guess to the actual checkerboard. That last step is known in the parlance of the game as "testing the hypothesis."

Of course, there is another name to this game: It's called "science." All we've done is describe what real scientists do to understand nature, albeit in a highly idealized context. In the case of physics, a good theory has three ingredients: a specification of the *stuff* that makes up the universe, the *arena* through which the stuff

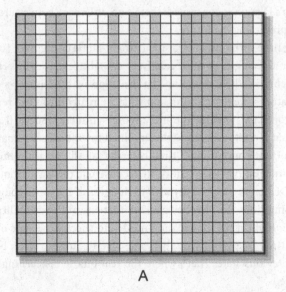

A

Figure 32: An example of checkerboard world, featuring a simple pattern within each vertical column.

is distributed, and a set of *rules* that the stuff obeys. For example, the stuff might be elementary particles or quantum fields, the arena might be four-dimensional spacetime, and the rules might be the laws of physics. Checkerboard world is a model for exactly that: The stuff is a set of bits (o's and 1's, white and gray squares), the arena through which they are distributed is the checkerboard itself, and the rules—the laws of nature in this toy world—are the patterns we discern in the behavior of the squares. When we play this game, we're putting ourselves in the position of imaginary physicists who live in one of these imaginary checkerboard worlds, and who spend their time looking for patterns in the appearance of the squares as they attempt to formulate laws of nature.[107]

In Figure 32 we have a particularly simple example of the game, labeled "checkerboard A." Clearly there is some pattern going on here, which should be pretty evident. One way of putting it would be, "every square in any particular column is in the same state." We should be careful to check that there aren't any *other* patterns lurking around—if someone else finds more patterns than we do, we lose the game, and they go collect the checkerboard Nobel Prize. From the looks of checkerboard A, there doesn't seem to be any obvious pattern as we go across a row that would allow us to make further simplifications, so it looks like we're done.

As simple as it is, there are some features of checkerboard A that are extremely relevant for the real world. For one thing, notice that the pattern distinguishes between "time," running vertically up the columns, and "space," running horizontally across the rows. The difference is that anything can happen within a row—as far as we can tell, knowing the state of one particular square tells us nothing about the state of nearby squares. In the real world, analogously, we believe that we can start with any configuration of matter in space that we like. But once that configuration is chosen, the "laws of physics" tell us exactly what must happen through time. If there is a cat sitting on our lap, we can be fairly sure that the same cat will be nearby a moment later; but knowing there's a cat doesn't tell us very much about what else is in the room.

Starting completely from scratch in inventing the universe, it's not at all obvious that there *must* be a distinction of this form between time and space. We could imagine a world in which things changed just as sharply and unpredictably from moment to moment in time as they do from place to place in space. But the real universe in which we live does seem to feature such a distinction. The idea of "time," through which things in the universe evolve, isn't a logically necessary part of the world; it's an idea that happens to be extremely useful when thinking about the reality in which we actually find ourselves.

We characterized the rule exhibited by checkerboard A as "every square in a

Figure 33: The laws of physics can be thought of as a machine that tells us, given what the world is like right now, what it will evolve into a moment later.

column is in the same state." That's a global description, referring to the entire column all at once. But we could rephrase the rule in another way, more local in character, which works by starting from some particular row (a "moment in time") and working up or down. We can express the rule as "given the state of any particular square, the square immediately above it must be in the same state." In other words, we can express the patterns we see in terms of *evolution through time*—starting with whatever state we find at some particular moment, we can march forward (or backward), one row at a time. That's a standard way of thinking about the laws of physics, as illustrated schematically in Figure 33. You tell me what is going on in the world (say, the position and velocity of every single particle in the universe) at one moment of time, and the laws of physics are a black box that tells us what the world will evolve into just one moment later.[108] By repeating the process, we can build up the entire future. What about the past?

FLIPPING TIME UPSIDE DOWN

A checkerboard is a bit sterile and limiting, as imaginary worlds go. It would be hard to imagine these little squares throwing a party, or writing epic poetry. Nevertheless, if there were physicists living in the checkerboards, they would find interesting things to talk about once they were finished formulating the laws of time evolution.

For example, the physics of checkerboard A seems to have a certain degree of symmetry. One such symmetry is *time-translation invariance*—the simple idea that the laws of physics don't change from moment to moment. We can shift our point of view forward or backward in time (up or down the columns) and the rule "the square immediately above this one must be in the same state" remains true.[109] Symmetries are always like that: If you do a certain thing, it doesn't matter; the rules still work in the same way. As we've discussed, the real world is also invariant under time shifts; the laws of physics don't seem to be changing as time passes.

Another kind of symmetry is lurking in checkerboard A: *time-reversal invariance*. The idea behind time reversal is relatively straightforward—just make time run backward. If the result "looks the same"—that is, looks like it's obeying the same laws of physics as the original setup—then we say that the rules are time-reversal invariant. To apply this to a checkerboard, just pick some particular row of the board, and reflect the squares vertically around that row. As long as the rules of the checkerboard are also invariant under time shifts, it doesn't matter which row we choose, since all rows are created equal. If the rules that described the original pattern also describe the new pattern, the checkerboard is said to be time-reversal invariant. Example A, featuring straight vertical columns of the same color squares, is clearly invariant under time reversal—not only does the reflected pattern satisfy the same rules; it is precisely the same as the original pattern.

Let's look at a more interesting example to get a better feeling for this idea. In Figure 34 we show another checkerboard world, labeled "B." Now there are two different kinds of patterns of gray squares running from bottom to top—diagonal series of squares running in either direction. (They kind of look like light cones, don't they?) Once again, we can express this pattern in terms of evolution from moment to moment in time, with one extra thing to keep in mind: Along any

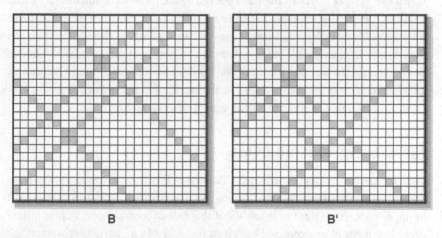

B   B'

Figure 34: Checkerboard B, on the left, has slightly more elaborate dynamics than checkerboard A, with diagonal lines of gray squares in both directions. Checkerboard B', on the right, is what happens when we reverse the direction of time by reflecting B about the middle row.

single row, it's not enough to keep track of whether a particular square is white or gray. We also need to keep track of what kinds of diagonal lines of gray squares, if any, are passing through that point. We could choose to label each square in one of four different states: "white," "diagonal line of grays going up and to the right," "diagonal line of grays going up and to the left," or "diagonal lines of grays going in both directions." If, on any particular row, we simply listed a bunch of o's and 1's, that wouldn't be enough to figure out what the next row up should look like.[110] It's as if we had discovered that there were two different kinds of "particles" in this universe, one always moving left and one always moving right, but that they didn't interact or interfere with each other in any way.

What happens to checkerboard B under time reversal? When we reverse the direction of time in this example, the result looks similar in form, but the actual configuration of white and black squares has certainly changed (in contrast to checkerboard A, where flipping time just gave us precisely the set of whites and grays we started with). The second panel in Figure 34, labeled B', shows the results of reflecting about some row in checkerboard B. In particular, the diagonal lines that were extending from lower left to upper right now extend from upper left to lower right, and vice versa.

Is the checkerboard world portrayed in example B invariant under time reversal? Yes, it is. It doesn't matter that the individual distribution of white and gray squares is altered when we reflect time around some particular row; what matters is that the "laws of physics," the rules obeyed by the patterns of squares, are unaltered. In the original example B, before reversing time, the rules were that there were two kinds of diagonal lines of gray squares, going in either direction; the same is true in example B'. The fact that the two kinds of lines switched identities doesn't change the fact that the same two kinds of lines could be found before and after. So imaginary physicists living in the world of checkerboard B would certainly proclaim that the laws of nature were time-reversal invariant.

THROUGH THE LOOKING GLASS

Well, then, what about checkerboard C, shown in Figure 35? Once again, the rules seem to be pretty simple: We see nothing but diagonal lines going from lower left to upper right. If we want to think about this rule in terms of one-step-at-a-time evolution, it could be expressed as "given the state of any particular square, the square one step above and one step to the right must be in the same state." It is certainly invariant under shifts in time, since that rule doesn't care about what row you start from.

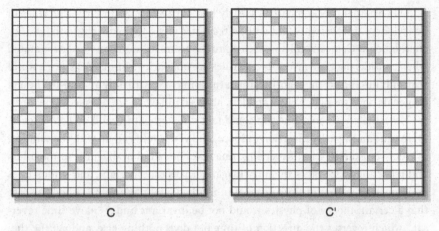

C   C'

Figure 35: Checkerboard world C only has diagonal lines of gray squares running from lower left to upper right. If we reverse the direction of time to obtain C', we only have lines running from bottom right to top left. Strictly speaking, checkerboard C is not time-reversal invariant, but it is invariant under simultaneous reflection in space and reversal in time.

If we reverse the direction of time in checkerboard C, we get something like the checkerboard C' shown in the figure. Clearly this is a different situation than before. The rules obeyed in C' are not those obeyed in C—diagonal lines stretching from lower left to upper right have been replaced by diagonal lines stretching the other way. Physicists who lived in one of these checkerboards would say that time reversal was *not* a symmetry of their observed laws of nature. We can tell the difference between "forward in time" and "backward in time"—forward is the direction in which the diagonal lines move to the right. It is completely up to us which direction we choose to label "the future," but once we make that choice it's unambiguous.

However, that's surely not the end of the story. While checkerboard C might not be, strictly speaking, invariant under time reversal as we have defined it, there does seem to be something "reversible" about it. Let's see if we can't put our fingers on it.

In addition to time reversal, we could also consider "space reversal," which would be obtained by flipping the checkerboard *horizontally* around some given column. In the real world, that's the kind of thing we get by looking at something in a mirror; we can think of space reversal as just taking the mirror image of something. In physics, it usually goes by the name of "parity," which (when we have three dimensions of space rather than just the one of the checkerboard) can be obtained by simultaneously inverting every spatial direction. Let's call it parity, so that we can sound like physicists when the occasion demands it.

Our original checkerboard A clearly had parity as a symmetry—the rules of behavior we uncovered would still be respected if we flipped right with left. For checkerboard C, meanwhile, we face a situation similar to the one we encountered when considering time reversal—the rules are not parity symmetric, since a world with only up-and-to-the-right diagonals turns into one with only up-and-to-the-left diagonals once we switch right and left, just as it did when we reversed time.

Nevertheless, it looks like you could take checkerboard C and do *both* a reversal in time and a parity inversion in space, and you would end up with the same set of rules you started with. Reversing time takes one kind of diagonal to the other, and reflecting space takes them back again. That's exactly right, and it illustrates an important feature of time reversal in fundamental physics: It is often the case that a certain theory of physics would not be invariant under "naïve time reversal," which reverses the direction of time but does nothing else, and yet the theory is invariant under an appropriately generalized symmetry transformation that reverses the direction of time and also does some other things. The way this works in the real world is a tiny bit subtle and becomes enormously more confusing the way it is often discussed in physics books. So let's leave the blocky world of checkerboards and take a look at the actual universe.

THE STATE-OF-THE-SYSTEM ADDRESS

The theories that physicists often use to describe the real world share the underlying framework of a "state" that "evolves with time." That's true for classical mechanics as put together by Newton, or for general relativity, or for quantum mechanics, all the way up to quantum field theory and the Standard Model of particle physics. On one of our checkerboards, a state is a horizontal row of squares, each of which is either white or gray (with perhaps some additional information). In different approaches to real-world physics, what counts as a "state" will be different. But in each case we can ask an analogous set of questions about time reversal and other possible symmetries.

A "state" of a physical system is "all of the information about the system, at some fixed moment in time, that you need to specify its future evolution,"[111] given the laws of physics." In particular, we have in mind isolated systems—those that aren't subject to unpredictable external forces. (If there are predictable external forces, we can simply count those as part of the "laws of physics" relevant to that system.) So we might be thinking of the whole universe, which is isolated by hypothesis, or some spaceship far away from any planets or stars.

First consider classical mechanics—the world of Sir Isaac Newton.[112] What information do we need to predict the future evolution of a system in Newtonian mechanics? I've already alluded to the answer: the position and velocity of every component of the system. But let's creep up on it gradually.

When someone brings up "Newtonian mechanics," you know sooner or later you'll be playing with billiard balls.[113] But let's imagine a game that is not precisely like conventional eight ball; it's a unique, hypothetical setup, which we might call "physicist's billiards." In our eagerness to strip away complications and get to the essence of a thing, physicists imagine games of billiards in which there is no noise or friction, so that perfectly round spheres roll along the table and bounce off one another without losing any energy. Real billiard balls don't quite behave this way—there is some dissipation and sound as they knock into one another and roll along the felt. That's the arrow of time at work, as noise and friction create entropy—so we're putting those complications aside for the moment.

Start by considering a *single* billiard ball moving alone on a table. (Generalization to many billiard balls is not very hard.) We imagine that it never loses energy, and bounces cleanly off of the bumper any time it hits. For purposes of this problem, "bounces cleanly off the bumper" is part of the "laws of physics" of our closed system, the billiard ball. So what counts as the state of that single ball?

You might guess that the state of the ball at any one moment of time is simply its position on the table. That is, after all, what would show up if we took a picture of the table—you would see where the ball was. But we defined the state to consist of all the information you would need to predict the future evolution of the system, and just specifying the position clearly isn't enough. If I tell you that the ball is precisely in the middle of the table (and nothing else), and ask you to predict where it will be one second later, you would be at a loss, since you wouldn't know whether the ball is moving.

Of course, to predict the motion of the ball from information defined at a single moment in time, you need to know *both the position and the velocity* of the ball. When we say "the state of the ball," we mean the position and the velocity, and—crucially—nothing else. We don't need to know (for example) the acceleration of the ball, the time of day, what the ball had for breakfast that morning, or any other pieces of information.

We often characterize the motion of particles in classical mechanics in terms of *momentum* rather than velocity. The concept of "momentum" goes all the way back to the great Persian thinker Ibn Sina (often Latinized as Avicenna) around the year 1000. He proposed a theory of motion in which "inclination"—weight times velocity—remained constant in the absence of outside influences. The momentum tells us how much oomph an object has, and the direction in which it is moving[114];

in Newtonian mechanics it is equal to mass times velocity, and in relativity the formula is slightly modified so that the momentum goes to infinity as the velocity approaches the speed of light. For any object with a fixed mass, when you know the momentum you know the velocity, and vice versa. We can therefore specify the state of a single particle by giving its position and its momentum.

Figure 36: A lone billiard ball, moving on a table, without friction. Three different moments in time are shown. The arrows denote the momentum of the ball; it remains constant until the ball rebounds off a wall.

Once you know the position and momentum of the billiard ball, you can predict the entire trajectory as it rattles around the table. When the ball is moving freely without hitting any walls, the momentum stays the same, while the position changes with a constant velocity along a straight line. When the ball does hit a wall, the momentum is suddenly reflected with respect to the line defined by the wall, after which the ball continues on at constant velocity. That is to say, it bounces. I'm making simple things sound complicated, but there's a method behind the madness.

All of Newtonian mechanics is like that. If you have many billiard balls on the same table, the complete state of the system is simply a list of the positions and momenta of each ball. If it's the Solar System you are interested in, the state is the position and momentum of each planet, as well as of the Sun. Or, if we want to be even more comprehensive and realistic, we can admit that the state is really the position and momentum of every single particle constituting these objects. If it's your boyfriend or girlfriend you are interested in, all you need to do is precisely specify the position and momentum of every atom in his or her body. The rules of classical mechanics give unambiguous predictions for how the system will involve, using only the information of its current state. Once you specify that list, Laplace's Demon takes over, and the rest of history is determined. You are not as smart as Laplace's Demon, nor do you have access to the same amount of information, so boyfriends and girlfriends are going to remain mysterious. Besides, they are open systems, so you would have to know about the rest of the world as well.

It will often be convenient to think about "every possible state the system could conceivably be in." That is known as the *space of states* of the system. Note that *space* is being used in two somewhat different senses. We have "space," the physical arena through which actual objects move in the universe, and a more abstract notion of "a space" as any kind of mathematical collection of objects (almost the same as "set," but with the possibility of added structure). The space of states is *a* space, which will take different forms depending on the laws of physics under consideration.

In Newtonian mechanics, the space of states is called "phase space," for reasons that are pretty mysterious. It's just the collection of all possible positions and momenta of every object in the system. For our checkerboards, the space of states consists of all possible sequences of white and gray squares along one row, possibly with some extra information when diagonal lines ran into one another. Once we get to quantum mechanics, the space of states will consist of all possible wave functions describing the quantum system; the technical term is *Hilbert space*. Any good theory of physics has a space of states, and then some rule describing how a particular state evolves in time.

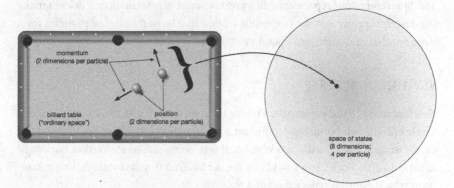

Figure 37: Two balls on a billiard table, and the corresponding space of states. Each ball requires two numbers to specify its position on the table, and two numbers to specify its momentum. The complete state of both particles is a point in an eight-dimensional space, on the right. We can't draw eight dimensions, but you should imagine they are there. Every extra ball on the table adds four more dimensions to the space of states.

The space of states can have a huge number of dimensions, even when ordinary space is just three-dimensional. In this abstract context, a "dimension" is just "a number you need to specify a point in the space." The space of states has one dimension for each component of position, and one dimension for each component of momentum, for *every particle* in the system. For a billiard ball confined to move on a flat two-dimensional table, we need to give two numbers to specify the position (because the table itself is two-dimensional), and also two numbers to specify the momentum, which has a magnitude and a direction. So the space of states of a single billiard ball confined to a two-dimensional table is *four*-dimensional: Two numbers fix the position, two fix the momentum. If we had nine balls on the table, we would have to specify two numbers for the position of each ball and two numbers for the momentum of each ball, so the total phase space would be thirty-six-dimensional. There are always an equal number of dimensions for position and momentum, since there can be momentum along every direction of real space. For a baseball flying through the air, which can be thought of as a single particle moving freely in three-dimensional space, the space of states would be six-dimensional; for 1,000 particles, it would be 6,000-dimensional.

In realistic cases, the space of states is very big indeed. An actual billiard ball consists of approximately 10^{25} atoms, and the space of states is a list of the position and momentum of each one of them. Instead of thinking of the evolution through time of all those atoms moving through three-dimensional space with their individual momenta, we can equally well think of the evolution of the entire system as the motion of a single point (the state) through a giant-dimensional space of states. This is a tremendous repackaging of a great amount of information; it doesn't make the description any simpler (we've just traded in a large number of particles for a large number of dimensions), but it provides a different way of looking at things.

NEWTON IN REVERSE

Newtonian mechanics is invariant under time reversal. If you made a movie of our single billiard ball bouncing around on a table, nobody to whom you showed the movie would be able to discern whether it was being played forward or backward in time. In either case, you would just see the ball moving in a straight line at constant velocity until it reflected off of a wall.

But that's not quite the whole story. Back in checkerboard world, we defined time-reversal invariance as the idea that we could reverse the time ordering of the sequence of states of the system, and the result would still obey the laws of physics. On the checkerboard, a state was a row of white and gray squares; for our billiard ball, it's a point in the space of states—that is, the position and momentum of the ball.

Take a look at the first part of the trajectory of the ball shown in Figure 36. The ball is moving uniformly up and to the right, and the momentum is fixed at a constant value, pointing up and to the right. So the time-reverse of that would be a series of positions of the ball moving from upper right to lower left, and a series of fixed momenta *pointing up and to the right*. But that's crazy. If the ball is moving along a time-reversed trajectory, from upper right to lower left, the momentum should surely be pointing in that direction, along the velocity of the ball. Clearly the simple recipe of taking the original set of states, ordered in time, and playing exactly the same states backward in time, does *not* give us a trajectory that obeys the laws of physics. (Or, apparently, of common sense—how can the momentum point oppositely to the velocity? It's equal to the velocity times the mass.[115])

The solution to this ginned-up dilemma is simple enough. In classical mechanics, we *define* the operation of time reversal to not simply play the original set of states backward, but also to *reverse the momenta*. And then, indeed, classical mechanics is perfectly invariant under time reversal. If you give me some evolution of a system through time, consisting of the position and momentum of each component at every moment, then I can reverse the momentum part of the state at every point, play it backward in time, and get a new trajectory that is also a perfectly good solution to the Newtonian equations of motion.

This is more or less common sense. If you think of a planet orbiting the Sun, and decide that you would like to contemplate that process in reverse, you imagine the planet reversing its course and orbiting the other way. And if you watched that for a while, you would conclude that the result still looked like perfectly reasonable behavior. But that's because your brain automatically reversed the momenta, without even thinking about it—the planet was obviously moving in the opposite direction. We don't make a big deal about it, because we don't *see* momentum in the same way that we see position, but it is just as much part of the state as the position is.

It is, therefore, *not true* that Newtonian mechanics is invariant under the most naïve definition of time reversal: Take an allowed sequence of states through time, reverse their ordering, and ask whether the new sequence is allowed by the laws of physics. And nobody is bothered by that, even a little bit. Instead, they simply define a more sophisticated version of time-reversal: Take an allowed sequence of states through time, *transform* each individual state in some simple and specific way, then reverse their ordering. By "transform" we just mean to change each state according to a predefined rule; in the case of Newtonian mechanics, the relevant transformation is "reverse the momentum." If we are able to find a sufficiently simple way to transform each individual state so that the time-reversed sequence of states is allowed by the laws of physics, we declare with a great sense of achievement that those laws are invariant under time reversal.

It's all very reminiscent (or should be, if my master plan has been successful) of the diagonal lines from checkerboard C. There we found that if you simply reversed the time ordering of states, as shown on checkerboard C', the result did not conform to the original pattern, so checkerboard C is not naïvely time-reversal invariant. But if we first flipped the checkerboard from right to left, and only then reversed the direction of time, the result *would* obey the original rules. So there does exist a well-defined procedure for transforming the individual states (rows of squares) so that checkerboard C really is time-reversal invariant, in this more sophisticated sense.

This notion of time reversal, which involves transforming states around as well as literally reversing time, might seem a little suspicious, but it is what physicists do all the time. For example, in the theory of electricity and magnetism, time-reversal leaves the electric field unchanged, but reverses the direction of the magnetic field. That's just a part of the necessary transformation; the magnetic field and the momentum both get reversed before we run time backward.[116]

The lesson of all this is that the statement "this theory is invariant under time reversal" does not, in common parlance, mean "you can reverse the direction of time and the theory is just as good." It means something like "you can transform the state at every time in some simple way, and then reverse the direction of time, and the theory is just as good." Admittedly, it sounds a bit fishy when we start including phrases like *in some simple way* into the definitions of fundamental physical concepts. Who is to say what counts as sufficiently simple?

At the end of the day, it doesn't matter. If there exists *some* transformation that you can do to the state of some system at every moment of time, so that the time-reversed evolution obeys the original laws of physics, you are welcome to define that as "invariance under time reversal." Or you are welcome to call it some other symmetry, related to time reversal but not precisely the same. The names don't matter; what matters is understanding all of the various symmetries that are respected or violated by the laws. In the Standard Model of particle physics, in fact, we are faced precisely with a situation where it's possible to transform the states in such a way that they can be run backward in time and obey the original equations of motion, but physicists choose *not* to call that "time-reversal invariance." Let's see how that works.

RUNNING PARTICLES BACKWARD

Elementary particles don't really obey the rules of classical mechanics; they operate according to quantum mechanics. But the basic principle is still the same: We can transform the states in a particular way, so that reversing the direction of time after that transformation gives us a perfectly good solution to the original theory. You will often hear that particle physics is *not* invariant under time reversal, and occasionally it will be hinted darkly that this has something to do with the arrow of time. That's misleading; the behavior of elementary particles under time reversal has nothing whatsoever to do with the arrow of time. Which doesn't stop it from being an interesting subject in its own right.

Let's imagine that we wanted to do an experiment to investigate whether elementary particle physics is time-reversal invariant. You might consider some particular process involving particles, and run it backward in time. For example, two particles could interact with each other and create other particles (as in a particle accelerator), or one particle could decay into several others. If it took a different amount of time for such a process to happen going forward and backward, that would be evidence for a violation of time-reversal invariance.

Atomic nuclei are made of neutrons and protons, which are in turn made of quarks. Neutrons can be stable if they are happily surrounded by protons and other neutrons within a nucleus, but left all alone they will decay in a number of minutes. (The neutron is a bit of a drama queen.) The problem is that a neutron will decay into a combination of a proton, an electron, and a neutrino (a very light, neutral particle).[117] You could imagine running that backward, by shooting a proton, an electron, and a neutrino at one another in precisely the right way as to make a neutron. But even if this interaction were likely to reveal anything interesting about time reversal, the practical difficulties would be extremely difficult to overcome; it's too much to ask that we arrange those particles exactly right to reproduce the time reversal of a neutron decay.

But sometimes we get lucky, and there are specific contexts in particle physics where a single particle "decays" into a single other particle, which can then "decay" right back into the original. That's not really a decay at all, since only one particle is involved—instead, such processes are known as *oscillations*. Clearly, oscillations can happen only under very special circumstances. A proton can't oscillate into a neutron, for example; their electrical charges are different. Two particles can oscillate into each other only if they have the same electric charge, the same number of quarks, and the same mass, since an oscillation shouldn't create or destroy energy. Note that a quark is different from an antiquark, so neutrons cannot oscillate into antineutrons. Basically, they have to be almost the same particle, but not quite.

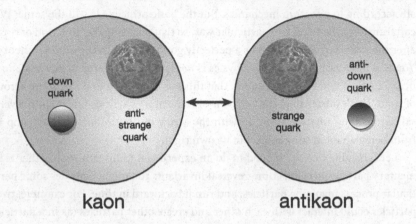

Figure 38: A neutral kaon and a neutral antikaon. Since they both have zero electric charge, and the net number of quarks is also zero, the kaon and antikaon can oscillate into each other, even though they are different particles.

Nature hands us the perfect candidate for such oscillations in the form of the *neutral kaon*. A kaon is a type of meson, which means it consists of one quark and one antiquark. If we want the two types of quarks to be different, and for the total charge to add up to zero, the easiest such particle to make will consist of one down quark and one strange antiquark, or vice versa.[118] By convention, we refer to the down/anti-strange combination as "the neutral kaon," and the strange/anti-down combination as "the neutral antikaon." They have precisely the same mass, about half the mass of a proton or neutron. It's natural to look for oscillations between kaons and antikaons, and indeed it's become something of an industry within experimental particle physics. (There are also electrically charged kaons, combinations of up quarks with strange quarks, but those aren't useful for our purposes; even if we drop the *neutral* for simplicity, we will always be referring to neutral kaons.)

So you'd like to make a collection of kaons and antikaons, and keep track as they oscillate back and forth into each other. If time-reversal invariance is violated, we would expect one process to take just a bit longer sthan the other; as a result, on average your collection would have a bit more kaons than antikaons, or vice versa. Unfortunately, the particles themselves don't come with little labels telling us which kind they are. They do, however, eventually decay into other particles entirely—the kaon decays into a negatively charged pion, an antielectron, and a neutrino, while the antikaon decays into a positively charged pion, an electron, and an antineutrino. If you measure how often one kind of decay happens compared to the other kind, you can figure out whether the original particle spends more time as a kaon than an antikaon.

Even though the theoretical predictions had been established for a while, this experiment wasn't actually carried out until 1998, by the CPLEAR experiment at the CERN laboratory in Geneva, Switzerland.[119] They found that their beam of particles, after oscillating back and forth between kaons and antikaons, decayed slightly more frequently (about 2/3 of 1 percent) like a kaon than like an antikaon; the oscillating beam was spending slightly more time as kaons than as antikaons. In other words, the process of going from a kaon to an antikaon took slightly longer than the time-reversed process of going from an antikaon to a kaon. Time reversal is *not* a symmetry of elementary particle physics in the real world.

At least, not "naïve" time reversal, as I defined it above. Is it possible to include some additional transformations that preserve some kind of time-reversal invariance in the world of elementary particles? Indeed it is, and that's worth discussing.

THREE REFLECTIONS OF NATURE

When you dig deeply into the guts of how particle physics works, it turns out that there are three different kinds of possible symmetries that involve "inverting" a physical property, each of which is denoted by a capital letter. We have time reversal T, which exchanges past and future. We also have parity P, which exchanges right and left. We discussed parity in the context of our checkerboard worlds, but it's just as relevant to three-dimensional space in the real world. Finally, we have "charge conjugation" C, which is a fancy name for the process of exchanging particles with their antiparticles. The transformations C, P, and T all have the property that when you repeat them twice in a row you simply return to the state you started with.

In principle, we could imagine a set of laws of physics that were invariant under each of these three transformations separately. Indeed, the real world superficially looks that way, as long as you don't probe it too carefully (for example, by studying decays of neutral kaons). If we made an anti-hydrogen atom by combining an antiproton with an antielectron, it would have almost exactly the same properties as an ordinary hydrogen atom—except that, if it were to touch an ordinary hydrogen atom, they would mutually annihilate into radiation. So *C* seems at first blush like a good symmetry, and likewise for *P* and *T*.

It therefore came as quite a surprise in the 1950s when one of these transformations—parity—was shown *not* to be a symmetry of nature, largely through the efforts of three Chinese-born American physicists: Tsung-Dao Lee, Chen Ning Yang, and Chien-Shiung Wu. The idea of parity violation had been floating around for a while, suggested by various people but never really taken seriously. In physics, credit accrues not just to someone who makes an offhand suggestion, but to someone who takes that suggestion seriously enough to put in the work and turn it into a respectable theory or a decisive experiment. In the case of parity violation, it was Lee and Yang who sat down and performed a careful analysis of the problem. They discovered that there was ample experimental evidence that electromagnetism and the strong nuclear force both were invariant under *P*, but that the question was open as far as the weak nuclear force was concerned.

Lee and Yang also suggested a number of ways that one could search for parity violation in the weak interactions. They finally convinced Wu, who was an experimentalist specializing in the weak interactions and Lee's colleague at Columbia, that this was a project worth tackling. She recruited physicists at the National Bureau of Standards to join her in performing an experiment on cobalt-60 atoms in magnetic fields at very low temperatures.

As they designed the experiment, Wu became convinced of the project's fundamental importance. In a later recollection, she explained vividly what it is like to be caught up in the excitement of a crucial moment in science:

> Following Professor Lee's visit, I began to think things through. This was a golden opportunity for a beta-decay physicist to perform a crucial test, and how could I let it pass?—That Spring, my husband, Chia-Liu Yuan, and I had planned to attend a conference in Geneva and then proceed to the Far East. Both of us had left China in 1936, exactly twenty years earlier. Our passages were booked on the Queen Elizabeth before I suddenly realized that I had to do the experiment immediately, before the rest of the Physics Community recognized the importance of this experiment and did it first. So I asked Chia-Liu to let me stay and go without me.
>
> As soon as the Spring semester ended in the last part of May, I started work in earnest in preparing for the experiment. In the middle of September, I finally went to Washington, D.C., for my first meeting with Dr. Ambler. . . . Between experimental runs in Washington,

I had to dash back to Columbia for teaching and other research activities. On Christmas eve, I returned to New York on the last train; the airport was closed because of heavy snow. There I told Professor Lee that the observed asymmetry was reproducible and huge. The asymmetry parameter was nearly -1. Professor Lee said that this was very good. This result is just what one should expect for a two-component theory of the neutrino. [120]

Your spouse and a return to your childhood home will have to wait—Science is calling! Lee and Yang were awarded the Nobel Prize in Physics in 1957; Wu should have been included among the winners, but she wasn't.

Once it was established that the weak interactions violated parity, people soon noticed that the experiments seemed to be invariant if you combined a parity transformation with charge conjugation C, exchanging particles with antiparticles. Moreover, this seemed to be a prediction of the theoretical models that were popular at the time. Therefore, people who were surprised that P is violated in nature took some solace in the idea that combining C and P appeared to yield a good symmetry.

It doesn't. In 1964, James Cronin and Val Fitch led a collaboration that studied our friend the neutral kaon. They found that the kaon decayed in a way that violated parity, and that the antikaon decayed in a way that violated parity slightly differently. In other words, the combined transformation of reversing parity and trading particles for antiparticles is *not* a symmetry of nature. [121] Cronin and Fitch were awarded the Nobel Prize in 1980.

At the end of the day, all of the would-be symmetries C, P, and T are violated in Nature, as well as any combination of two of them together. The obvious next step is to inquire about the combination of all three: CPT. In other words, if we take some process observed in nature, switch all the particles with their antiparticles, flip right with left, and run it backward in time, do we get a process that obeys the laws of physics? At this point, with everything else being violated, we might conclude that a stance of suspicion toward symmetries of this form is a healthy attitude, and guess that even CPT is violated.

Wrong again! (It's good to be the one both asking and answering the questions.) As far as any experiment yet performed can tell, CPT is a perfectly good symmetry of Nature. And it's more than that; under certain fairly reasonable assumptions about the laws of physics, you can *prove* that CPT must be a good symmetry—this result is known imaginatively as the "CPT Theorem." Of course, even reasonable assumptions might be wrong, and neither experimentalists nor theorists have shied away from exploring the possibility of *CPT* violation. But as far as we can tell, this particular symmetry is holding up.

I argued previously that it was often necessary to fix up the operation of time reversal to obtain a transformation that was respected by nature. In the case of the Standard Model of particle physics, the requisite fixing-up involves adding charge conjugation and parity inversion to our time reversal. Most physicists find it more convenient to distinguish between the hypothetical world in which *C*, *P*, and *T* were all individually invariant, and the real world, in which only the combination *CPT* is invariant, and therefore proclaim that the real world is not invariant under time reversal. But it's important to appreciate that there is a way to fix up time reversal so that it does appear to be a symmetry of Nature.

CONSERVATION OF INFORMATION

We've seen that "time reversal" involves not just reversing the evolution of a system, playing each state in the opposite order in time, but also doing some sort of transformation on the states at each time—maybe just reversing the momentum or flipping a row on our checkerboards, or maybe something more sophisticated like exchanging particles with antiparticles.

In that case, is *every* sensible set of laws of physics invariant under some form of "sophisticated time reversal"? Is it always possible to find some transformation on the states so that the time-reversed evolution obeys the laws of physics?

No. Our ability to successfully define "time reversal" so that some laws of physics are invariant under it depends on one other crucial assumption: *conservation of information*. This is simply the idea that two different states in the past always evolve into two distinct states in the future—they never evolve into the same state. If that's true, we say that "information is conserved," because knowledge of the future state is sufficient to figure out what the appropriate state in the past must have been. If that feature is respected by some laws of physics, the laws are *reversible*, and there will exist some (possibly complicated) transformations we can do to the states so that time-reversal invariance is respected.[122]

To see this idea in action, let's return to checkerboard world. Checkerboard D, portrayed in Figure 39, looks fairly simple. There are some diagonal lines, and one vertical column of gray squares. But something interesting happens here that didn't happen in any of our previous examples: The different lines of gray squares are "interacting" with one another. In particular, it would appear that diagonal lines can approach the vertical column from either the right or the left, but when they get there they simply come to an end.

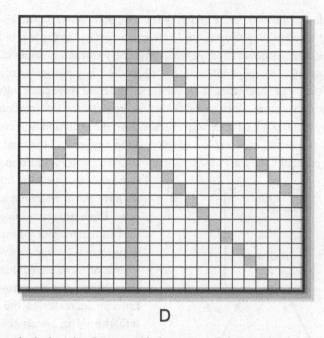

D

Figure 39: A checkerboard with irreversible dynamics. Information about the past is not preserved into the future.

That is a fairly simple rule and makes for a perfectly respectable set of "laws of physics." But there is a radical difference between checkerboard D and our previous ones: This one is not reversible. The space of states is, as usual, just a list of white and gray squares along any one row, with the additional information that the square is part of a right-moving diagonal, a left-moving diagonal, or a vertical column. And given that information, we have no problem at all in evolving the state forward in time—we know exactly what the next row up will look like, and the row after that, and so on.

But if we are told the state along one row, we cannot evolve it *backward* in time. The diagonal lines would keep going, but from the time-reversed point of view, the vertical column could spit out diagonal lines at completely random intervals (corresponding, from the point of view portrayed in the figure, to a diagonal hitting the vertical column of grays and being absorbed). When we say that a physical process is irreversible, we mean that we cannot construct the past from knowledge of the current state, and this checkerboard is a perfect example of that.

In a situation like this, information is lost. Knowing the state at one time, we can't be completely sure what the earlier states were. We have a space of states—a specification of a row of white and gray squares, with labels on the gray squares indicating whether they move up and to the right, up and to the left, or vertically. That space of states doesn't change with time; every row is a member of the same space of states, and any possible state is allowed on any particular row. But the unusual feature of checkerboard D is that two different rows can evolve into the same row in the future. Once we get to that future state, the information of which past configurations got us there is irrevocably lost; the evolution is irreversible.

In the real world, *apparent* loss of information happens all the time. Consider two different states of a glass of water. In one state, the water is uniform and at the same cool temperature; in the other, we have warm water but also an ice cube. These two states can evolve into the future into what appears to be the same state: a glass of cool water.

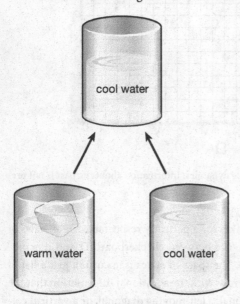

We've encountered this phenomenon before: It's the arrow of time. Entropy increases as the ice melts into the warm water; that's a process that can happen but will never un-happen. The puzzle is that the motion of the individual molecules making up the water is perfectly invariant under time reversal, while the macroscopic description in terms of ice and liquid is not. To understand how reversible underlying laws give rise to macroscopic irreversibility, we must return to Boltzmann and his ideas about entropy.

Figure 40: Apparent loss of information in a glass of water. A future state of a glass of cool water could have come either from the same state of cool water, or from warm water with an ice cube.

8

ENTROPY AND DISORDER

Nobody can imagine in physical terms the act of reversing the order of time. Time is not reversible.

—Vladimir Nabokov, *Look at the Harlequins!*

Why is it that discussions of entropy and the Second Law of Thermodynamics so often end up being about food? Here are some popular (and tasty) examples of the increase of entropy in irreversible processes:

- Breaking eggs and scrambling them.
- Stirring milk into coffee.
- Spilling wine on a new carpet.
- The diffusion of the aroma of a freshly baked pie into a room.
- Ice cubes melting in a glass of water.

To be fair, not all of these are equally appetizing; the ice-cube example is kind of bland, unless you replace the water with gin. Furthermore, I should come clean about the scrambled-eggs story. The truth is that the act of cooking the eggs in your skillet isn't a straightforward demonstration of the Second Law; the cooking is a chemical reaction that is caused by the introduction of heat, which wouldn't happen if the eggs weren't an open system. Entropy comes into play when we break the eggs and whisk the yolks together with the whites; the point of cooking the resulting mixture is to avoid salmonella poisoning, not to illustrate thermodynamics.

The relationship between entropy and food arises largely from the ubiquity of *mixing*. In the kitchen, we are often interested in combining together two things that had been kept separate—either two different forms of the same substance (ice and liquid water) or two altogether different ingredients (milk and coffee, egg whites and yolks). The original nineteenth-century thermodynamicists were extremely interested in the dynamics of heat, and the melting ice cube would have been of foremost concern to them; they would have been less fascinated by processes where all the ingredients were at the same temperature, such as spilling wine onto a carpet. But clearly there is some underlying similarity in what is going on; an initial state in which substances are kept separate evolves into a final state in which they are mixed together. It's easy to mix things and hard to unmix them—the arrow of time looms over everything we do in the kitchen.

Why is mixing easy and unmixing hard? When we mix two liquids, we see them swirl together and gradually blend into a uniform texture. By itself, that process doesn't offer much clue into what is really going on. So instead let's visualize what happens when we mix together two different kinds of colored sand. The important thing about sand is that it's clearly made of discrete units, the individual grains. When we mix together, for example, blue sand and red sand, the mixture as a whole begins to look purple. But it's not that the individual grains turn purple; they maintain their identities, while the blue grains and the red grains become jumbled together. It's only when we look from afar ("macroscopically") that it makes sense to think of the mixture as being purple; when we peer closely at the sand ("microscopically") we see individual blue and red grains.

The great insight of the pioneers of kinetic theory—Daniel Bernoulli in Switzerland, Rudolf Clausius in Germany, James Clerk Maxwell and William Thomson in Great Britain, Ludwig Boltzmann in Austria, and Josiah Willard Gibbs in the United States—was to understand all liquids and gases in the same way we think of sand: as collections of very tiny pieces with persistent identities. Instead of grains, of course, we think of liquids and gases as composed of atoms and molecules. But the principle is the same. When milk and coffee mix, the individual milk molecules don't combine with the individual coffee molecules to make some new kind of molecule; the two sets of molecules simply intermingle. Even heat is a property of atoms and molecules, rather than constituting some kind of fluid in its own right—the heat contained in an object is a measure of the energy of the rapidly moving molecules within it. When an ice cube melts into a glass of water, the molecules remain the same, but they gradually bump into one another and distribute their energy evenly throughout the molecules in the glass.

Without (yet) being precise about the mathematical definition of "entropy," the example of blending two kinds of colored sand illustrates why it is easier to mix things than to unmix them. Imagine a bowl of sand, with all of the blue grains on one side of the bowl and the red grains on the other. It's pretty clear that this arrangement is somewhat *delicate*—if we disturb the bowl by shaking it or stirring with a spoon, the two colors will begin to mix together. If, on the other hand, we start with the two colors completely mixed, such an arrangement is *robust*—if we disturb the mixture, it will stay mixed. The reason is simple: To separate out two kinds of sand that are mixed together requires a much more precise operation than simply shaking or stirring. We would have to reach in carefully with tweezers and a magnifying glass to move all of the red grains to one side of the bowl and all of the blue grains to the other. It takes much more care to create the delicate unmixed state of sand than to create the robust mixed state.

That's a point of view that can be made fearsomely quantitative and scientific, which is exactly what Boltzmann and others managed to do in the 1870s. We're going to dig into the guts of what they did, and explore what it explains and what it doesn't, and how it can be reconciled with underlying laws of physics that are perfectly reversible. But it should already be clear that a crucial role is played by the *large numbers* of atoms that we find in macroscopic objects in the real world. If we had only one grain of red sand and one grain of blue sand, there would be no distinction between "mixed" and "unmixed." In the last chapter we discussed how the underlying laws of physics work equally well forward or backward in time (suitably defined). That's a microscopic description, in which we keep careful track of each and every constituent of a system. But very often in the real world, where large numbers of atoms are involved, we don't keep track of nearly that much information. Instead, we make simplifications—thinking about the average color or temperature or pressure, rather than the specific position and momentum of each atom. When we think macroscopically, we forget (or ignore) detailed information about every particle—and that's where entropy and irreversibility begin to come into play.

SMOOTHING OUT

The basic idea we want to understand is "how do macroscopic features of a system made of many atoms evolve as a consequence of the motion of the individual atoms?" (I'll use "atoms" and "molecules" and "particles" more or less interchangeably, since all we care is that they are tiny things that obey reversible laws of physics, and that you need a lot of them to make something macroscopic.) In that spirit, consider a sealed box divided in two by a wall with a hole in it. Gas molecules can bounce around on one side of the box and will usually bounce right off the central wall, but every once in a while they will sneak through to the other side. We might imagine, for example, that the molecules bounce off the central wall 995 times out of 1,000, but one-half of 1 percent of the time (each second, let's say) they find the hole and move to the other side.

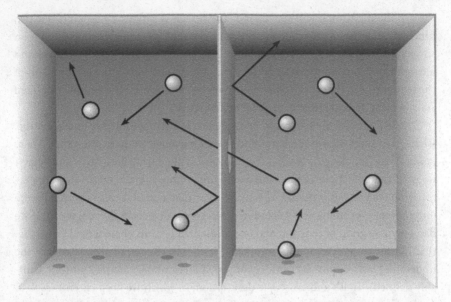

Figure 41: A box of gas molecules, featuring a central partition with a hole. Every second, each molecule has a tiny chance to go through the hole to the other side.

This example is pleasingly specific; we can examine a particular instance in detail and see what happens.[123] Every second, each molecule on the left side of the box has a 99.5 percent chance of staying on that side, and a 0.5 percent chance of moving to the other side; likewise for the right side of the box. This rule is perfectly time-reversal invariant; if you made a movie of the motion of just one particle obeying this rule, you couldn't tell whether it was being run forward or backward in time. At the level of individual particles, we can't distinguish the past from the future.

In Figure 42 we have portrayed one possible evolution of such a box; time moves upward, as always. The box has 2,000 "air molecules" in it, and starts at time $t = 1$ with 1,600 molecules on the left-hand side and only 400 on the right. (You're not supposed to ask *why* it starts that way—although later, when we replace "the box" with "the universe," we will start asking such questions.) It's not very surprising what happens as we sit there and let the molecules bounce around inside the box. Every second, there is a small chance that any particular molecule will switch sides; but, because we started with a much larger number of molecules on the one side, there is a general tendency for the numbers to even out. (Exactly like temperature, in Clausius's formulation of the Second Law.) When there are more molecules

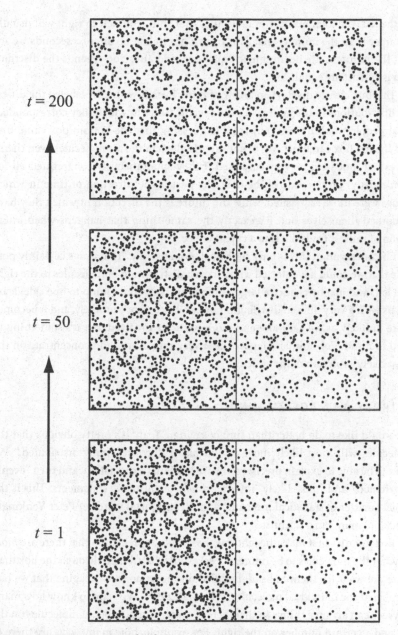

Figure 42: Evolution of 2,000 molecules in a divided box of gas. We start with 1,600 molecules on the left, 400 on the right. After 50 seconds, there are about 1,400 on the left and 600 on the right; by the time 200 seconds have passed, the molecules are distributed equally between the two sides.

on the left, the total number of molecules that shift from left to right will usually be larger than the number that shift from right to left. So after 50 seconds we see that the numbers are beginning to equal out, and after 200 seconds the distribution is essentially equal.

This box clearly displays an arrow of time. Even if we hadn't labeled the different distributions in the figure with the specific times to which they corresponded, most people wouldn't have any trouble guessing that the bottom box came first and the top box came last. We're not surprised when the air molecules even themselves out, but we'd be very surprised if they spontaneously congregated all (or even mostly) on one side of the box. The past is the direction of time in which things were more segregated, while the future is the direction in which they have smoothed themselves out. It's exactly the same thing that happens when a teaspoon of milk spreads out into a cup of coffee.

Of course, all of this is only statistical, not absolute. That is, it's certainly possible that we could have started with an even distribution of molecules to the right and left, and just by chance a large number of them could jump to one side, leaving us with a very uneven distribution. As we'll see, that's unlikely, and it becomes more unlikely as we get more and more particles involved; but it's something to keep in mind. For now, let's ignore these very rare events and concentrate on the most likely evolution of the system.

ENTROPY À LA BOLTZMANN

We would like to do better than simply saying, "Yeah, it's pretty obvious that the molecules will most likely move around until they are evenly distributed." We want to be able to explain precisely why we have that expectation, and turn "evenly distributed" and "most likely" into rigorously quantitative statements. This is the subject matter of statistical mechanics. In the immortal words of Peter Venkman: "Back off, man, I'm a scientist."

Boltzmann's first major insight was the simple appreciation that there are *more ways* for the molecules to be (more or less) evenly distributed through the box than there are ways for them to be all huddled on the same side. Imagine that we had numbered the individual molecules, 1 through 2,000. We want to know how many ways we can arrange things so that there are a certain number of molecules on the left and a certain number on the right. For example, how many ways are there to arrange things so that all 2,000 molecules are on the left, and zero on the right? There is only one way. We're just keeping track of whether each molecule is on the left or on the right, not any details about its specific position or momentum, so we simply put every molecule on the left side of the box.

But now let's ask how many ways there are for there to be 1,999 molecules on the left and exactly 1 on the right. The answer is: 2,000 different ways—one for each of the specific molecules that could be the lucky one on the right side. If we ask how many ways there are to have 2 molecules on the right side, we find 1,999,000 possible arrangements. And when we get bold and consider 3 molecules on the right, with the other 1,997 on the left, we find 1,331,334,000 ways to make it happen.[124]

It should be clear that these numbers are growing rapidly: 2,000 is a lot bigger than 1, and 1,999,000 is a lot bigger than 2,000, and 1,331,334,000 is bigger still. Eventually, as we imagine moving more and more molecules to the right and emptying out the left, they would begin to go down again; after all, if we ask how many ways we can arrange things so that all 2,000 are on the right and zero are on the left, we're back to only one unique way.

The situation corresponding to the largest number of different possible arrangements is, unsurprisingly, when things are exactly balanced: 1,000 molecules on the left and 1,000 molecules on the right. In that case, there are—well, a really big number of ways to make that happen. We won't write out the whole thing, but it's approximately 2×10^{600} different ways; a 2 followed by 600 zeroes. And that's with only 2,000 total particles. Imagine the number of possible arrangements of atoms we could find in a real roomful of air or even a glass of water. (Objects you can hold in your hand typically have about 6×10^{23} molecules in them—Avogadro's Number.) The age of the universe is only about 4×10^{17} seconds, so you are welcome to contemplate how quickly you would have to move molecules back and forth before you explored every possible allowed combination.

This is all very suggestive. There are relatively few ways for all of the molecules to be hanging out on the same side of the box, while there are very many ways for them to be distributed more or less equally—*and* we expect that a highly uneven distribution will evolve easily into a fairly even one, but not vice versa. But these statements are not quite the same. Boltzmann's next step was to suggest that, if we didn't know any better, we should expect systems to evolve from "special" configurations into "generic" ones—that is, from situations corresponding to a relatively small number of arrangements of the underlying particles, toward arrangements corresponding to a larger number of such arrangements.

Boltzmann's goal in thinking this way was to provide a basis in atomic theory for the Second Law of Thermodynamics, the statement that the entropy will always increase (or stay constant) in a closed system. The Second Law had already been formulated by Clausius and others, but Boltzmann wanted to *derive* it from some simple set of underlying principles. You can see how this statistical thinking leads us in the right direction—"systems tend to evolve from uncommon arrangements into common ones" bears a family resemblance to "systems tend to evolve from low-entropy configurations into high-entropy ones."

So we're tempted to define "entropy" as "the number of ways we can rearrange the microscopic components of a system that will leave it macroscopically unchanged." In our divided-box example, that would correspond to the number of ways we could rearrange individual molecules that would leave the total number on each side unchanged.

That's almost right, but not quite. The pioneers of thermodynamics actually knew more about entropy than simply "it tends to go up." For example, they knew that if you took two different systems and put them into contact next to each other, the total entropy would simply be the *sum* of the individual entropies of the two systems. Entropy is additive, just like the number of particles (but not, for example, like the temperature). But the number of rearrangements is certainly not additive; if you combine two boxes of gas, the number of ways you can rearrange the molecules between the two boxes is enormously larger than the number of ways you can rearrange them *within* each box.

Boltzmann was able to crack the puzzle of how to define entropy in terms of microscopic rearrangements. We use the letter W—from the German *Wahrscheinlichkeit*, meaning "probability" or "likelihood"—to represent the number of ways we can rearrange the microscopic constituents of a system without changing its macroscopic appearance. Boltzmann's final step was to take the *logarithm* of W and proclaim that the result is proportional to the entropy.

The word *logarithm* sounds very highbrow, but it's just a way to express how many digits it takes to express a number. If the number is a power of 10, its logarithm is just that power.[125] So the logarithm of 10 is 1, the logarithm of 100 is 2, the logarithm of 1,000,000 is 6, and so on.

In the Appendix, I discuss some of the mathematical niceties in more detail. But those niceties aren't crucial to the bigger picture; if you just glide quickly past any appearance of the word *logarithm*, you won't be missing much. You only really need to know two things:

- As numbers get bigger, their logarithms get bigger.
- But not very fast. The logarithm of a number grows slowly as the number itself gets bigger and bigger. One billion is much greater than

1,000, but 9 (the logarithm of 1 billion) is not much greater than 3 (the logarithm of 1,000).

That last bit is a huge help, of course, when it comes to the gigantic numbers we are dealing with in this game. The number of ways to distribute 2,000 particles equally between two halves of a box is 2×10^{600}, which is an unimaginably enormous quantity. But the logarithm of that number is just 600.3, which is relatively manageable.

Boltzmann's formula for the entropy, which is traditionally denoted by S (you wouldn't have wanted to call it E, which usually stands for energy), states that it is equal to some constant k, cleverly called "Boltzmann's constant," times the logarithm of W, the number of microscopic arrangements of a system that are macroscopically indistinguishable.[126] That is:

$$S = k \log W.$$

This is, without a doubt, one of the most important equations in all of science—a triumph of nineteenth-century physics, on a par with Newton's codification of dynamics in the seventeenth century or the revolutions of relativity and quantum mechanics in the twentieth. If you visit Boltzmann's grave in Vienna, you will find this equation engraved on his tombstone (see Chapter Two).[127]

Taking the logarithm does the trick, and Boltzmann's formula leads to just the properties we think something called "entropy" should have—in particular, when you combine two systems, the total entropy is just the sum of the two entropies you started with. This deceptively simple equation provides a quantitative connection between the microscopic world of atoms and the macroscopic world we observe.[128]

BOX OF GAS REDUX

As an example, we can calculate the entropy of the box of gas with a small hole in a divider that we illustrated in Figure 42. Our macroscopic observable is simply the total number of molecules on the left side or the right side. (We don't know which particular molecules they are, nor do we know their precise coordinates and momenta.) The quantity W in this example is just the number of ways we could distribute the 2,000 total particles without changing the numbers on the left and right. If there are 2,000 particles on the left, W equals 1, and $\log W$ equals 0. Some of the other possibilities are listed in Table 1.

PARTICLES ON LEFT, RIGHT SIDES	W	LOG W
2,000, 0	1	0
1,999, 1	2,000	3.3
1,998, 2	1,999,000	6.3
1,997, 3	1,331,334,000	9.1
...	...	...
1,000, 1,000	2×10^{600}	600.3
...	...	...
3, 1,997	1,331,334,000	9.1
2, 1,998	1,999,000	6.3
1, 1,999	2,000	3.3
0, 2,000	1	0

Table 1: The number of arrangements W, and the logarithm of that number, corresponding to a divided box of 2,000 particles with some on the left side and some on the right side.

In Figure 43 we see how the entropy, as defined by Boltzmann, changes in our box of gas. I've scaled things so that the maximum possible entropy of the box is equal to 1. It starts out relatively low, corresponding to the first configuration in Figure 42, where 1,600 molecules were on the left and only 400 on the right. As molecules gradually slip through the hole in the central divider, the entropy tends to increase. This is one particular example of the evolution; because our "law of physics" (each particle has a 0.5 percent chance of switching sides every second) involved probabilities, the details of any particular example will be slightly different. But it is overwhelmingly likely that the entropy will increase, as the system tends to wander into macroscopic configurations that correspond to larger numbers of microscopic arrangements. The Second Law of Thermodynamics in action.

So this is the origin of the arrow of time, according to Boltzmann and his friends. We start with a set of microscopic laws of physics that are time-reversal invariant: They don't distinguish between past and future. But we deal with systems featuring large numbers of particles, where we don't keep track of every detail necessary to fully specify the state of the system; instead, we keep track of some observable macroscopic features. The entropy characterizes (by which we mean, "is proportional to the logarithm of") the number of microscopic states that are macroscopically indistinguishable. Under the reasonable assumption that the system will tend to evolve toward the macroscopic configurations that correspond to a large number of possible states, it's natural that entropy will increase with time.

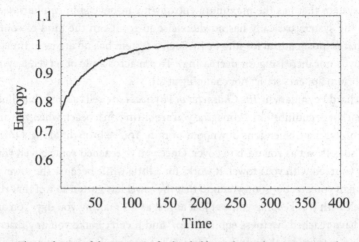

Figure 43: The evolution of the entropy of a divided box of gas. The gas starts with most molecules on the left, and the distribution evens out in time, as we saw in Figure 42. The entropy correspondingly rises, as there are more ways for the molecules to be distributed evenly than to be mostly on one side or the other. For convenience we have plotted the entropy in terms of the maximum entropy, so the maximum value attainable on this plot is 1.

In particular, it would be very surprising if it spontaneously decreased. The arrow of time arises because the system (or the universe) naturally evolves from rare configurations into more common configurations as time goes by.

All of this seems superficially plausible and will turn out to be basically true. But along the way we made some "reasonable" leaps of logic, which deserve more careful examination. For the rest of this chapter we will bring to light the various assumptions that go into Boltzmann's way of thinking about entropy, and try to decide just how plausible they are.

USEFUL AND USELESS ENERGY

One interesting feature of this box-of-gas example is that the arrow of time is only temporary. After the gas has had a chance to even itself out (at around time 150 in Figure 43), nothing much happens anymore. Individual molecules will continue to bounce between the right and left sides of the box, but these will tend to average out, and the system will spend almost all of its time with approximately equal numbers of molecules on each side. Those are the kinds of configurations that correspond to the largest number of rearrangements of the individual molecules, and correspondingly have the highest entropy the system can possibly have.

A system that has the maximum entropy it can have is in *equilibrium*. Once there, the system basically has nowhere else to go; it's in the kind of configuration that is most natural for it to be in. Such a system has no arrow of time, as the entropy is not increasing (or decreasing). To a macroscopic observer, a system in equilibrium appears static, not changing at all.

Richard Feynman, in *The Character of Physical Law*, tells a story that illustrates the concept of equilibrium.[129] Imagine you are sitting on a beach when you are suddenly hit with a tremendous downpour of rain. You've brought along a towel, but that also gets wet as you dash to cover. Once you've reached some cover, you start to dry yourself with your towel. It works for a little while because the towel is a bit drier than you are, but soon you find that the towel has gotten so wet that rubbing yourself with it is keeping you wet just as fast as it's making you dry. You and the towel have reached "wetness equilibrium," and it can't make you any drier. Your situation maximizes the number of ways the water molecules can arrange themselves on you and the towel.[130]

Once you've reached equilibrium, the towel is no longer useful for its intended purpose (drying you off). Note that the total amount of water doesn't change as you dry yourself off; it is simply transferred from you to the towel. Similarly, the total *energy* doesn't change in a box of gas that is isolated from the rest of the world; energy is conserved, at least in circumstances where we can neglect the expansion of space. But energy can be arranged in more or less useful ways. When energy is arranged in a low-entropy configuration, it can be harnessed to perform useful work, like propelling a vehicle. But the same amount of energy, when it's in an equilibrium configuration, is completely useless, just like a towel that is in wetness equilibrium with you. Entropy measures the uselessness of a configuration of energy.[131]

Consider our divided box once again. But instead of the divider being a fixed wall with a hole in it, passively allowing molecules to move back and forth, imagine that the divider is movable, and hooked up to a shaft that reaches outside the box. What we've constructed is simply a piston, which can be used to do work under the right circumstances.

In Figure 44 we've depicted two different situations for our piston. The top row shows a piston in the presence of a low-entropy configuration of some gas—all the molecules on one side of the divider—while the bottom row shows a high-entropy configuration—equal amounts of gas on both sides. The total number of molecules, and the total amount of energy, is assumed to be the same in both cases; the only difference is the entropy. But it's clear that what happens in the two cases is very different. In the top row, the gas is all on the left side of the piston, and the force of the molecules bumping into it exerts pressure that pushes the piston to the right until the gas fills the container. The moving piston shaft can be used to do useful work—run a flywheel or some such thing, at least for a little while. That extracts energy from the gas; at the end of the process, the gas will have a lower temperature. (The pistons in your car engine operate in exactly this way, expanding and cooling the hot vapor created by igniting vaporized gasoline, performing the useful work of moving your car.)

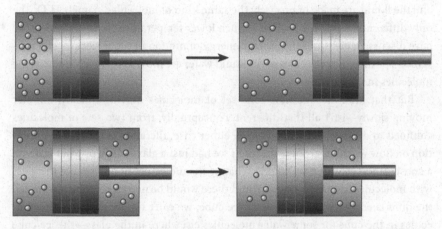

Figure 44: Gas in a divided box, used to drive a cylinder. On the top, gas in a low-entropy state pushes the cylinder to the right, doing useful work. On the bottom, gas in a high-entropy state doesn't push the cylinder in either direction.

On the bottom row in the figure, meanwhile, we imagine starting with the same amount of energy in the gas but in an initial state with a much higher entropy—an equal number of particles on each side of the divider. High entropy implies equilibrium, which implies that the energy is useless, and indeed we see that our piston isn't going anywhere. The pressure from gas on one side of the divider is exactly canceled by pressure coming from the other side. The gas in this box has the same total energy as the gas in the upper left box, but in this case we can't harness that energy to make the piston move to do something useful.

This helps us understand the relationship between Boltzmann's viewpoint on entropy and that of Rudolf Clausius, who first formulated the Second Law. Remember that Clausius and his predecessors didn't think of entropy in terms of atoms at all; they thought of it as an autonomous substance with its own dynamics. Clausius's original version of the Second Law didn't even mention entropy; it was the simple statement that "heat never flows spontaneously from a colder object to a hotter one." If we put two objects with different temperatures into contact with each other, they will both evolve toward a common middle temperature; if we put two objects with the same temperature into contact with each other, they will simply stay that way. (They're in thermal equilibrium.)

From the point of atoms, this all makes sense. Consider the classic example of two objects at different temperatures in contact with each other: an ice cube in a glass of warm water, discussed at the end of the previous chapter. Both the ice cube and the liquid are made of precisely the same kind of molecules, namely H_2O. The only difference is that the ice is at a much lower temperature. Temperature, as we have discussed, measures the average energy of motion in the molecules of a substance. So while the molecules of the liquid water are moving relatively quickly, the molecules in the ice are moving slowly.

But that kind of condition—one set of molecules moving quickly, another moving slowly—isn't all that different, conceptually, from two sets of molecules confined to different sides of a box. In either case, there is a broad-brush limitation on how we can rearrange things. If we had just a glass of nothing but water at a constant temperature, we could exchange the molecules in one part of the glass with molecules in some other part, and there would be no macroscopic way to tell the difference. But when we have an ice cube, we can't simply exchange the molecules in the cube for some water molecules elsewhere in the glass—the ice cube would move, and we would certainly notice that even from our everyday macroscopic perspective. The division of the water molecules into "liquid" and "ice" puts a serious constraint on the number of rearrangements we can do, so that configuration has a low entropy. As the temperature between the water molecules that started out as ice equilibrates with that of the rest of the glass, the entropy goes up. Clausius's rule that temperatures tend to even themselves out, rather than spontaneously flowing from cold to hot, is precisely equivalent to the statement that the entropy as defined by Boltzmann never decreases in a closed system.

None of this means that it's impossible to cool things down, of course. But in everyday life, where most things around us are at similar temperatures, it takes a bit more ingenuity than heating them up. A refrigerator is a more complicated machine than a stove. (Refrigerators work on the same basic principle as the piston in Figure 44, expanding a gas to extract energy and cool it off.) When Grant Achatz, chef of Chicago's Alinea restaurant, wanted a device that would rapidly freeze food in the same way a frying pan rapidly heats food up, he had to team with culinary technologist Philip Preston to create their own. The result is the "anti-griddle," a microwave-oven-sized machine with a metallic top that attains a temperature of –34 degrees Celsius. Hot purees and sauces, poured on the anti-griddle, rapidly freeze on the bottom while remaining soft on the top. We have understood the basics of thermodynamics for a long time now, but we're still inventing new ways to put them to good use.

DON'T SWEAT THE DETAILS

You're out one Friday night playing pool with your friends. We're talking about real-world pool now, not "physicist pool" where we can ignore friction and noise.[132] One of your pals has just made an impressive break, and the balls have scattered thoroughly across the table. As they come to a stop and you're contemplating your next shot, a stranger walks by and exclaims, "Wow! That's incredible!"

Somewhat confused, you ask what is so incredible about it. "Look at these balls at those *exact positions* on the table! What are the chances that you'd be able to put all the balls in precisely those spots? You'd never be able to repeat that in a million years!"

The mysterious stranger is a bit crazy—probably driven slightly mad by reading too many philosophical tracts on the foundations of statistical mechanics. But she does have a point. With several balls on the table, any particular configuration of them is extremely unlikely. Think of it this way: If you hit the cue ball into a bunch of randomly placed balls, which rattled around before coming to rest in a perfect arrangement as if they had just been racked, you'd be astonished. But that particular arrangement (all balls perfectly arrayed in the starting position) is no more or less unusual than any other precise arrangement of the balls.[133] What right do we have to single out certain configurations of the billiard balls as "astonishing" or "unlikely," while others seem "unremarkable" or "random"?

This example pinpoints a question at the heart of Boltzmann's definition of entropy and the associated understanding of the Second Law of Thermodynamics: Who decides when two specific microscopic states of a system look the same from our macroscopic point of view?

Boltzmann's formula for entropy hinges on the idea of the quantity W, which we defined as "the number of ways we can rearrange the microscopic constituents of a system without changing its macroscopic appearance." In the last chapter we defined the "state" of a physical system to be a complete specification of all the information required to uniquely evolve it in time; in classical mechanics, it would be the position and momentum of every single constituent particle. Now that we are considering statistical mechanics, it's useful to use the term *microstate* to refer to the precise state of a system, in contrast with the *macrostate*, which specifies only those features that are macroscopically observable. Then the short-hand definition of W is "the number of microstates corresponding to a particular macrostate."

For the box of gas separated in two by a divider, the microstate at any one time is the position and momentum of every single molecule in the box. But all we were keeping track of was how many molecules were on the left, and how many were on the right. Implicitly, every division of the molecules into a certain number on the left and a certain number on the right defined a "macrostate" for the box. And our calculation of W simply counted the number of microstates per macrostate.[134]

The choice to just keep track of how many molecules were in each half of the box seemed so innocent at the time. But we could imagine keeping track of much more. Indeed, when we deal with the atmosphere in an actual room, we keep track of a lot more than simply how many molecules are on each side of the room. We might, for example, keep track of the temperature, and density, and pressure of the atmosphere at every point, or at least at some finite number of places. If there were more than one kind of gas in the atmosphere, we might separately keep track of the density and so on for each different kind of gas. That's still enormously less information than the position and momentum of every molecule in the room, but the choice of which information to "keep" as a macroscopically measurable quantity and which information to "forget" as an irrelevant part of the microstate doesn't seem to be particularly well defined.

The process of dividing up the space of microstates of some particular physical system (gas in a box, a glass of water, the universe) into sets that we label "macroscopically indistinguishable" is known as *coarse-graining*. It's a little bit of black magic that plays a crucial role in the way we think about entropy. In Figure 45 we've portrayed how coarse-graining works; it simply divides up the space of all states of a system into regions (macrostates) that are indistinguishable by macroscopic observations. Every point within one of those regions corresponds to a different microstate, and the entropy associated with a given microstate is proportional to the logarithm of the area (or really volume, as it's a very high-dimensional space) of the macrostate to which it belongs. This kind of figure makes it especially clear why entropy tends to go up: Starting from a state with low entropy, corresponding to a very tiny part of the space of states, it's only to be expected that an ordinary system will tend to evolve to states that are located in one of the large-volume, high-entropy regions.

Figure 45 is not to scale; in a real example, the low-entropy macrostates would be much smaller compared to the high-entropy macrostates. As we saw with the divided-box example, the number of microstates corresponding to high-entropy

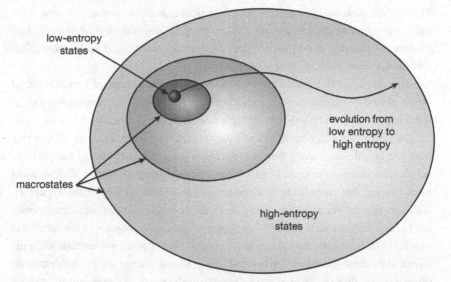

Figure 45: The process of coarse-graining consists of dividing up the space of all possible microstates into regions considered to be macroscopically indistinguishable, which are called macrostates. Each macrostate has an associated entropy, proportional to the logarithm of the volume it takes up in the space of states. The size of the low-entropy regions is exaggerated for clarity; in reality, they are fantastically smaller than the high-entropy regions.

macrostates is enormously larger than the number associated with low-entropy macrostates. Starting with low entropy, it's certainly no surprise that a system should wander into the roomier high-entropy parts of the space of states; but starting with high entropy, a typical system can wander for a very long time without ever bumping into a low-entropy condition. That's what equilibrium is like; it's not that the microstate is truly static, but that it never leaves the high-entropy macrostate it's in.

This whole business should strike you as just a little bit funny. Two microstates belong to the same macrostate when they are macroscopically indistinguishable. But that's just a fancy way of saying, "when we can't tell the difference between them on the basis of macroscopic observations." It's the appearance of "we" in that statement that should make you nervous. Why should our powers of observation be involved in any way at all? We like to think of entropy as a feature of *the world*, not as a feature of *our ability to perceive the world*. Two glasses of water are in the same macrostate if they have the same temperature throughout the glass, even if the exact distribution of positions and momenta of the water molecules are different, because we can't directly measure all of that information. But what if we ran across a race of superobservant aliens who could peer into a glass of liquid and observe the position and momentum of every molecule? Would such a race think that there was no such thing as entropy?

There are several different answers to these questions, none of which is found satisfactory by everyone working in the field of statistical mechanics. (If any of them were, you would need only that one answer.) Let's look at two of them.

The first answer is, it really doesn't matter. That is, it might matter a lot to you how you bundle up microstates into macrostates for the purposes of the particular physical situation in front of you, but it ultimately doesn't matter if all we want to do is argue for the validity of something like the Second Law. From Figure 45, it's clear why the Second Law should hold: There is a lot more room corresponding to high-entropy states than to low-entropy ones, so if we start in the latter it is natural to wander into the former. But that will hold true no matter how we actually do the coarse-graining. The Second Law is robust; it depends on the definition of entropy as the logarithm of a volume within the space of states, but not on the precise way in which we choose that volume. Nevertheless, in practice we do make certain choices and not others, so this transparent attempt to avoid the issue is not completely satisfying.

The second answer is that the choice of how to coarse-grain is not *completely* arbitrary and socially constructed, even if some amount of human choice does come into the matter. The fact is, we coarse-grain in ways that seem physically natural, not just chosen at whim. For example, when we keep track of the temperature and pressure in a glass of water, what we're really doing is throwing away all information that we could measure only by looking through a microscope. We're looking at average properties within relatively small regions of space because that's what our senses are actually able to do. Once we choose to do that, we are left with a fairly well-defined set of macroscopically observable quantities.

Averaging within small regions of space isn't a procedure that we hit upon randomly, nor is it a peculiarity of our human senses as opposed to the senses of a hypothetical alien; it's a very natural thing, given how the laws of physics work.[135] When I look at cups of coffee and distinguish between cases where a teaspoon of milk has just been added and ones where the milk has become thoroughly mixed, I'm not pulling a random coarse-graining of the states of the coffee out of my hat; that's how the coffee *looks* to me, immediately and phenomenologically. So even though in principle our choice of how to coarse-grain microstates into macro-states seems absolutely arbitrary, in practice Nature hands us a very sensible way to do it.

RUNNING ENTROPY BACKWARD

A remarkable consequence of Boltzmann's statistical definition of entropy is that the Second Law is not absolute—it just describes behavior that is overwhelmingly likely. If we start with a medium-entropy macrostate, almost all microstates within it will evolve toward higher entropy in the future, but a small number will actually evolve toward lower entropy.

It's easy to construct an explicit example. Consider a box of gas, in which the gas molecules all happened to be bunched together in the middle of the box in a low-entropy configuration. If we just let it evolve, the molecules will move around, colliding with one another and with the walls of the box, and ending up (with overwhelming probability) in a much higher-entropy configuration.

Now consider a particular microstate of the above box of gas at some moment after it has become high-entropy. From there, construct a new state by keeping all

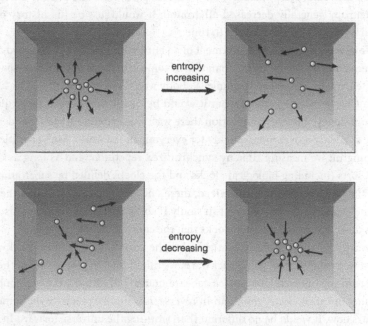

Figure 46: On the top row, ordinary evolution of molecules in a box from a low-entropy initial state to a high-entropy final state. At the bottom, we carefully reverse the momentum of every particle in the final state from the top, to obtain a time-reversed evolution in which entropy decreases.

of the molecules at exactly the same positions, but precisely reversing all of the velocities. The resulting state still has a high entropy—it's contained within the same macrostate as we started with. (If someone suddenly reversed the direction of motion of every single molecule of air around you, you'd never notice; on average there are equal numbers moving in every direction.) Starting in this state, the motion of the molecules will exactly retrace the path that they took from the previous low-entropy state. To an external observer, it will look as if the entropy is spontaneously decreasing. The fraction of high-entropy states that have this peculiar property is astronomically small, but they certainly exist.

We could even imagine an entire universe that was like that, if we believe that the fundamental laws are reversible. Take our universe today: It is described by some particular microstate, which we don't know, although we know something about the macrostate to which it belongs. Now simply reverse the momentum of every single particle in the universe and, moreover, do whatever extra transformations (changing particles to antiparticles, for example) are needed to maintain the integrity of time reversal. Then let it go. What we would see would be an evolution toward the "future" in which the universe collapsed, stars and planets unformed, and entropy generally decreased all around; it would just be the history of our actual universe played backward in time.

However—the thought experiment of an entire universe with a reversed arrow of time is much less interesting than that of some subsystem of the universe with a reversed arrow. The reason is simple: Nobody would ever notice.

In Chapter One we asked what it would be like if time passed more quickly or more slowly. The crucial question there was: Compared to what? The idea that "time suddenly moves more quickly for everyone in the world" isn't operationally meaningful; we measure time by synchronized repetition, and as long as clocks of all sorts (including biological clocks and the clocks defined by subatomic processes) remain properly synchronized, there's no way you could tell that the "rate of time" was in any way different. It's only if some particular clock speeds up or slows down compared to other clocks that the concept makes any sense.

Exactly the same problem is attached to the idea of "time running backward." When we visualize time going backward, we might imagine some part of the universe running in reverse, like an ice cube spontaneously forming out of a cool glass of water. But if the *whole thing* ran in reverse, it would be precisely the same as it appears now. It would be no different than running the universe forward in time, but choosing some perverse time coordinate that ran in the opposite direction.

The arrow of time isn't a consequence of the fact that "entropy increases to the future"; it's a consequence of the fact that "entropy is very different in one direction of time than the other." If there were some other part of the universe, which didn't interact with us in any way, where entropy decreased toward what we now call the future, the people living in that reversed-time world wouldn't notice anything out of the ordinary. They would experience an ordinary arrow of time and claim that entropy was lower in their past (the time of which they have memories) and grew to the future. The difference is that what they *mean* by "the future" is what we call "the past," and vice versa. The direction of the time coordinate on the universe is completely arbitrary, set by convention; it has no external meaning. The convention we happen to prefer is that "time" increases in the direction that entropy increases. The important thing is that entropy increases in the same temporal direction for everyone within the observable universe, so that they can agree on the direction of the arrow of time.

Of course, everything changes if two people (or other subsets of the physical universe) who can actually communicate and interact with each other disagree on the direction of the arrow of time. Is it possible for my arrow of time to point in a different direction than yours?

THE DECONSTRUCTION OF BENJAMIN BUTTON

We opened Chapter Two with a few examples of incompatible arrows of time in literature—stories featuring some person or thing that seemed to experience time backward. The homunculus narrator of *Time's Arrow* remembered the future but not the past; the White Queen experienced pain just before she pricked her finger; and the protagonist of F. Scott Fitzgerald's "The Curious Case of Benjamin Button" grew physically younger as time passed, although his memories and experiences accumulated in the normal way. We now have the tools to explain why none of those things happen in the real world.

As long as the fundamental laws of physics are perfectly reversible, given the precise state of the entire universe (or any closed system) at any one moment in time, we can use those laws to determine what the state will be at any future time, or what it was at any past time. We usually take that time to be the "initial" time, but in principle we could choose any moment—and in the present context, when we're worried about arrows of time pointing in different directions, there is no time that is initial for everything. So what we want to ask is: Why is it difficult/impossible to choose a state of the universe with the property that, as we evolve it forward in time, some parts of it have increasing entropy and some parts have decreasing entropy?

At first it would seem simple enough. Take two boxes of gas molecules. Prepare one of them in some low-entropy state, as in the top left of Figure 46; once the molecules are let go, their entropy will go up as expected. Prepare the other box by taking a high-entropy state that has just evolved from a low-entropy state, and reversing all of the velocities, as at the bottom left. That second box is delicately constructed so that the entropy will decrease with time. So, starting from that initial condition in both boxes, we will see the entropy evolve in opposite directions.

But we want more than that. It's not very interesting to have two completely separate systems with oppositely directed arrows of time. We would like to have systems that *interact*—one system can somehow communicate with the other.

And that ruins everything.[136] Imagine we started with these two boxes, one of which had an entropy that was ready to go up and the other ready to go down. But then we introduced a tiny interaction that connected the boxes—say, a few photons moving between the boxes, bouncing off a molecule in one before returning to the other. Certainly the interaction of Benjamin Button's body with the rest of the world is much stronger than that. (Likewise the White Queen, or Martin Amis's narrator in *Time's Arrow*.)

That extra little interaction will slightly alter the velocities of the molecules with which it interacts. (Momentum is conserved, so it has no choice.) That's no problem for the box that starts with low entropy, as there is no delicate tuning required to make the entropy go up. But it completely ruins our attempt to set up conditions in the other box so that entropy goes down. Just a tiny change in velocity will quickly propagate through the gas, as one affected molecule hits another molecule, and then they hit two more, and so on. It was necessary for all of the velocities to be very precisely aligned to make the gas miraculously conspire to decrease its entropy, and any interaction we might want to introduce will destroy the required conspiracy. The entropy in the first box will very sensibly go up, while the entropy in the other will just stay high; that subsystem will basically stay in equilibrium. You can't have incompatible arrows of time among interacting subsystems of the universe.[137]

ENTROPY AS DISORDER

We often say that entropy measures disorder. That's a shorthand translation of a very specific concept into somewhat sloppy language—perfectly adequate as a quick gloss, but there are ways in which it can occasionally go wrong. Now that we know the real definition of entropy given by Boltzmann, we can understand how close this informal idea comes to the truth.

The question is, what do you mean by "order"? That's not a concept that can easily be made rigorous, as we have done with entropy. In our minds, we associate "order" with a condition of purposeful arrangement, as opposed to a state of randomness. That certainly bears a family resemblance to the way we've been talking about entropy. An egg that has not yet been broken seems more orderly than one that we have split apart and whisked into a smooth consistency.

Entropy seems naturally to be associated with disorder because, more often than not, there are more ways to be disordered than to be ordered. A classic example of the growth of entropy is the distribution of papers on your desk. You can put them into neat piles—orderly, low entropy—and over time they will tend to get scattered across the desktop—disorderly, high entropy. Your desk is not a closed system, but the basic idea is on the right track.

But if we push too hard on the association, it doesn't quite hold up. Consider the air molecules in the room you're sitting in right now—presumably spread evenly throughout the room in a high-entropy configuration. Now imagine those molecules were instead collected into a small region in the center of the room, just a few centimeters across, taking on the shape of a miniature replica of the Statue of Liberty. That would be, unsurprisingly, much lower entropy—and we would all agree that it also seemed to be more orderly. But now imagine that all the gas in the room was collected into an extremely tiny region, only 1 millimeter across, in the shape of an amorphous blob. Because the region of space covered by the gas is even smaller now, the entropy of that configuration is lower than in the Statue of Liberty example. (There are more ways to rearrange the molecules within a medium-sized statuette than there are within a very tiny blob.) But it's hard to argue that an amorphous blob is more "orderly" than a replica of a famous monument, even if the blob is really small. So in this case the correlation between orderliness and low entropy seems to break down, and we need to be more careful.

That example seems a bit contrived, but we actually don't have to work that hard to see the relationship between entropy and disorder break down. In keeping with our preference for kitchen-based examples, consider oil and vinegar. If you shake oil and vinegar together to put on a salad, you may have noticed that they tend to spontaneously unmix themselves if you set the mixture down and leave it to its own devices. This is not some sort of spooky violation of the Second Law of Thermodynamics. Vinegar is made mostly of water, and water molecules tend to stick to oil molecules—and, due to the chemical properties of oil and water, they stick in very particular configurations. So when oil and water (or vinegar) are thoroughly mixed, the water molecules cling to the oil molecules in specific arrangements, corresponding to a relatively *low*-entropy state. Whereas, when the two substances are largely segregated, the individual molecules can move freely among the other molecules of similar type. At room temperature, it turns out that oil and water have a higher entropy in the unmixed state than in the mixed state.[138] Order appears spontaneously at the macroscopic level, but it's ultimately a matter of disorder at the microscopic level.

Things are also subtle for really big systems. Instead of the gas in a room, consider an astronomical-sized cloud of gas and dust—say, an interstellar nebula. That seems pretty disorderly and high-entropy. But if the nebula is big enough, it will contract under its own gravity and eventually form stars, perhaps with planets orbiting around them. Because such a process obeys the Second Law, we can be sure that the entropy goes up along the way (as long as we keep careful track of all the radiation emitted during the collapse and so forth). But a star with several orbiting planets seems, at least informally, to be more orderly than a dispersed interstellar cloud of gas. The entropy went up, but so did the amount of order, apparently.

The culprit in this case is gravity. We're going to have a lot to say about how gravity wreaks havoc with our everyday notions of entropy, but for now suffice it to say that the interaction of gravity with other forces seems to be able to create order while still making the entropy go up—temporarily, anyway. That is a deep clue to something important about how the universe works; sadly, we aren't yet sure what that clue is telling us.

For the time being, let's recognize that the association of entropy with disorder is imperfect. It's not bad—it's okay to explain entropy informally by invoking messy desktops. But what entropy really is telling us is how many microstates are macroscopically indistinguishable. Sometimes that has a simple relationship with orderliness, sometimes not.

THE PRINCIPLE OF INDIFFERENCE

There are a couple of other nagging worries about Boltzmann's approach to the Second Law that we should clean up, or at least bring out into the open. We have this large set of microstates, which we divide up into macrostates, and declare that the entropy is the logarithm of the number of microstates per macrostate. Then we are asked to swallow another considerable bite: The proposition that each microstate within a macrostate is "equally likely."

Following Boltzmann's lead, we want to argue that the reason why entropy tends to increase is simply that there are more ways to be high-entropy than to be low-entropy, just by counting microstates. But that wouldn't matter the least bit if a typical system spent a lot more time in the relatively few low-entropy microstates than it did in the many high-entropy ones. Imagine if the microscopic laws of physics had the property that almost all high-entropy microstates tended to naturally evolve toward a small number of low-entropy states. In that case, the fact that there were more high-entropy states wouldn't make any difference; we would still expect to find the system in a low-entropy state if we waited long enough.

It's not hard to imagine weird laws of physics that behave in exactly this way. Consider the billiard balls once again, moving around according to perfectly normal billiard-ball behavior, with one crucial exception: Every time a ball bumps into a particular one of the walls of the table, it sticks there, coming immediately to rest. (We're not imagining that someone has put glue on the rail or any such thing that could ultimately be traced to reversible behavior at the microscopic level, but contemplating an entirely new law of fundamental physics.) Note that the space of states for these billiard balls is exactly what it would be under the usual rules: Once we specify the position and momentum of every ball, we can precisely predict the future evolution. It's just that the future evolution, with overwhelming probability, ends up with all of the balls stuck on one wall of the table. That's a very low-entropy configuration; there aren't many microstates like that. In such a world, entropy would spontaneously decrease even for the closed system of the pool table.

It should be clear what's going on in this concocted example: The new law of physics is not reversible. It's much like checkerboard D from the last chapter, where diagonal lines of gray squares would run into a particular vertical column and simply come to an end. Knowing the positions and momenta of all the balls on this funky table is sufficient to predict the future, but it is not good enough to reconstruct the past. If a ball is stuck to the wall, we have no idea how long it has been there.

The real laws of physics seem to be reversible at a fundamental level. This is, if we think about it a bit, enough to guarantee that high-entropy states don't evolve preferentially into low-entropy states. Remember that reversibility is based on conservation of information: The information required to specify the state at one time is preserved as it evolves through time. That means that two different states now will always evolve into two different states some given amount of time in the future; if they evolved into the same state, we wouldn't be able to reconstruct the past of that state. So it's just impossible that high-entropy states all evolve preferentially into low-entropy states, because there aren't enough low-entropy states to allow it to happen. This is a technical result called *Liouville's Theorem*, after French mathematician Joseph Liouville.

That's almost what we want, but not quite. And what we want (as so often in life) is not something we can really get. Let's say that we have some system, and we know what macrostate it is in, and we would like to say something about what will happen next. It might be a glass of water with an ice cube floating in it. Liouville's Theorem says that *most* microstates in that macrostate will have to increase in entropy or stay the same, just as the Second Law would imply—the ice cube is likely to melt. But the system is in some particular microstate, even if we don't know which one. How can we be sure that the microstate isn't one of the very tiny number that is going to dramatically decrease in entropy any minute now? How can we guarantee that the ice cube isn't actually going to grow a bit, while the water around it heats up?

The answer is: We can't. There is bound to be some particular microstate, very rare in the ice-cube-and-water macrostate we are considering, that actually evolves toward an even lower-entropy microstate. Statistical mechanics, the version of thermodynamics based on atoms, is essentially *probabilistic*—we don't know for sure what is going to happen; we can only argue that certain outcomes are overwhelmingly likely. At least, that's what we'd like to be able to argue. What we can honestly argue is that most medium-entropy states evolve into higher-entropy states rather than lower-entropy ones. But you'll notice a subtle difference between "most microstates within this macrostate evolve to higher entropy" and "a microstate within this macrostate is likely to evolve to higher entropy." The first statement is just about counting the relative number of microstates with different properties ("ice cube melts" vs. "ice cube grows"), but the second statement is a claim about the probability of something happening in the real world. Those are not quite the same thing. There are more Chinese people in the world than there are Lithuanians; but that doesn't mean that you are more likely to run into a Chinese person than a Lithuanian, if you just happen to be walking down the streets of Vilnius.

Conventional statistical mechanics, in other words, makes a crucial assumption: Given that we know we are in a certain macrostate, and that we understand the complete set of microstates corresponding to that macrostate, we can assume that *all such microstates are equally likely*. We can't avoid invoking some assumption along these lines; otherwise there's no way of making the leap from counting states to assigning probabilities. The equal-likelihood assumption has a name that makes it sound like a dating strategy for people who prefer to play hard to get: the "Principle of Indifference." It was championed in the context of probability theory, long before statistical mechanics even came on the scene, by our friend Pierre-Simon Laplace. He was a die-hard determinist, but understood as well as anyone that we usually don't have access to all possible facts, and wanted to understand what we can say in situations of incomplete knowledge.

And the Principle of Indifference is basically the best we can do. When all we know is that a system is in a certain macrostate, we assume that every microstate within that macrostate is equally likely. (With one profound exception—the Past Hypothesis—to be discussed at the end of this chapter.) It would be nice if we could *prove* that this assumption should be true, and people have tried to do that. For example, if a system were to evolve through every possible microstate (or at least, through a set of microstates that came very close to every possible microstate) in a reasonable period of time, and we didn't know where it was in that evolution, there would be some justification for treating all microstates as equally likely. A system that wanders all over the space of states and covers every possibility (or close to it) is known as "ergodic." The problem is, even if a system is ergodic (and not all systems are), it would take forever to actually evolve close to every possible state. Or, if not forever, at least a horrifically long time. There are just too many states for a macroscopic system to sample them all in a time less than the age of the universe.

The real reason we use the Principle of Indifference is that we don't know any better. And, of course, because it seems to work.

OTHER ENTROPIES, OTHER ARROWS

We've been pretty definitive about what we mean by "entropy" and "the arrow of time." Entropy counts the number of macroscopically indistinguishable states, and the arrow of time arises because entropy increases uniformly throughout the observable universe. The real world being what it is, however, other people often use these words to mean slightly different things.

The definition of entropy we have been working with—the one engraved on Boltzmann's tombstone—associates a specific amount of entropy with each individual microstate. A crucial part of the definition is that we first decide on what counts as "macroscopically measurable" features of the state, and then use those to coarse-grain the entire space of states into a set of macrostates. To calculate the entropy of a microstate, we count the total number of microstates that are macroscopically indistinguishable from it, then take the logarithm.

But notice something interesting: As a state evolving through time moves from a low-entropy condition to a high-entropy condition, if we choose to forget everything other than the macrostate to which it belongs, we end up knowing less and less about which state we actually have in mind. In other words, if we are told that a system belongs to a certain macrostate, the probability that it is any particular microstate within that macrostate decreases as the entropy increases, just because there are more possible microstates it could be. Our *information* about the state—how accurately we have pinpointed which microstate it is—goes down as the entropy goes up.

This suggests a somewhat different way of defining entropy in the first place, a way that is most closely associated with Josiah Willard Gibbs. (Boltzmann actually investigated similar definitions, but it's convenient for us to associate this approach with Gibbs, since Boltzmann already has his.) Instead of thinking of entropy as something that characterizes individual states—namely, the number of other states that look macroscopically similar—we could choose to think of entropy as characterizing *what we know* about the state. In the Boltzmann way of thinking about entropy, the knowledge of which macrostate we are in tells us less and less about the microstate as entropy increases; the Gibbs approach inverts this perspective and defines entropy in terms of how much we know. Instead of starting with a coarse-graining on the space of states, we start with a probability distribution: the percentage chance, for each possible microstate, that the system is actually in that microstate right now. Then Gibbs gives us a formula, analogous to Boltzmann's, for calculating the entropy associated with that probability distribution.[139] Coarse-graining never comes into the game.

Neither the Boltzmann formula nor the Gibbs formula for entropy is the "right" one. They both are things you can choose to define, and manipulate, and use to help understand the world; each comes with its advantages and disadvantages. The Gibbs formula is often used in applications, for one very down-to-Earth reason: It's easy to calculate with. Because there is no coarse-graining, there is no discontinuous jump in entropy when a system goes from one macrostate to another; that's a considerable benefit when solving equations.

But the Gibbs approach also has two very noticeable disadvantages. One is epistemic: It associates the idea of "entropy" with our knowledge of the system, rather than with the system itself. This has caused all kinds of mischief among the community of people who try to think carefully about what entropy really means. Arguments go back and forth, but the approach I have taken in this book, which treats entropy as a feature of the state rather than a feature of our knowledge, seems to avoid most of the troublesome issues.

The other disadvantage is more striking: If you know the laws of physics and use them to study how the Gibbs entropy evolves with time, you find that it never changes. A bit of reflection convinces us that this must be true. The Gibbs entropy characterizes how well we know what the state is. But under the influence of reversible laws, that's a quantity that doesn't change—information isn't created or destroyed. For the entropy to go up, we would have to know less about the state in the future than we know about it now; but we can always run the evolution backward to see where it came from, so that can't happen. To derive something like the Second Law from the Gibbs approach, you have to "forget" something about the evolution. When you get right down to it, that's philosophically equivalent to the coarse-graining we had to do in the Boltzmann approach; we've just moved the "forgetting" step to the equations of motion, rather than the space of states.

Nevertheless, there's no question that the Gibbs formula for entropy is extremely useful in certain applications, and people are going to continue to take advantage of it. And that's not the end of it; there are several other ways of thinking about entropy, and new ones are frequently being proposed in the literature. There's nothing wrong with that; after all, Boltzmann and Gibbs were proposing definitions to supercede Clausius's perfectly good definition of entropy, which is still used today under the rubric of "thermodynamic" entropy. After quantum mechanics came on the scene, John von Neumann proposed a formula for entropy that is specifically adapted to the quantum context. As we'll discuss in the next chapter, Claude Shannon suggested a definition of entropy that was very similar in spirit to Gibbs's, but in the framework of information theory rather than physics. The point is not to find the one true definition of entropy; it's to come up with concepts that serve useful functions in the appropriate contexts. Just don't let anyone bamboozle you by pretending that one definition or the other is the uniquely correct meaning of entropy.

Just as there are many definitions of entropy, there are many different "arrows of time," another source of potential bamboozlement. We've been dealing with the thermodynamic arrow of time, the one defined by entropy and the Second Law. There is also the cosmological arrow of time (the universe is expanding), the psychological arrow of time (we remember the past and not the future), the radiation arrow of time (electromagnetic waves flow away from moving charges, not toward them), and so on. These different arrows fall into different categories. Some, like the cosmological arrow, reflect facts about the evolution of the universe but are nevertheless completely reversible. It might end up being true that the ultimate explanation for the thermodynamic arrow also explains the cosmological arrow (in fact it seems quite plausible), but the expansion of the universe doesn't present any puzzle with respect to the microscopic laws of physics in the same way the increase of entropy does. Meanwhile, the arrows that reflect true irreversibilities—the psychological arrow, radiation arrow, and even the arrow defined by quantum mechanics we will investigate later—all seem to be reflections of the same underlying state of affairs, characterized by the evolution of entropy. Working out the details of how they are all related is undeniably important and interesting, but I will continue to speak of "the" arrow of time as the one defined by the growth of entropy.

PROVING THE SECOND LAW

Once Boltzmann had understood entropy as a measure of how many microstates fit into a given macrostate, his next goal was to derive the Second Law of Thermodynamics from that perspective. I've already given the basic reasons why the Second Law works—there are more ways to be high-entropy than low-entropy, and distinct starting states evolve into distinct final states, so most of the time (with truly overwhelming probability) we would expect entropy to go up. But Boltzmann was a good scientist and wanted to do better than that; he wanted to *prove* that the Second Law followed from his formulation.

It's hard to put ourselves in the shoes of a late-nineteenth-century thermodynamicist. Those folks felt that the inability of entropy to decrease in a closed system was not just a good idea; it was a *Law*. The idea that entropy would "probably" increase wasn't any more palatable than a suggestion that energy would "probably" be conserved would have been. In reality, the numbers are just so overwhelming that the probabilistic reasoning of statistical mechanics might as well be absolute, for all intents and purposes. But Boltzmann wanted to prove something more definite than that.

In 1872, Boltzmann (twenty-eight years old at the time) published a paper in which he purported to use kinetic theory to prove that entropy would always increase or remain constant—a result called the "*H*-Theorem," which has been the subject of countless debates ever since. Even today, some people think that the *H*-Theorem explains why the Second Law holds in the real world, while others think of it as an amusing relic of intellectual history. The truth is that it's an interesting result for statistical mechanics but falls short of "proving" the Second Law.

Boltzmann reasoned as follows. In a macroscopic object such as a room full of gas or a cup of coffee with milk, there are a tremendous number of molecules—more than 10^{24}. He considered the special case where the gas is relatively dilute, so that two particles might bump into each other, but we can ignore those rare events when three or more particles bump into one another at the same time. (That really is an unobjectionable assumption.) We need some way of characterizing the macrostate of all these particles. So instead of keeping track of the position and momentum of every molecule (which would be the whole microstate), let's keep track of the average number of particles that have any particular position and momentum. In a box of gas in equilibrium at a certain temperature, for example, the average number of particles is equal at every position in the box, and there will be a certain distribution of momenta, so that the average energy per particle gives the right temperature. Given just that information, you can calculate the entropy of the gas. And then you could prove (if you were Boltzmann) that the entropy of a gas that is not in equilibrium will go up as time goes by, until it reaches its maximum value, and then it will just stay there. The Second Law has, apparently, been derived.[140]

But there is clearly something fishy going on. We started with microscopic laws of physics that are perfectly time-reversal invariant—they work equally well running forward or backward in time. And then Boltzmann claimed to derive a result from them that is manifestly *not* time-reversal invariant—one that demonstrates a clear arrow of time, by saying that entropy increases toward the future. How can you possibly get irreversible conclusions from reversible assumptions?

This objection was put forcefully in 1876 by Josef Loschmidt, after similar concerns had been expressed by William Thomson (Lord Kelvin) and James Clerk Maxwell. Loschmidt was close friends with Boltzmann and had served as a mentor to the younger physicist in Vienna in the 1860s. And he was no skeptic of atomic theory; in fact Loschmidt was the first scientist to accurately estimate the physical sizes of molecules. But he couldn't understand how Boltzmann could have derived time asymmetry without sneaking it into his assumptions.

The argument behind what is now known as "Loschmidt's reversibility objection" is simple. Consider some specific microstate corresponding to a low-entropy macrostate. It will, with overwhelming probability, evolve toward higher entropy. But time-reversal invariance guarantees that for every such evolution, there is another allowed evolution—the time reversal of the original—that starts in the high-entropy state and evolves toward the low-entropy state. In the space of all things that can happen over time, there are precisely as many examples of entropy starting high and decreasing as there are examples of entropy starting low and increasing. In Figure 45, showing the space of states divided up into macrostates, we illustrated a trajectory emerging from a very low-entropy macrostate; but trajectories don't just pop into existence. That history had to come from somewhere, and that somewhere had to have higher entropy—an explicit example of a path along which entropy decreased. It is manifestly impossible to prove that entropy always increases, if you believe in time-reversal-invariant dynamics (as they all did).[141]

But Boltzmann had proven *something*—there were no mathematical or logical errors in his arguments, as far as anyone could tell. It would appear that he must have smuggled in some assumption of time asymmetry, even if it weren't explicitly stated.

And indeed he had. A crucial step in Boltzmann's reasoning was the assumption of *molecular chaos*—in German, the *Stosszahlansatz*, translated literally as "collision number hypothesis." It amounts to assuming that there are no sneaky conspiracies in the motions of individual molecules in the gas. But a sneaky conspiracy is precisely what is required for the entropy to decrease! So Boltzmann had effectively proven that entropy could increase only by dismissing the alternative possibilities from the start. In particular, he had assumed that the momenta of every pair of particles were uncorrelated *before* they collided. But that "before" is an explicitly time-asymmetric step; if the particles really were uncorrelated before a collision, they would generally be correlated afterward. That's how an irreversible assumption was sneaked into the proof.

If we start a system in a low-entropy state and allow it to evolve to a high-entropy state (let an ice cube melt, for example), there will certainly be a large number of correlations between the molecules in the system once all is said and done. Namely, there will be correlations that guarantee that if we reversed all the momenta, the system would evolve back to its low-entropy beginning state. Boltzmann's analysis didn't account for this possibility. He proved that entropy would never decrease, if we neglected those circumstances under which entropy would decrease.

WHEN THE LAWS OF PHYSICS AREN'T ENOUGH

Ultimately, it's perfectly clear what the resolution to these debates must be, at least within our observable universe. Loschmidt is right in that the set of all possible evolutions has entropy decreasing as often as it is increasing. But Boltzmann is also right, that statistical mechanics explains why low-entropy conditions will evolve into high-entropy conditions with overwhelming probability. The conclusion should be obvious: In addition to the dynamics controlled by the laws of physics, we need to assume that the universe began in a low-entropy state. That is a *boundary condition*, an extra assumption, not part of the laws of physics themselves. (At least, not until we start talking about what happened before the Big Bang, which is not a discussion one could have had in the 1870s.) Unfortunately, that conclusion didn't seem sufficient to people at the time, and subsequent years have seen confusions about the status of the *H*-Theorem proliferate beyond reason.

In 1876, Boltzmann wrote a response to Loschmidt's reversibility objection, which did not really clarify the situation. Boltzmann certainly understood that Loschmidt had a point, and admitted that there must be something undeniably probabilistic about the Second Law; it couldn't be absolute, if kinetic theory were true. At the beginning of his paper, he makes this explicit:

> Since the entropy would decrease as the system goes through this
> sequence in reverse, we see that the fact that entropy actually increases

in all physical processes in our own world cannot be deduced solely from the nature of the forces acting between the particles, but must be a consequence of the initial conditions.

We can't ask for a more unambiguous statement than that: "the fact that entropy increases in our own world . . . must be a consequence of the initial conditions." But then, still clinging to the idea of proving something without relying on initial conditions, he immediately says this:

> Nevertheless, we do not have to assume a special type of initial condition in order to give a mechanical proof of the Second Law, if we are willing to accept a statistical viewpoint.

"Accepting a statistical viewpoint" presumably means that he admits we can argue only that increasing entropy is overwhelmingly likely, not that it always happens. But what can he mean by now saying that we don't have to assume a special type of initial condition? The next sentences confirm our fears:

> While any individual non-uniform state (corresponding to low entropy) has the same probability as any individual uniform state (corresponding to high entropy), there are many more uniform states than non-uniform states. Consequently, if the initial state is chosen at random, the system is almost certain to evolve into a uniform state, and entropy is almost certain to increase.

That first sentence is right, but the second is surely wrong. If an initial state is chosen at random, it is not "almost certain to evolve into a uniform state"; rather, it is almost certain to *be* in a uniform (high-entropy) state. Among the small number of low-entropy states, almost all of them evolve toward higher-entropy states. In contrast, only a very tiny fraction of high-entropy states will evolve toward low-entropy states; however, there are a fantastically larger number of high-entropy states to begin with. The total number of low-entropy states that evolve to high entropy is equal, as Loschmidt argued, to the total number of high-entropy states that evolve to low entropy.

Reading through Boltzmann's papers, one gets a strong impression that he was several steps ahead of everyone else—he saw the ins and outs of all the arguments better than any of his interlocutors. But after zooming through the ins and outs, he didn't always stop at the right place; moreover, he was notoriously inconsistent about the working assumptions he would adopt from paper to paper. We should cut him some slack, however, since here we are 140 years later and we still don't agree on the best way of talking about entropy and the Second Law.

THE PAST HYPOTHESIS

Within our observable universe, the consistent increase of entropy and the corresponding arrow of time cannot be derived from the underlying reversible laws of physics alone. They require a boundary condition at the beginning of time. To understand why the Second Law works in our real world, it is not sufficient to simply apply statistical reasoning to the underlying laws of physics; we must also assume that the observable universe began in a state of very low entropy. David Albert has helpfully given this assumption a simple name: the *Past Hypothesis*.[142]

The Past Hypothesis is the one profound exception to the Principle of Indifference that we alluded to above. The Principle of Indifference would have us imagine that, once we know a system is in some certain macrostate, we should consider every possible microstate within that macrostate to have an equal probability. This assumption turns out to do a great job of predicting the *future* on the basis of statistical mechanics. But it would do a terrible job of reconstructing the *past*, if we really took it seriously.

Boltzmann has told us a compelling story about why entropy increases: There are more ways to be high entropy than low entropy, so most microstates in a low-entropy macrostate will evolve toward higher-entropy macrostates. But that argument makes no reference to the direction of time. Following that logic, most microstates within some macrostate will increase in entropy toward the future but will also have evolved from a higher-entropy condition in the past.

Consider all the microstates in some medium-entropy macrostate. The overwhelming majority of those states have come from prior states of *high* entropy. They must have, because there aren't that many low-entropy states from which they could have come. So with high probability, a typical medium-entropy microstate appears as a "statistical fluctuation" from a higher-entropy past. This argument is exactly the same argument that entropy should increase into the future, just with the time direction reversed.

As an example, consider the divided box of gas with 2,000 particles. Starting from a low-entropy condition (80 percent of the particles on one side), the entropy tends to go up, as plotted in Figure 43. But in Figure 47 we show how the entropy evolves to the past as well as to the future. Since the underlying dynamical rule ("each particle has a 0.5 percent chance of changing sides per second") doesn't distinguish between directions of time, it's no surprise that the entropy is higher in the past of that special moment just as it is in the future.

You may object, thinking that it's very unlikely that a system would start out in equilibrium and then dive down to a low-entropy state. That's certainly true; it would be much more likely to remain at or near equilibrium. But given that we insist on having a low-entropy state at all, it is overwhelmingly likely that such a state represents a minimum on the entropy curve, with higher entropy both to the past and to the future.

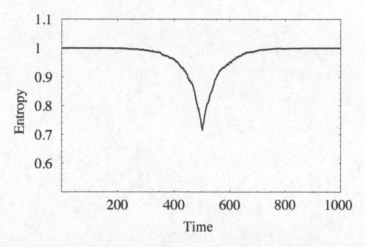

Figure 47: The entropy of a divided box of gas. The "boundary" condition is set at time = 500, where 80 percent of the particles are on one side and 20 percent on the other (a low-entropy macrostate). Entropy increases both to the future and to the past of that moment.

At least, it would be overwhelmingly likely, if all we had to go on were the Principle of Indifference. The problem is, no one in the world thinks that the entropy of the real universe behaves as shown in Figure 47. Everyone agrees that the entropy will be higher tomorrow than it is today, but nobody thinks it was higher yesterday than it is today. There are good reasons for that agreement, as we'll discuss in the next chapter—if we currently live at a minimum of the entropy curve, all of our memories of the past are completely unreliable, and we have no way of making any kind of sense of the universe.

So if we care about what actually happens in the world, we have to supplement the Principle of Indifference with the Past Hypothesis. When it comes to picking out microstates within our macrostate, we do not assign every one equal probability:

We choose only those microstates that are compatible with a much lower-entropy past (a very tiny fraction), and take all of *those* to have equal probability.[143]

But this strategy leaves us with a question: Why is the Past Hypothesis true? In Boltzmann's time, we didn't know anything about general relativity or the Big Bang, much less quantum mechanics or quantum gravity. But the question remains with us, only in a more specific form: Why did the universe have a low entropy near the Big Bang?

9

INFORMATION AND LIFE

You should call it entropy, for two reasons. In the first place, your uncertainty function has been used in statistical mechanics under that name, so it already has a name. In the second place, and more important, no one knows what entropy really is, so in a debate you will always have the advantage.

—John von Neumann, to Claude Shannon[144]

In a celebrated episode in *Swann's Way,* Marcel Proust's narrator is feeling cold and somewhat depressed. His mother offers him tea, which he reluctantly accepts. He is then pulled into an involuntary recollection of his childhood by the taste of a traditional French teatime cake, the madeleine.

> And suddenly the memory appeared. That taste was the taste of the little piece of madeleine which on Sunday mornings at Combray . . . when I went to say good morning to her in her bedroom, my aunt Léonie would give me after dipping it in her infusion of tea or lime blossom . . . And as soon as I had recognized the taste of the piece of madeleine dipped in lime-blossom tea that my aunt used to give me . . . immediately the old gray house on the street, where her bedroom was, came like a stage set to attach itself to the little wing opening onto a garden that had been built for my parents behind it . . . ; and with the house the town, from morning to night and in all weathers, the Square, where they sent me before lunch, the streets where I went on errands, the paths we took if the weather was fine.[145]

Swann's Way is the first of the seven volumes of *À la recherche du temps perdu,* which translates into English as *In Search of Lost Time.* But C. K. Scott Moncrieff, the original translator, borrowed a line from Shakespeare's Sonnet 30 to render Proust's novel as *Remembrance of Things Past.*

The past, of course, is a natural thing to have remembrances of. What else would we be remembering, anyway? Surely not the future. Of all the ways in which the arrow of time manifests itself, memory—and in particular, the fact that it applies to the past but not the future—is the most obvious, and the most central to our lives. Perhaps the most important difference between our experience of one moment and our experience of the next is the accumulation of memories, propelling us forward in time.

My stance so far has been that all the important ways in which the past differs from the future can be traced to a single underlying principle, the Second Law of Thermodynamics. This implies that our ability to remember the past but not the future must ultimately be explained in terms of entropy, and in particular by recourse to the Past Hypothesis that the early universe was in a very low-entropy state. Examining how that works will launch us on an exploration of the relationship between entropy, information, and life.

PICTURES AND MEMORIES

One of the problems in talking about "memory" is that there's a lot we don't understand about how the human brain actually works, not to mention the phenomenon of consciousness.[146] For our present purposes, however, that's not a significant handicap. When we talk about remembering the past, we're interested not specifically in the human experience of memory, but in the general notion of reconstructing past events from the present state of the world. We don't lose anything by considering well-understood mechanical recording devices, or even such straightforward artifacts as photographs or history books. (We are making an implicit assumption that human beings are part of the natural world, and in particular that our minds can in principle be understood in terms of our brains, which obey the laws of physics.)

So let's imagine you have in your possession something you think of as a reliable record of the past: for example, a photograph taken of your tenth birthday party. You might say to yourself, "I can be confident that I was wearing a red shirt at my tenth birthday party, because this photograph of that event shows me wearing a red shirt." Put aside any worries that you might have over whether the photo has been tampered with or otherwise altered. The question is, what right do we have to conclude something about the past from the existence of this photo in the present?

In particular, let's imagine that we did not buy into this Past Hypothesis business. All we have is some information about the current macrostate of the universe, including the fact that it has this particular photo, and we have certain memories and so on. We certainly don't know the current *microstate*—we don't know the position and momentum of every particle in the world—but we can invoke the Principle of Indifference to assign equal probability to every microstate compatible with the macrostate. And, of course, we know the laws of physics—maybe not the complete Theory of Everything, but enough to give us a firm handle on our everyday world. Are those—the present macrostate including the photo, plus the Principle of Indifference, plus the laws of physics—enough to conclude with confidence that we really were wearing a red shirt at our tenth birthday party?

Not even close. We tend to think that they are, without really worrying about the details too much as we get through our lives. Roughly speaking, we figure that a photograph like that is a highly specific arrangement of its constituent molecules. (Likewise for a memory in our brain of the same event.) It's not as if those molecules are just going to randomly assemble themselves into the form of that particular photo—that's astronomically unlikely. If, however, there really was an event in the past corresponding to the image portrayed in the photo, and someone was there with a camera, then the existence of the photo becomes relatively likely. It's therefore very reasonable to conclude that the birthday party really did happen in the way seen in the photo.

All of those statements are reasonable, but the problem is that they are not nearly enough to justify the final conclusion. The reason is simple, and precisely analogous to our discussion of the box of gas at the end of the last chapter. Yes, the photograph is a very specific and unlikely arrangement of molecules. However, the story we are telling to "explain" it—an elaborate reconstruction of the past, involving birthday parties and cameras and photographs surviving essentially undisturbed to the present day—is even *less* likely than the photo all by itself. At least, if "likely" is judged by assuming that all possible microstates consistent with our current macrostate have an equal probability—which is precisely what we assumed.

Think of it this way: You would never think to appeal to some elaborate story in the *future* in order to explain the existence of a particular artifact in the present. If we ask about the future of our birthday photo, we might have some plans to frame it or whatnot, but we'll have to admit to a great deal of uncertainty—we could lose it, it could fall into a puddle and decay, or it could burn in a fire. Those are all perfectly plausible extrapolations of the present state into the future, even with the specific anchor point provided by the photo here in the present. So why are we so confident about what the photo implies concerning the past?

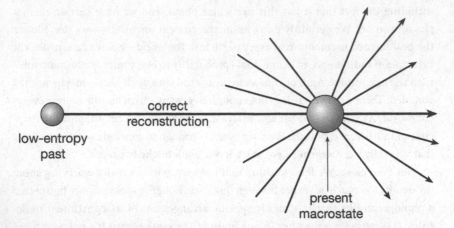

low-entropy
past

correct
reconstruction

present
macrostate

Figure 48: Trajectories through (part of) state space, consistent with our present macrostate. We can reconstruct the past accurately only by assuming a Past Hypothesis, in addition to knowledge of our current macrostate.

The answer, of course, is the Past Hypothesis. We don't really apply the Principle of Indifference to the current macrostate of the world—we only consider those microstates that are compatible with a very low-entropy past. And that makes all the difference when drawing inferences about the meaning of photographs or memories or other sorts of records. If we ask, "What is the most likely way, in the space of all possible evolutions of the universe, to get this particular photograph?" the answer is that it is most likely to evolve as a random fluctuation from a higher-entropy past—by exactly the same arguments that convince us it is likely to evolve toward a high-entropy future. But if instead we ask, "What is the most likely way, in the space of all evolutions of the universe from a very low-entropy beginning, to get this particular photograph?" then we find very naturally that it is most likely to go through the intermediate steps of an actual birthday party, a red shirt, a camera, and all the rest. Figure 48 illustrates the general principle—by demanding that our history stretch from a low-entropy beginning to here, we dramatically restrict the space of allowed trajectories, leaving us with those for which our records are (for the most part) reliable reflections of the past.

COGNITIVE INSTABILITY

I know from experience that not everyone is convinced by this argument. One stumbling block is the crucial assertion that what we start with is knowledge of our present macrostate, including some small-scale details about a photograph or a history book or a memory lurking in our brains. Although it seems like a fairly innocent assumption, we have an intuitive feeling that we don't know something only about the present; we *know* something about the past, because we see it, in a way that we don't see the future. Cosmology is a good example, just because the speed of light plays an important role, and we have a palpable sense of "looking at an event in the past." When we try to reconstruct the history of the universe, it's tempting to look at (for example) the cosmic microwave background and say, "I can *see* what the universe was like almost 14 billion years ago; I don't have to appeal to any fancy Past Hypothesis to reason my way into drawing any conclusions."

That's not right. When we look at the cosmic microwave background (or light from any other distant source, or a photograph of any purported past event), we're not looking at the past. We're observing what certain photons are doing right here and now. When we scan our radio telescope across the sky and observe a bath of radiation at about 2.7 Kelvin that is very close to uniform in every direction, we've learned something about the radiation passing through our *present* location, which we then need to extrapolate backward to infer something about the past. It's conceivable that this uniform radiation came from a past that was actually highly non-uniform, but from which a set of finely tuned conspiracies between temperatures and Doppler shifts and gravitational effects produced a very smooth-looking set of photons arriving at us today. You may say that's very unlikely, but the time-reverse of that is exactly what we would expect if we took a typical microstate within our present macrostate and evolved it toward a Big Crunch. The truth is, we don't have any more direct empirical access to the past than we have to the future, unless we allow ourselves to assume a Past Hypothesis.

Indeed, the Past Hypothesis is more than just "allowed"; it's completely necessary, if we hope to tell a sensible story about the universe. Imagine that we simply refused to invoke such an idea and stuck solely with the data given to us by our current macrostate, including the state of our brains and our photographs and our history books. We would then predict with strong probability that the past as well as the future was a higher-entropy state, and that all of the low-entropy features of our present condition arose as random fluctuations. That sounds bad enough, but the reality is worse. Under such circumstances, among the things that randomly fluctuated into existence are all of the pieces of information we traditionally use to justify our understanding of the laws of physics, or for that matter all of the mental states (or written-down arguments) we traditionally use to justify mathematics and logic and the scientific method. Such assumptions, in other words, give us absolutely no reason to believe we have justified anything, including those assumptions themselves.

David Albert has referred to such a conundrum as *cognitive instability*—the condition we face when a set of assumptions undermines the reasons we might have used to justify those very assumptions.[147] It is a kind of helplessness that can't be escaped without reaching beyond the present moment. Without the Past Hypothesis, we simply can't tell any intelligible story about the world; so we seem to be stuck with it, or stuck with trying to find a theory that actually explains it.

CAUSE AND EFFECT

There is a dramatic temporal asymmetry in this story of how we use memories and records: We invoke a Past Hypothesis but not a future one. In making predictions, we do not throw away any microstates consistent with our current macrostate on the grounds that they are incompatible with any particular future boundary condition. What if we did? In Chapter Fifteen we will examine the Gold cosmology, in which the universe eventually stops expanding and begins to re-collapse, while the arrow of time reverses itself and entropy begins to decrease as we approach the Big Crunch. In that case there would be no overall difference between the collapsing phase and the expanding phase we find ourselves in today—they are identical (at least statistically). Observers who lived in the collapsing phase wouldn't think anything was funny about their universe, any more than we do; they would think that *we* were evolving backward in time.

It's more illuminating to consider the ramifications of a minor restriction on allowed trajectories into our nearby future. This is essentially the situation we would face if we had a reliable *prophecy* of future events. When Harry Potter learns that either he will kill Voldemort or Voldemort will kill him, that places a very tight restriction on the allowed space of states.[148]

Craig Callender tells a vivid story about what a future boundary condition would be like. Imagine that an oracle with an impeccable track record (much better than Professor Trelawney from the Harry Potter books) tells you that all of the world's Imperial Fabergé eggs will end up in your dresser drawer, and that when they get there your life will end. Not such a believable prospect, really—you're not even especially fond of Russian antiques, and now you know better than to let any into your bedroom. But somehow, through a series of unpredictable and unlikely fluke occurrences, those eggs keep finding a way into your drawer. You lock it, but the lock jiggles open; you inform the eggs' owners to keep them where they are, but thieves and random accidents conspire to gradually collect them all in your room. You get a package that was mistakenly delivered to your address—it was supposed to go to the museum—and you open it to find an egg inside. In a panic, you throw it out the window, but the egg bounces off a street lamp at a crazy angle and careens back into your room to land precisely in your dresser drawer. And then you have a heart attack and die.[149]

Throughout this chain of events, no laws of physics are broken along the way. At every step, events occur that are not impossible, just extremely unlikely. As a result, our conventional notions of cause and effect are overturned. We operate in our daily lives with a deep-seated conviction that causes precede effects: "There is a broken egg on the floor because I just dropped it," not "I just dropped that egg because there was going to be a broken egg on the floor." In the social sciences, where the causal relationship between different features of the social world can be hard to ascertain, this intuitive feeling has been elevated to the status of a principle. When two properties are highly correlated with each other, it's not always obvious which is the cause and which is the effect, or whether both are caused by a different effect altogether. If you find that people who are happier in their marriages tend to eat more ice cream, is that because ice cream improves marriage, or happiness leads to more ice-cream eating? But there is one case where you know for sure: When one of the properties comes before the other one in time. Your grandparents' level of educational attainment may affect your adult income, but your income doesn't change your grandparents' education.[150]

Future boundary conditions overturn this understanding of cause and effect by insisting that some specific otherwise-unlikely things are necessarily going to happen. The same holds for the idea of free will. Ultimately, our ability to "choose" how to act in the future is a reflection of our ignorance concerning the specific microstate of the universe; if Laplace's Demon were around, he would know exactly how we are going to act. A future boundary condition is a form of predestination.

All of which may seem kind of academic and not worth dwelling on, for the basic reason that we don't think there is any kind of future boundary condition that restricts our current microstate, and therefore we believe that causes precede effects. But we have no trouble believing in a *past* condition that restricts our current microstate. The microscopic laws of physics draw no distinction between past and future, and the idea that one event "causes" another or that we can "choose" different actions in the future in a way that we can't in the past is nowhere to be found therein. The Past Hypothesis is necessary to make sense of the world around us, but it has a lot to answer for.

MAXWELL'S DEMON

Let's shift gears a bit to return to the thought-experiment playground of nineteenth-century kinetic theory. Ultimately this will lead us to the connection between entropy and information, which will circle back to illuminate the question of memory.

Perhaps the most famous thought experiment in all of thermodynamics is Maxwell's Demon. James Clerk Maxwell proposed his Demon—more famous than Laplace's, and equally menacing in its own way—in 1867, when the atomic hypothesis was just beginning to be applied to problems of thermodynamics. Boltzmann's first work on the subject wasn't until the 1870s, so Maxwell didn't have recourse to the definition of entropy in the context of kinetic theory. But he did know about Clausius's formulation of the Second Law: When two systems are in contact, heat will tend to flow from the hotter to the cooler, bringing both temperatures closer to equilibrium. And Maxwell knew enough about atoms to understand that "temperature" measures the average kinetic energy of the atoms. But with his Demon, he seemed to come up with a way to increase the difference in temperature between two systems, without injecting any energy—in apparent violation of the Second Law.

The setup is simple: the same kind of box of gas divided into two sides that we're very familiar with by now. But instead of a small opening that randomly lets molecules pass back and forth, there's a small opening with a very tiny door—one that can be opened and closed without exerting a noticeable amount of energy. At the door sits a Demon, who monitors all of the molecules on either side of the box. If a fast-moving molecule approaches from the right, the Demon lets it through to the left side of the box; if a slow-moving molecule approaches from the left, the Demon lets it through to the right. But if a slow-moving molecule approaches from the right, or a fast-moving one from the left, the Demon shuts the door so they stay on the side they're on.

It's clear what will happen: Gradually, and without any energy being exerted, the high-energy molecules will accumulate on the left, and the low-energy ones on the right. If the temperatures on both sides of the box started out equal, they will gradually diverge—the left will get hotter, and the right will get cooler. But that's in direct violation of Clausius's formulation of the Second Law. What's going on?

If we started in a high-entropy state, with the gas at equal temperature throughout the box, and we evolve reliably (for any beginning state, not just some finely tuned ones) into a lower-entropy state, we've gone from a situation where a large

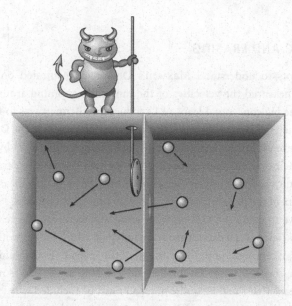

Figure 49: By letting high-energy molecules move from the right half of the box to the left, and slow-moving molecules move from the left to the right, Maxwell's Demon lets heat flow from a cold system to a hotter one, in apparent violation of the Second Law.

number of initial states all evolve into a small number of final states. But that simply can't happen, if the dynamical laws are information conserving and reversible. There's no room for all of those initial states to be squeezed into the smaller number of final states. So clearly there has to be a compensating increase in entropy somewhere, if the entropy in the gas goes down. And there's only one place that entropy could go: into the Demon.

The question is, how does that work? It doesn't look like the Demon increased in entropy; at the start of the experiment it's sitting there peacefully, waiting for the right molecules to come along, and at the end of the experiment it's still sitting there, just as peacefully. The embarrassing fact is that it took a long time—more than a century—for scientists to really figure out the right way to think about this problem. Hungarian-American physicist Leó Szilárd and French physicist Léon Brillouin— both of whom were pioneers in applying the new science of quantum mechanics to problems of practical interest—helped pinpoint the crucial relationship between the information gathered by the Demon and its entropy. But it wasn't until the contributions of two different physicist/computer scientists who worked for IBM, Rolf Landauer in 1961 and Charles Bennett in 1982, that it finally became clear why exactly the Demon's entropy must always increase in accordance with the Second Law.[151]

RECORDING AND ERASING

Many attempts to understand Maxwell's Demon concentrated on the means by which it measured the velocities of the molecules zooming around its vicinity. One of the big conceptual leaps of Landauer and Bennett was to focus on the means by which the Demon *recorded* that information. After all, the Demon has to remember—even if just for a microsecond—which molecules to let by, and which to keep on their original sides. Indeed, if the Demon simply knew from the start which molecules had which velocities, it wouldn't have to do any measurements at all; so the crux of the problem can't be in the measurement process.

So we have to equip the Demon with some way to record the velocities of all the molecules—perhaps it carries around a notepad, which for convenience we can imagine has just enough room to record all of the relevant information. (Nothing changes if we consider larger or smaller pads, as long as the pad is not infinitely big.) That means that the state of the notepad must be included when we calculate the entropy of the combined gas/Demon system. In particular, the notepad must start out blank, in order to be ready to record the velocities of the molecules.

But a blank notepad is, of course, nothing other than a low-entropy past boundary condition. It's just the Maxwell's Demon version of the Past Hypothesis, sneaked in under another guise. If that's the case, the entropy of the combined gas/Demon system clearly wasn't as high as it could have been. The Demon doesn't lower the entropy of the combined system; it simply transfers the entropy from the state of the gas to the state of the notepad.

You might be suspicious of this argument. After all, you might think, can't the Demon just *erase* the notepad when all is said and done? And wouldn't that return the notepad to its original state, while the gas went down in entropy?

This is the crucial insight of Landauer and Bennett: No, you can't just erase the notepad. At least, you can't erase information if you are part of a closed system operating under reversible dynamical laws. When phrased that way, the result is pretty believable: If you were able to erase the information entirely, how would you ever be able to reverse the evolution to its previous state? If erasure is possible, either the fundamental laws are irreversible—in which case it's not at all surprising that the Demon can lower the entropy—or you're not really in a closed system. The act of erasing information necessarily transfers entropy to the outside world. (In the case of real-world erasing of actual pencil markings, this entropy comes mostly in the form of heat, dust, and tiny flecks of rubber.)

So you have two choices. Either the Demon starts with a blank low-entropy notepad, in a demonic version of the Past Hypothesis, and simply transfers entropy from the gas to the notepad; or the Demon needs to erase information from the notepad, in which case entropy gets transferred to the outside world. In either case, the Second Law is safe. But along the way, we've opened the door to the fascinating connection between information and entropy.

INFORMATION IS PHYSICAL

Even though we've tossed around the word *information* a lot in discussing dynamical laws of physics—reversible laws conserve information—the concept still seems a bit abstract compared to the messy world of energy and heat and entropy. One of the lessons of Maxwell's Demon is that this is an illusion: *Information is physical.* More concretely, possessing information allows us to extract useful work from a system in ways that would have otherwise been impossible.

Leó Szilárd showed this explicitly in a simplified model of Maxwell's Demon. Imagine that our box of gas contained just a single molecule; the "temperature" would just be the energy of that one gas molecule. If that's all we know, there's no way to use that molecule to do useful work; the molecule just rattles around like a pebble in a can. But now imagine that we have a single bit of information: whether the molecule is on the left side of the box or the right. With that, plus some clever thought-experiment-level manipulation, we can use the molecule to do work. All we have to do is quickly insert a piston into the other half of the box. The molecule will bump into it, pushing the piston, and we can use the external motion to do something useful, like turn a flywheel.[152]

Note the crucial role played by information in Szilárd's setup. If we didn't know which half of the box the molecule was in, we wouldn't know where to insert the piston. If we inserted it randomly, half the time it would be pushed out and half the time it would be pulled in; on average, we wouldn't be getting any useful work at all. The information in our possession allowed us to extract energy from what appeared to be a system at maximal entropy.

To be clear: In the final analysis, none of these thought experiments are letting us violate the Second Law. Rather, they provide ways that we could appear to violate the Second Law, if we didn't properly account for the crucial role played by information. The information collected and processed by the Demon must somehow be accounted for in any consistent story of entropy.

The concrete relationship between entropy and information was developed in the 1940s by Claude Shannon, an engineer/mathematician working for Bell Labs.[153] Shannon was interested in finding efficient and reliable ways of sending signals across noisy channels. He had the idea that some messages carry more effective information than others, simply because the message is more "surprising" or unexpected. If I tell you that the Sun is going to rise in the East tomorrow morning, I'm not actually conveying much information, because you already expected that was going to happen. But if I tell you the peak temperature tomorrow is going to be exactly 25 degrees Celsius, my message contains more information, because without the message you wouldn't have known precisely what temperature to expect.

Shannon figured out how to formalize this intuitive idea of the effective information content of a message. Imagine that we consider the set of all possible messages we could receive of a certain type. (This should remind you of the "space of states" we considered when talking about physical systems rather than messages.) For example, if we are being told the outcome of a coin flip, there are only two possible messages: "heads" or "tails." Before we get the message, either alternative is equally likely; after we get the message, we have learned precisely one bit of information.

If, on the other hand, we are being told what the high temperature will be tomorrow, there are a large number of possible messages: say, any integer between –273 and plus infinity, representing the temperature in degrees Celsius. (Minus 273 degrees Celsius is absolute zero.) But not all of those are equally likely. If it's summer in Los Angeles, temperatures of 27 or 28 degrees Celsius are fairly common, while temperatures of –13 or +4,324 degrees Celsius are comparatively rare. Learning that the temperature tomorrow would be one of those unlikely numbers would convey a great deal of information indeed (presumably related to some global catastrophe).

Roughly speaking, then, the information content of a message goes *up* as the probability of a given message taking that form goes *down*. But Shannon wanted to be a little bit more precise than that. In particular, he wanted it to be the case that if we receive two messages that are completely independent of each other, the total information we get is equal to the sum of the information contained in each individual message. (Recall that, when Boltzmann was inventing his entropy formula, one of the properties he wanted to reproduce was that the entropy of a combined system was the sum of the entropies of the individual systems.) After some playing around, Shannon figured out that the right thing to do was to take the *logarithm* of the probability of receiving a given message. His final result is this: The "self-information" contained in a message is equal to minus the logarithm of the probability that the message would take that particular form.

If many of these words sound familiar, it's not an accident. Boltzmann associated the entropy with the logarithm of the number of microstates in a certain macrostate. But given the Principle of Indifference, the number of microstates in a macrostate is clearly proportional to the probability of picking one of them randomly in the entire space of states. A low-entropy state is like a surprising, information-filled message, while knowing that you're in a high-entropy state doesn't tell you much at all. When all is said and done, if we think of the "message" as a specification of which macrostate a system is in, the relationship between entropy and information is very simple: The information is the difference between the maximum possible entropy and the actual entropy of the macrostate.[154]

DOES LIFE MAKE SENSE?

It should come as no surprise that these ideas connecting entropy and information come into play when we start thinking about the relationship between thermodynamics and life. Not that this relationship is very straightforward; although there certainly is a close connection, scientists haven't even yet agreed on what "life" really means, much less understood all its workings. This is an active research area, one that has seen an upsurge in recent interest, drawing together insights from biology, physics, chemistry, mathematics, computer science, and complexity studies.[155]

Without yet addressing the question of how "life" should be defined, we can ask what sounds like a subsequent question: Does life make thermodynamic sense? The answer, before you get too excited, is "yes." But the opposite has been claimed—not by any respectable scientists, but by creationists looking to discredit Darwinian natural selection as the correct explanation for the evolution of life on Earth. One of their arguments relies on a misunderstanding of the Second Law, which they read as "entropy always increases," and then interpret as a universal tendency toward decay and disorder in all natural processes. Whatever life is, it's pretty clear that life is complicated and orderly—how, then, can it be reconciled with the natural tendency toward disorder?

There is, of course, no contradiction whatsoever. The creationist argument would equally well imply that refrigerators are impossible, so it's clearly not correct. The Second Law doesn't say that entropy always increases. It says that entropy always increases (or stays constant) in a closed system, one that doesn't interact noticeably with the external world. It's pretty obvious that life is not like that; living organisms interact very strongly with the external world. They are the quint essential examples of open systems. And that is pretty much that; we can wash our hands of the issue and get on with our lives.

But there's a more sophisticated version of the creationist argument, which is not quite as silly—although it's still wrong—and it's illuminating to see exactly how it fails. The more sophisticated argument is quantitative: Sure, living beings are open systems, so in principle they can decrease entropy somewhere as long as it increases somewhere else. But how do you know that the increase in entropy in the outside world is really enough to account for the low entropy of living beings?

As I mentioned back in Chapter Two, the Earth and its biosphere are systems that are very far away from thermal equilibrium. In equilibrium, the temperature is the same everywhere, whereas when we look up we see a very hot Sun in an otherwise very cold sky. There is plenty of room for entropy to increase, and that's exactly what's happening. But it's instructive to run the numbers.[156]

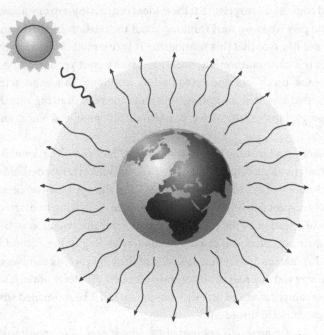

Figure 50: We receive energy from the Sun in a concentrated, low-entropy form, and radiate it back to the universe in a diffuse, high-entropy form. For every 1 high-energy photon we receive, the Earth radiates about 20 low-energy photons.

The energy budget of the Earth, considered as a single system, is pretty simple. We get energy from the Sun via radiation; we lose the same amount of energy to empty space, also via radiation. (Not exactly the same; processes such as nuclear decays also heat up the Earth and leak energy into space, and the rate at which energy is radiated is not strictly constant. Still, it's an excellent approximation.) But while the amount is the same, there is a big difference in the *quality* of the energy we get and the energy we give back. Remember back in the pre-Boltzmann days, entropy was understood as a measurement of the uselessness of a certain amount of energy; low-entropy forms of energy could be put to useful work, such as powering an engine or grinding flour, while high-entropy forms of energy just sat there.

The energy we get from the Sun is of a low-entropy, useful form, while the energy we radiate back out into space has a much higher entropy. The temperature of the Sun is about 20 times the average temperature of the Earth. For radiation, the temperature is just the average energy of the photons of which it is made, so the Earth needs to radiate 20 low-energy (long-wavelength, infrared) photons for every 1 high-energy (short-wavelength, visible) photon it receives. It turns out, after a bit of math, that 20 times as many photons directly translates into 20 times the entropy. The Earth emits the same amount of energy as it receives, but with 20 times higher entropy.

The hard part is figuring out just what we mean when we say that the life forms here on Earth are "low-entropy." How exactly do we do the coarse-graining? It is possible to come up with reasonable answers to that question, but it's complicated. Fortunately, there is a dramatic shortcut we can take. Consider the entire biomass of the Earth—all of the molecules that are found in living organisms of any type. We can easily calculate the maximum entropy that collection of molecules could have, if it were in thermal equilibrium; plugging in the numbers (the biomass is 10^{15} kilograms; the temperature of the Earth is 255 Kelvin), we find that its maximum entropy is 10^{44}. And we can compare that to the minimum entropy it could possibly have—if it were in an exactly unique state, the entropy would be precisely zero.

So the largest conceivable change in entropy that would be required to take a completely disordered collection of molecules the size of our biomass and turn them into absolutely any configuration at all—including the actual ecosystem we currently have—is 10^{44}. If the evolution of life is consistent with the Second Law, it must be the case that the Earth has generated more entropy over the course of life's evolution by converting high-energy photons into low-energy ones than it has decreased entropy by creating life. The number 10^{44} is certainly an overly generous estimate—we don't have to generate nearly that much entropy, but if we can generate that much, the Second Law is in good shape.

How long does it take to generate that much entropy by converting useful solar energy into useless radiated heat? The answer, once again plugging in the temperature of the Sun and so forth, is: about 1 year. Every year, if we were really efficient, we could take an undifferentiated mass as large as the entire biosphere and arrange it in a configuration with as small an entropy as we can imagine. In reality, life has evolved over billions of years, and the total entropy of the "Sun + Earth (including life) + escaping radiation" system has increased by quite a bit. So the Second Law is perfectly consistent with life as we know it—not that you were ever in doubt.

LIFE IN MOTION

It's good to know that life doesn't violate the Second Law of Thermodynamics. But it would also be nice to have a well-grounded understanding of what "life" actually means. Scientists haven't yet agreed on a single definition, but there are a number of features that are often associated with living organisms: complexity, organization, metabolism, information processing, reproduction, response to stimuli, aging. It's difficult to formulate a set of criteria that clearly separates living beings—algae, earthworms, house cats—from complex nonliving objects—forest fires, galaxies, personal computers. In the meantime, we are able to analyze some of life's salient features, without drawing a clear distinction between their appearance in living and nonliving contexts.

One famous attempt to grapple with the concept of life from a physicist's perspective was the short book *What Is Life?* written by none other than Erwin Schrödinger. Schrödinger was one of the inventors of quantum theory; it's his equation that replaces Newton's laws of motion as the dynamical description of the world when we move from classical mechanics to quantum mechanics. He also originated the Schrödinger's Cat thought experiment to highlight the differences between our direct perceptions of the world and the formal structure of quantum theory.

After the Nazis came to power, Schrödinger left Germany, but despite winning the Nobel Prize in 1933 he had difficulty in finding a permanent position elsewhere, largely because of his colorful personal life. (His wife Annemarie knew that he had mistresses, and she had lovers of her own; at the time Schrödinger was involved with Hilde March, wife of one of his assistants, who would eventually bear a child with him.) He ultimately settled in Ireland, where he helped establish an Institute for Advanced Studies in Dublin.

In Ireland Schrödinger gave a series of public lectures, which were later published as *What Is Life?* He was interested in examining the phenomenon of life from the perspective of a physicist, and in particular an expert on quantum mechanics and statistical mechanics. Perhaps the most remarkable thing about the book is Schrödinger's deduction that the stability of genetic information over time is best explained by positing the existence of some sort of "aperiodic crystal" that stored the information in its chemical structure. This insight helped inspire Francis Crick to leave physics in favor of molecular biology, eventually leading to his discovery with James Watson of the double-helix structure of DNA.[157]

But Schrödinger also mused on how to define "life." He made a specific proposal in that direction, which comes across as somewhat casual and offhand, and perhaps hasn't been taken as seriously as it might have been:

> What is the characteristic feature of life? When is a piece of matter said to be alive? When it goes on 'doing something', exchanging material with its environment, and so forth, and that for a much longer period than we would expect an inanimate piece of matter to 'keep going' under similar circumstances. [158]

Admittedly, this is a bit vague; what exactly does it mean to "keep going," how long should we "expect" it to happen, and what counts as "similar circumstances"? Furthermore, there's nothing in this definition about organization, complexity, information processing, or any of that.

Nevertheless, Schrödinger's idea captures something important about what distinguishes life from non-life. In the back of his mind, he was certainly thinking of Clausius's version of the Second Law: objects in thermal contact evolve toward a common temperature (thermal equilibrium). If we put an ice cube in a glass of warm water, the ice cube melts fairly quickly. Even if the two objects are made of very different substances—say, if we put a plastic "ice cube" in a glass of water— they will still come to the same temperature. More generally, nonliving physical objects tend to wind down and come to rest. A rock may roll down a hill during an avalanche, but before too long it will reach the bottom, dissipate energy through the creation of noise and heat, and come to a complete halt.

Schrödinger's point is simply that, for living organisms, this process of coming to rest can take much longer, or even be put off indefinitely. Imagine that, instead of an ice cube, we put a goldfish into our glass of water. Unlike the ice cube (whether water or plastic), the goldfish will not simply equilibrate with the water— at least, not within a few minutes or even hours. It will stay alive, doing something, swimming, exchanging material with its environment. If it's put into a lake or a fish tank where food is available, it will keep going for much longer.

And that, suggests Schrödinger, is the essence of life: staving off the natural tendency toward equilibration with one's surroundings. At first glance, most of the features we commonly associate with life are nowhere to be found in this definition. But if we start thinking about *why* organisms are able to keep doing something long after nonliving things would wind down—why the goldfish is still swimming long after the ice cube would have melted—we are immediately drawn to the complexity of the organism and its capacity for processing information. The outward sign of life is the ability of an organism to keep going for a long time, but the mechanism behind that ability is a subtle interplay between numerous levels of hierarchical structure.

We would like to be a little more specific than that. It's nice to say, "living beings are things that keep going for longer than we would otherwise expect, and the reason they can keep going is because they're complex," but surely there is more to the story. Unfortunately, it's not a simple story, nor one that scientists understand very well. Entropy certainly plays a big role in the nature of life, but there are important aspects that it doesn't capture. Entropy characterizes individual states at a single moment in time, but the salient features of life involve processes that evolve through time. By itself, the concept of entropy has only very crude implications for evolution through time: It tends to go up or stay the same, not go down. The Second Law says nothing about *how fast* entropy will increase, or the particular methods by which entropy will grow—it's all about Being, not Becoming.[159]

Nevertheless, even without aspiring to answer all possible questions about the meaning of "life," there is one concept that undoubtedly plays an important role: *free energy*. Schrödinger glossed over this idea in the first edition of *What Is Life?*, but in subsequent printings he added a note expressing his regret for not giving it greater prominence. The idea of free energy helps to tie together entropy, the Second Law, Maxwell's Demon, and the ability of living organisms to keep going longer than nonliving objects.

FREE ENERGY, NOT FREE BEER

The field of biological physics has witnessed a dramatic rise in popularity in recent years. That's undoubtedly a good thing—biology is important, and physics is important, and there are a great number of interesting problems at the interface of the two fields. But it's also no surprise that the field lay relatively fallow for as long as it did. If you pick up an introductory physics textbook and compare it with a biological physics text, you'll notice a pronounced shift in vocabulary.[160] Conventional introductory physics books are filled with words like *force* and *momentum* and *conservation*, while biophysics books feature words like *entropy* and *information* and *dissipation*.

This difference in terminology reflects an underlying difference in philosophy.

Ever since Galileo first encouraged us to ignore air resistance when thinking about how objects fall in a gravitational field, physics has traditionally gone to great lengths to minimize friction, dissipation, noise, and anything else that would detract from the unimpeded manifestation of simple microscopic dynamical laws. In biological physics, we can't do that; once you start ignoring friction, you ignore life itself. Indeed, that's an alternative definition worth contemplating: Life is organized friction.

But, you are thinking, that doesn't sound right at all. Life is all about maintaining structure and organization, whereas friction creates entropy and disorder. In fact, both perspectives capture some of the underlying truth. What life does is to create entropy somewhere, in order to maintain structure and organization somewhere else. That's the lesson of Maxwell's Demon.

Let's examine what that might mean. Back when we first talked about the Second Law in Chapter Two, we introduced the distinction between "useful" and "useless" energy: Useful energy can be converted into some kind of work, while useless energy is useless. One of the contributions of Josiah Willard Gibbs was to formalize these concepts, by introducing the concept of "free energy." Schrödinger didn't use that term in his lectures because he worried that the connotations were confusing: The energy isn't really "free" in the sense that you can get it for nothing; it's "free" in the sense that it's available to be used for some purpose.[161] (Think "free speech," not "free beer," as free-software guru Richard Stallman likes to say.) Gibbs realized that he could use the concept of entropy to cleanly divide the total amount of energy into the useful part, which he called "free," and the useless part:[162]

total energy = free energy + useless (high-entropy) energy.

When a physical process creates entropy in a system with a fixed total amount of energy, it uses up free energy; once all the free energy is gone, we've reached equilibrium.

That's one way of thinking about what living organisms do: They maintain order in their local environment (including their own bodies) by taking advantage of free energy, degrading it into useless energy. If we put a goldfish in an otherwise empty container of water, it can maintain its structure (far from equilibrium with its surroundings) for a lot longer than an ice cube can; but eventually it will die from starvation. But if we feed the goldfish, it can last for a much longer time even than that. From a physics point of view, food is simply a supply of free energy, which a living organism can take advantage of to power its metabolism.

From this perspective, Maxwell's Demon (along with his box of gas) serves as an illuminating paradigm for how life works. Consider a slightly more elaborate version of the Demon story. Let's take the divided box of gas and embed it in an "environment," which we model by an arbitrarily large collection of stuff at a constant temperature—what physicists call a "heat bath." (The point is that the environment is so large that its own temperature won't be affected by interactions with the smaller system in which we are interested, in this case the box of gas.) Even though the molecules of gas stay inside the walls of the box, thermal energy can pass in and out; therefore, even if the Demon were to segregate the gas effectively into one cool half and one hot half, the temperature would immediately begin to even out through interactions with the surrounding environment.

We imagine that the Demon would really like to keep its particular box far from equilibrium—it wants to do its best to keep the left side of the box at a high temperature and the right side at a low temperature. (Note that we have turned the Demon into a protagonist, rather than a villain.) So it has to do its traditional sorting of molecules according to their velocities, but now it has to keep doing that in perpetuity, or otherwise each side will equilibrate with its environment. By our previous discussion, the Demon can't do its sorting without affecting the outside world; the process of erasing records will inevitably generate entropy. What the Demon requires, therefore, is a continual supply of free energy. It takes in the free energy ("food"), then takes advantage of that free energy to erase its records, generating entropy in the process and degrading the energy into uselessness; the useless

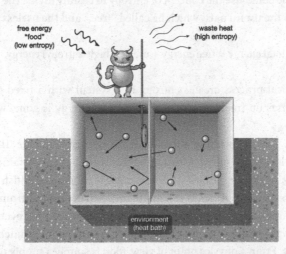

Figure 51: Maxwell's Demon as a paradigm for life. The Demon maintains order—a separation of temperatures—in the box, against the influence of the environment, by processing information through the transformation of free energy into high-entropy heat.

energy is then discarded as heat (or whatever). With its newly erased notepad, the Demon is ready to keep its box of gas happily displaced from equilibrium, at least until it fills the notepad once more, and the cycle repeats itself.

This charming vignette obviously fails to encapsulate everything we mean by the idea of "life," but it succeeds in capturing an essential part of the bigger picture. Life strives to maintain order in the face of the demands of the Second Law, whether it's the actual body of the organism, or its mental state, or the works of Ozymandias. And it does so in a specific way: by degrading free energy in the outside world in the cause of keeping itself far from thermal equilibrium. And that's an operation, as we have seen, that is tightly connected to the idea of information processing. The Demon carries out its duty by converting free energy into information about the molecules in its box, which it then uses to keep the temperature in the box from evening out. At some very basic level, the purpose of life boils down to survival—the organism wants to preserve the smooth operation of its own complex structure.[163] Free energy and information are the keys to making it happen.

From the point of view of natural selection, there are many reasons why a complex, persistent structure might be adaptively favored: An eye, for example, is a complex structure that clearly contributes to the fitness of an organism. But increasingly complex structures require that we turn increasing amounts of free energy into heat, just to keep them intact and functioning. This picture of the interplay of energy and information therefore makes a prediction: The more complex an organism becomes, the more *inefficient* it will be at using energy for "work" purposes—simple mechanical operations like running and jumping, as opposed to the "upkeep" purposes of keeping the machinery in good working condition. And indeed, that's true; in real biological organisms, the more complex ones are correspondingly less efficient in their use of energy.[164]

COMPLEXITY AND TIME

There are any number of fascinating topics at the interface of entropy, information, life, and the arrow of time that we don't have a chance to discuss here: aging, evolution, mortality, thinking, consciousness, social structures, and countless more. Confronting all of them would make this a very different book, and our primary goals are elsewhere. But before returning to the relatively solid ground of conventional statistical mechanics, we can close this chapter with one more speculative thought, the kind that may hopefully be illuminated by new research in the near future.

As the universe evolves, entropy increases. That is a very simple relationship:

At early times, near the Big Bang, the entropy was very low, and it has grown ever since and will continue to grow into the future. But apart from entropy, we can also characterize (at least roughly) the state of the universe at any one moment in time in terms of its *complexity*, or by the converse of complexity, its simplicity. And the evolution of complexity with time isn't nearly that straightforward.

There are a number of different ways we could imagine quantifying the complexity of a physical situation, but there is one measure that has become widely used, known as the *Kolmogorov complexity* or *algorithmic complexity*.[165] This idea formalizes our intuition that a simple situation is easy to describe, while a complex situation is hard to describe. The difficulty we have in describing a situation can be quantified by specifying the shortest possible computer program (in some given programming language) that would generate a description of that situation. The Kolmogorov complexity is just the length of that shortest possible computer program.

Consider two strings of numbers, each a million characters long. One string consists of nothing but 8's in every digit, while the other is some particular sequence of digits with no discernible pattern within them:

8888888888888888888 . . .
60462491123396078395 . . .

The first of these is simple—it has a low Kolmogorov complexity. That's because it can be generated by a program that just says, "Print the number 8 a million times." The second string, however, is complex. Any program that prints it out has to be at least one million characters long, because the only way to describe this string is to literally specify every single digit. This definition becomes helpful when we consider numbers like pi or the square root of two—they look superficially complex, but there is actually a short program in either case that can calculate them to any desired accuracy, so their Kolmogorov complexity is quite low.

The complexity of the early universe is low, because it's very easy to describe. It was a hot, dense state of particles, very smooth over large scales, expanding at a certain rate, with some (fairly simple to specify) set of tiny perturbations in density from place to place. From a coarse-grained perspective, that's the entire description of the early universe; there's nothing else to say. Far in the future, the complexity of the universe will also be very low: It will just be empty space, with an increasingly dilute gruel of individual particles. But in between—like right now—things look extremely complicated. Even after coarse-graining, there is no simple way of expressing the hierarchical structures described by gas, dust, stars, galaxies, and clusters, much less all of the interesting things going on much smaller scales, such as our ecosystem here on Earth.

So while the entropy of the universe increases straightforwardly from low to high as time goes by, the complexity is more interesting: It goes from low, to relatively high, and then back down to low again. And the question is: Why? Or perhaps: What are the ramifications of this form of evolution? There are a whole host of questions we can think to ask. Under what general circumstances does complexity tend to rise and then fall again? Does such behavior inevitably accompany the evolution of entropy from low to high, or are other features of the underlying dynamics necessary? Is the emergence of complexity (or "life") a generic feature of evolution in the presence of entropy gradients? What is the significance of the fact that our early universe was simple as well as low-entropy? How long can life survive as the universe relaxes into a simple, high-entropy future?[166]

Science is about answering hard questions, but it's also about pinpointing the right questions to ask. When it comes to understanding life, we're not even sure what the right questions are. We have a bunch of intriguing concepts that we're pretty sure will play some sort of role in an ultimate understanding—entropy, free energy, complexity, information. But we're not yet able to put them together into a unified picture. That's okay; science is a journey in which getting there is, without question, much of the fun.

10
RECURRENT NIGHTMARES

Nature is a big series of unimaginable catastrophes.

—Slavoj Žižek

In Book Four of *The Gay Science*, written in 1882, Friedrich Nietzsche proposes a thought experiment. He asks us to imagine a scenario in which everything that happens in the universe, including our lives down to the slightest detail, will eventually repeat itself, in a cycle that plays out for all eternity.

> What if some day or night a demon were to steal into your loneliest loneliness and say to you: "This life as you now live it and have lived it you will have to live once again and innumerable times again; and there will be nothing new in it, but every pain and every joy and every thought and sigh and everything unspeakably small or great in your life must return to you, all in the same succession and sequence— even this spider and this moonlight between the trees, and even this moment and I myself. The eternal hourglass of existence is turned over again and again, and you with it, speck of dust!"[167]

Nietzsche's interest in an eternally repeating universe was primarily an ethical one. He wanted to ask: How would you feel about knowing that your life would be repeated an infinite number of times? Would you feel dismayed—gnashing your teeth is mentioned—at the horrible prospect, or would you rejoice? Nietzsche felt that a successful life was one that you would be proud to have repeated in an endless cycle.[168]

The idea of a cyclic universe, or "eternal return," was by no means original with Nietzsche. It appears now and again in ancient religions—in Greek myths, Hinduism, Buddhism, and some indigenous American cultures. The Wheel of Life spins, and history repeats itself.

But soon after Nietzsche imagined his demon, the idea of eternal recurrence popped up in physics. In 1890 Henri Poincaré proved an intriguing mathematical theorem, showing that certain physical systems would necessarily return to any particular configuration infinitely often, if you just waited long enough. This result was seized upon by a young mathematician named Ernst Zermelo, who claimed that it was incompatible with Boltzmann's purported derivation of the Second Law of Thermodynamics from underlying reversible rules of atomic motion.

In the 1870s, Boltzmann had grappled with Loschmidt's "reversibility paradox." By comparison, the 1880s were a relatively peaceful time for the development of statistical mechanics—Maxwell had died in 1879, and Boltzmann concentrated on technical applications of the formalism he had developed, as well as on climbing the academic ladder. But in the 1890s controversy flared again, this time in the form of Zermelo's "recurrence paradox." To this day, the ramifications of these arguments have not been completely absorbed by physicists; many of the issues that Boltzmann and his contemporaries argued about are still being hashed out right now. In the context of modern cosmology, the problems suggested by the recurrence paradox are still very much with us.

POINCARÉ'S CHAOS

Oscar II, king of Sweden and Norway, was born on January 21, 1829. In 1887, the Swedish mathematician Gösta Mittag-Leffler proposed that the king should mark his upcoming sixtieth birthday in a somewhat unconventional way: by sponsoring a mathematical competition. Four different questions were proposed, and a prize would be given to whoever submitted the most original and creative solution to any of them.

One of these questions was the "three-body problem"—how three massive objects would move under the influence of their mutual gravitational pull. (For two bodies it's easy, and Newton had solved it: Planets move in ellipses.) This problem was tackled by Henri Poincaré, who in his early thirties was already recognized as one of the world's leading mathematicians. He did not solve it, but submitted an essay that seemed to demonstrate a crucial feature: that the orbits of the planets would be *stable*. Even without knowing the exact solutions, we could be confident that the planets would at least behave predictably. Poincaré's method was so ingenious that he was awarded the prize, and his paper was prepared for publication in Mittag-Leffler's new journal, *Acta Mathematica*.[169]

Figure 52: Henri Poincaré, pioneer of topology, relativity, and chaos theory, and later president of the Bureau of Longitude.

But there was a slight problem: Poincaré had made a mistake. Edvard Phragmén, one of the journal editors, had some questions about the paper, and in the course of answering them Poincaré realized that he had left out an important case in constructing his proof. Such tiny mistakes occur frequently in complicated mathematical writing, and Poincaré set about correcting his presentation. But as he tugged at the loose thread, the entire argument became completely unraveled. What Poincaré ended up proving was the opposite of his original claim—three-body orbits were not stable at all. Not only are orbits not periodic; they don't even approach any sort of regular behavior. Now that we have computers to run simulations, this kind of behavior is less surprising, but at the time it came as an utter shock. In his attempt to prove the stability of planetary orbits, Poincaré ended up doing something quite different— he invented chaos theory.

But the story doesn't quite end there. Mittag-Leffler, convinced that Poincaré would be able to fix things up in his prize essay, had gone ahead and printed it. By the time he heard from Poincaré that no such fixing-up would be forthcoming, the journal had already been mailed to leading mathematicians throughout Europe. Mittag-Leffler swung into action, telegraphing Berlin and Paris in an attempt to have all copies of the journal destroyed. He basically succeeded, but not without creating a minor scandal in elite mathematical circles across the continent.

In the course of revising his argument, Poincaré established a deceptively simple and powerful result, now known as the *Poincaré recurrence theorem*. Imagine you have some system in which all of the pieces are confined to some finite region of space, like planets orbiting the Sun. The recurrence theorem says that if we start with the system in a particular configuration and simply let it evolve according to Newton's laws, we are guaranteed that the system will return to its original configuration—again and again, infinitely often into the future.

That seems pretty straightforward, and perhaps unsurprising. If we have assumed from the start that all the components of our system (planets orbiting the Sun, or molecules bouncing around inside a box) are confined to a finite region, but we allow time to stretch on forever, it makes sense that the system is going to keep returning to the same state over and over. Where else can it go?

Things are a bit more subtle than that. The most basic subtlety is that there can be an infinite number of possible states, even if the objects themselves don't actually run away to infinity.[170] A circular orbit is confined to a finite region, but there are an infinite number of points along it; likewise, there are an infinite number of points inside a finite-sized box of gas. In that case, a system will typically not return to *precisely* the original state. What Poincaré realized is that this is a case where "almost" is good enough. If you decide ahead of time how close two states need to be so that you can't tell the difference between them, Poincaré proved that the system would return at least that close to the original state an infinite number of times.

Consider the three inner planets of the Solar System: Mercury, Venus, and Earth. Venus orbits the Sun once every 0.61520 years (about 225 days), while Mercury orbits once every 0.24085 years (about 88 days). As shown in Figure 53, imagine that we started in an arrangement where all three planets were arranged in a straight line. After 88 days have passed, Mercury will have returned to its starting point, but Venus and Earth will be at some other points in their orbits. But if we wait long enough, they will all line up again, or very close to it. After 40 years, for example, these three planets will be in almost the same arrangement as when they started.

Poincaré showed that all confined mechanical systems are like that, even ones with large numbers of moving parts. But notice that the amount of time we have to wait before the system returns close to its starting point keeps getting larger as we add more components. If we waited for all nine of the planets to line up,[171] we would have to wait much longer than 40 years; that's partly because the outer planets orbit more slowly, but in large part it simply takes longer for more objects to conspire in the right way to re-create any particular starting configuration.

This is worth emphasizing: As we consider more and more particles, the time it takes for a system to return close to its starting point—known, reasonably enough, as the *recurrence time*—quickly becomes unimaginably huge.[172] Consider the divided box of gas we played with in Chapter Eight, where individual particles had a small chance of hopping from one side of the box to the other every second. Clearly if there are only two or three particles, it won't take long for the system to return to where it started. But once we consider a box with 60 total particles, we

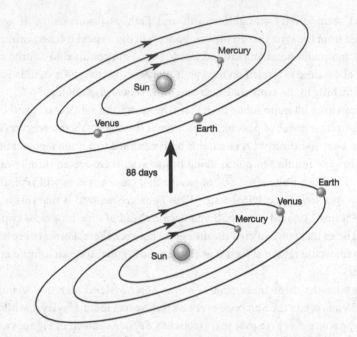

Figure 53: The inner Solar System in a configuration with Mercury, Venus, and Earth all aligned (bottom), and 88 days later (top). Mercury has returned to its original position, but Venus and Earth are somewhere else along their orbits.

find that the recurrence time has become as large as the current age of the observable universe.

Real objects usually have a lot more than 60 particles in them. For a typical macroscopic-sized object, the recurrence time would be at least

$$10^{1,000,000,000,000,000,000,000,000} \text{ seconds.}$$

That's a long time. For the total number of particles in the observable universe, the recurrence time would be even much longer—but who cares? The recurrence time for any interestingly large object is much longer than any time relevant to our experience. The observable universe is about 10^{18} seconds old. An experimental physicist who put in a grant proposal suggesting that they would pour a teaspoon of milk into a cup of coffee and then wait one recurrence time for the milk to unmix itself would have a very hard time getting funding.

But if we waited long enough, it would happen. Nietzsche's Demon isn't wrong; it's just thinking long-term.

ZERMELO VERSUS BOLTZMANN

Poincaré's original paper in which he proved the recurrence theorem was mainly concerned with the crisp, predictable world of Newtonian mechanics. But he was familiar with statistical mechanics, and within a short while realized that the idea of eternal recurrence might, at first blush, be incompatible with attempts to derive the Second Law of Thermodynamics. After all, the Second Law says that entropy only ever goes one way: It increases. But the recurrence theorem seems to imply that if a low-entropy state evolves to a higher-entropy state, all we have to do is wait long enough and the system will return to its low-entropy beginnings. That means it must decrease somewhere along the way.

In 1893, Poincaré wrote a short paper that examined this apparent contradiction more closely. He pointed out that the recurrence theorem implied that the entropy of the universe would eventually start decreasing:

> I do not know if it has been remarked that the English kinetic theories can extract themselves from this contradiction. The world, according to them, tends at first toward a state where it remains for a long time without apparent change; and this is consistent with experience; but it does not remain that way forever, if the theorem cited above is not violated; it merely stays that way for an enormously long time, a time which is longer the more numerous are the molecules. This state will not be the final death of the universe, but a sort of slumber, from which it will awake after millions of millions of centuries. According to this theory, to see heat pass from a cold body to a warm one, it will not be necessary to have the acute vision, the intelligence, and the dexterity of Maxwell's demon; it will suffice to have a little patience.[173]

By "the English kinetic theories," Poincaré was presumably thinking of the work of Maxwell and Thomson and others—no mention of Boltzmann (or for that matter Gibbs). Whether it was for that reason or just because he didn't come across the paper, Boltzmann made no direct reply to Poincaré.

But the idea would not be so easily ignored. In 1896, Zermelo made a similar argument to Poincaré's (referencing Poincaré's long 1890 paper that stated the recurrence theorem, but not his shorter 1893 paper), which is now known as "Zermelo's recurrence objection."[174] Despite Boltzmann's prominence, atomic theory and statistical mechanics were not nearly as widely accepted in the late-nineteenth-century German-speaking world as they were in the English-speaking world. Like many German scientists, Zermelo thought that the Second Law was an absolute rule of nature; the entropy of a closed system would *always* increase or stay constant, not merely most of the time. But the recurrence theorem clearly implied that if entropy initially went up, it would someday come down as the system returned to its starting configuration. The lesson drawn by Zermelo was that the edifice of statistical mechanics was simply wrong; the behavior of heat and entropy could not be reduced to the motions of molecules obeying Newton's laws.

Zermelo would later go on to great fame within mathematics as one of the founders of set theory, but at the time he was a student studying under Max Planck, and Boltzmann didn't take the young interloper's objections very seriously. He did bother to respond, although not with great patience.

> Zermelo's paper shows that my writings have been misunderstood; nevertheless it pleases me for it seems to be the first indication that these writings have been paid any attention in Germany. Poincaré's theorem, which Zermelo explains at the beginning of his paper, is clearly correct, but his application of it to the theory of heat is not.[175]

Oh, snap. Zermelo wrote another paper in response to Boltzmann, who replied again in turn.[176] But the two were talking past each other, and never seemed to reach a satisfactory conclusion.

Boltzmann, by this point, was completely comfortable with the idea that the Second Law was only statistical in nature, rather than absolute. The main thrust of his response to Zermelo was to distinguish between theory and practice. In theory, the whole universe could start in a low entropy state, evolve toward thermal equilibrium, and eventually evolve back to low entropy again; that's an implication of Poincaré's theorem, and Boltzmann didn't deny it. But the actual time you would have to wait is enormous, much longer than what we currently think of as "the age of the universe," and certainly much longer than any timescales that were contemplated by scientists in the nineteenth century. Boltzmann argued that we should accept the implications of the recurrence theorem as an interesting mathematical curiosity, but not one that was in any way relevant to the real world.

TROUBLES OF AN ETERNAL UNIVERSE

In Chapter Eight we discussed Loschmidt's reversibility objection to Boltzmann's *H*-Theorem: It is impossible to use reversible laws of physics to derive an irreversible result. In other words, there are just as many high-entropy states whose entropy will decrease as there are low-entropy states whose entropy will increase, because the former trajectories are simply the time-reverses of the latter. (And neither is anywhere near as numerous as high-entropy states that remain high-entropy.) The proper response to this objection, at least within our observable universe, is to accept the need for a Past Hypothesis—an additional postulate, over and above the dynamical laws of nature, to the effect that the early universe had an extremely low entropy.

In fact, by the time of his clash with Zermelo, Boltzmann himself had cottoned on to this realization. He called his version of the Past Hypothesis "assumption A," and had this to say about it:

> The second law will be explained mechanically by means of assump-
> tion A (which is of course unprovable) that the universe, considered
> as a mechanical system—or at least a very large part of it which sur-
> rounds us—started from a very improbable state, and is still in an
> improbable state.[177]

This short excerpt makes Boltzmann sound more definitive than he really is; in the context of this paper, he offers several different ways to explain why we see entropy increasing around us, and this is just one of them. But notice how careful he is—not only admitting up front that the assumption is unprovable, but even limiting consideration to "a very large part of [the universe] which surrounds us," not the whole thing.

Unfortunately, this strategy isn't quite sufficient. Zermelo's recurrence objec-tion is closely related to the reversibility objection, but there is an important dif-ference. The reversibility objection merely notes that there are an equal number of entropy-decreasing evolutions as entropy-increasing ones; the recurrence objec-tion points out that the entropy-decreasing processes *will eventually happen some time in the future*. It's not just that a system could decrease in entropy—if we wait long enough, it is eventually guaranteed to do so. That's a stronger statement and requires a better comeback.

We can't rely on the Past Hypothesis to save us from the problems raised by recurrence. Let's say we grant that, at some point in the relatively recent past—perhaps billions of years ago, but much more recently than one recurrence time—the universe found itself in a state of extremely low entropy. Afterward, as Boltzmann taught us, the entropy would increase, and the time it would take to do so is much shorter than one recurrence time. But if the universe truly lasts forever, that shouldn't matter. Eventually the entropy is going to go down again, even if we're not around to witness it. The question then becomes: Why do we find ourselves living in the particular part of the history of the universe in the relatively recent aftermath of the low-entropy state? Why don't we live in some more "natural" time in the history of the universe?

Something about that last question, especially the appearance of the word *natural*, opens a can of worms. The basic problem is that, according to Newtonian physics, the universe doesn't have a "beginning" or an "end." From our twenty-first-century post-Einsteinian perspective, the idea that the universe began at the Big Bang is a familiar one. But Boltzmann and Zermelo and contemporaries didn't know about general relativity or the expansion of the universe. As far as they were concerned, space and time were absolute, and the universe persisted forever. The option of sweeping these embarrassing questions under the rug of the Big Bang wasn't available to them.

That's a problem. If the universe truly lasts forever, having neither a beginning nor an end, what is the Past Hypothesis supposed to mean? There was some moment, earlier than the present, when the entropy was small. But what about before that? Was it always small—for an infinitely long time—until some transition occurred that allowed the entropy to grow? Or was the entropy also higher before that moment, and if so, why is there a special low-entropy moment in the middle of the history of the universe? We seem to be stuck: If the universe lasts forever, and the assumptions underlying the recurrence theorem are valid, entropy can't increase forever; it must go up and then eventually come back down, in an endless cycle.

There are at least three ways out of this dilemma, and Boltzmann alluded to all three of them.[178] (He was convinced he was right but kept changing his mind about the reason why.)

First, the universe might really have a beginning, and that beginning would involve a low-entropy boundary condition. This is implicitly what Boltzmann must have been imagining in the context of "assumption A" discussed above, although he doesn't quite spell it out. But at the time, it would have been truly dramatic to claim that time had a beginning, as it requires a departure from the basic rules of physics as Newton had established them. These days we have such a departure, in the form of general relativity and the Big Bang, but those ideas weren't on the table in the 1890s. As far as I know, no one at the time took the problem of the universe's low entropy at early times seriously enough to suggest explicitly that time must have had a beginning, and that something like the Big Bang must have occurred.

Second, the assumptions behind the Poincaré recurrence theorem might simply not hold in the real world. In particular, Poincaré had to assume that the space of states was somehow bounded, and particles couldn't wander off to infinity. That sounds like a technical assumption, but deep truths can be hidden under the guise of technical assumptions. Boltzmann also floats this as a possible loophole:

> If one first sets the number of molecules equal to infinity and allows the time of the motion to become very large, then in the overwhelming majority of cases one obtains a curve [for entropy as a function of time] which asymptotically approaches the abscissa axis. The Poincaré theorem is not applicable in this case, as can easily be seen.[179]

But he doesn't really take this option seriously. As well he shouldn't, as it avoids the strict implication of the recurrence theorem but not the underlying spirit. If the average density of particles through space is some nonzero number, you will still see all sorts of unlikely fluctuations, including into low-entropy states; it's just that the fluctuations will typically consist of different sets of particles each time, so that "recurrence" is not strictly occurring. That scenario has all of the problems of a truly recurring system.

The third way out of the recurrence objection is not a way out at all—it's a complete capitulation. Admit that the universe is eternal, and that recurrences happen, so that the universe witnesses moments when entropy is increasing and moments when it is decreasing. And then just say: That's the universe in which we live.

Let's put these three possibilities in the context of modern thinking. Many contemporary cosmologists subscribe, often implicitly, to something like the first option—conflating the puzzle of our low-entropy initial conditions with the puzzle of the Big Bang. It's a viable possibility but seems somewhat unsatisfying, as it requires that we specify the state of the universe at early times over and above the laws of physics. The second option, that there are an infinite number of things in the universe and the recurrence theorem simply doesn't apply, helps us wriggle out of the technical requirements of the theorem but doesn't give us much guidance concerning why our universe looks the particular way that it does. We could consider a slight variation on this approach, in which there were only a finite number of particles in the universe, but they had an infinite amount of space in which to evolve. Then recurrences would truly be absent; the entropy would grow without limit in the far past and far future. This is somewhat reminiscent of the multiverse scenario I will be advocating later in the book. But as far as I know, neither Boltzmann nor any of his contemporaries advocated such a picture.

The third option—that recurrences really do happen, and that's the universe we live in—can't be right, as we will see. But we can learn some important lessons from the way in which it fails to work.

FLUCTUATING AROUND EQUILIBRIUM

Recall the divided box of gas we considered in Chapter Eight. There is a partition between two halves of the box that occasionally lets gas molecules pass through and switch sides. We modeled the evolution of the unknown microstate of each particle by imagining that every molecule has a small, fixed chance of moving from one side of the box to the other. We can use Boltzmann's entropy formula to show how the entropy evolves with time; it has a strong tendency to increase, at least if we start the system by hand in a low-entropy state, with most of the molecules on one side. The natural tendency is for things to even out and approach an equilibrium state with approximately equal numbers of molecules on each side. Then the entropy reaches its maximum value, labeled as "1" on the vertical axis of the graph.

What if we *don't* start the system in a low-entropy state? What happens if it starts in equilibrium? If the Second Law were absolutely true, and entropy could never decrease, once the system reached equilibrium it would have to strictly stay there. But in Boltzmann's probabilistic world, that's not precisely right. With high probability, a system that is in equilibrium will stay in equilibrium or very close to it. But there will inevitably be random fluctuations away from the state, if we wait long enough. And if we wait very long, we could see some rather large fluctuations.

In Figure 54, we see the evolution of the entropy in a divided box of gas with 2,000 particles, but now at a later time, after it has reached equilibrium. Note that this is an extreme close-up on the change in entropy; whereas the plots in Chapter Eight showed the entropy evolving from about 0.75 up to 1, this plot shows the entropy ranging from between 0.997 and 1.

What we see are small fluctuations from the equilibrium value where the entropy is maximal and the molecules are equally divided. This makes perfect sense, the way we've set up the situation; most of the time, there will be equal numbers of particles on the right side of the box and the left side, but occasionally there will be a slight excess on one side or the other, corresponding to a slightly lower entropy. It's exactly the same idea as flipping a coin—on average, a sequence of many coin flips will average to half heads and half tails, but if we wait long enough, we will see sequences of the same result many times in a row.

The fluctuations seen here are very small, but on the other hand we didn't

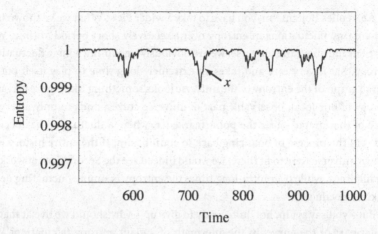

Figure 54: The evolution of the entropy of a divided box of gas, starting from equilibrium. The state spends most of its time near maximum entropy, but there are occasional fluctuations to lower-entropy states. Note from the vertical axis that we have zoomed up close; typical fluctuations are very small. The point x marks a return to equilibrium from a relatively large fluctuation.

wait very long. If we stretched out the plot to much longer times—and here we're talking *much* longer times—the entropy would eventually dip down to its original value, representing a state with 80 percent of the particles on one side and only 20 percent on the other. Keep in mind that this graph shows what happens with 2,000 particles; in the real world, with many more particles in any macroscopic object, fluctuations in entropy are correspondingly smaller and more rare. But they will be there; that's an inevitable consequence of the probabilistic nature of entropy.

So here is Boltzmann's final, dramatic suggestion: Maybe the universe is like that. Maybe time does last forever, and the underlying laws of physics are Newtonian and reversible, and maybe the assumptions underlying the recurrence theorem are valid.[180] And maybe, therefore, the plot of entropy versus time shown in Figure 54 is how the entropy of the real universe actually evolves.

THE ANTHROPIC APPEAL

But—you say—that can't be right. On that graph, entropy goes up half the time and goes down half the time. That's not at all like the real world, where entropy only ever goes up, as far as we can see.

Ah, replies Boltzmann, you have to take a wider view. What we've shown in the plot are tiny fluctuations in entropy over a relatively short period of time. When we're talking about the universe, we are obviously imagining a huge fluctuation in entropy that is very rare and takes an extremely long time to play itself out. The overall graph of the entropy of the universe looks something like Figure 54, but the entropy of our local, observable part of universe corresponds to only a very tiny piece of that graph—near the point marked x, where a fluctuation has occurred and is in the process of bouncing back to equilibrium. If the entire history of the known universe were to fit there, we would indeed see the Second Law at work over our lifetimes, while over ultra-long times the entropy is simply fluctuating near its maximum value.

But—you say again, not quite ready to give up—why should we live at that particular part of the curve, in the aftermath of a giant entropy fluctuation? We've already admitted that such fluctuations are exceedingly rare. Shouldn't we find ourselves at a more typical period in the history of the universe, where things basically look like they are in equilibrium?

Boltzmann, naturally, has foreseen your objection. And at this point he makes a startlingly modern move—he invokes the *anthropic principle*. The anthropic principle is basically the idea that any sensible account of the universe around us must take into consideration the fact that we exist. It comes in various forms, from the uselessly weak—"the fact that life exists tell us that the laws of physics must be compatible with the existence of life"—to the ridiculously strong—"the laws of physics had to take the form they do because the existence of life is somehow a necessary feature." Arguments over the status of the anthropic principle—Is it useful? Is it science?—grow quite heated and are rarely very enlightening.

Fortunately, we (and Boltzmann) need only a judicious medium-strength version of the anthropic principle. Namely, imagine that the real universe is much bigger (in space, or in time, or both) than the part we directly observe. And imagine further that different parts of this bigger universe exist in very different conditions. Perhaps the density of matter is different, or even something as dramatic as different local laws of physics. We can label each distinct region a "universe," and the whole collection is the "multiverse." The different universes within the multiverse may or may not be physically connected; for our present purposes it doesn't matter. Finally, imagine that some of these different regions are hospitable to the existence of life, and some are not. (That part is inevitably a bit fuzzy, given how little we know about "life" in a wider context.) Then—and this part is pretty much unimpeachable—we will always find ourselves existing in one of the parts of the universe where life is allowed to exist, and not in the other parts. That sounds completely empty, but it's not. It represents a *selection effect* that distorts our view of the universe as a whole—we don't see the entire thing; we see only one of the parts, and that part might not be representative.

Boltzmann appeals to exactly this logic. He asks us to imagine a universe consisting of some collection of particles moving through an absolute Newtonian spacetime that exists for all eternity. What would happen?

> There must then be in the universe, which is in thermal equilibrium as a whole and therefore dead, here and there relatively small regions of the size of our galaxy (which we call worlds), which during the relatively short time of eons deviate significantly from thermal equilibrium. Among these worlds the state probability [entropy] increases as often as it decreases. For the universe as a whole the two directions of time are indistinguishable, just as in space there is no up or down. However, just as at a certain place on the earth's surface we can call "down" the direction toward the centre of the earth, so a living being that finds itself in such a world at a certain period of time can define the time direction as going from less probable to more probable states (the former will be the "past" and the latter the "future") and by virtue of this definition he will find that this small region, isolated from the rest of the universe, is "initially" always in an improbable state.[181]

This is a remarkable paragraph, which would be right at home in a modern cosmology discussion, with just a few alterations in vocabulary. Boltzmann imagines that the universe (or the multiverse, if you prefer) is basically an infinitely big box of gas. Most of the time the gas is distributed uniformly through space, at constant temperature—thermal equilibrium. The thing is, we can't live in thermal equilibrium—it's "dead," as he bluntly puts it. From time to time there will be random fluctuations, and eventually one of these will create something like the universe we see around us. (He refers to "our galaxy," which at the time was synonymous with "the observable universe.") It's in those environments, the random fluctuations away from equilibrium, where we can possibly live, so it's not much surprise that we find ourselves there.

Even in the course of a fluctuation, of course, the entropy is only increasing half the time—in the other half it's decreasing, moving from equilibrium down to the minimum value it will temporarily reach. But this sense of "increasing" or "decreasing" describes the evolution of entropy with respect to some arbitrarily chosen time coordinate, which—as we discussed in the last chapter—is completely

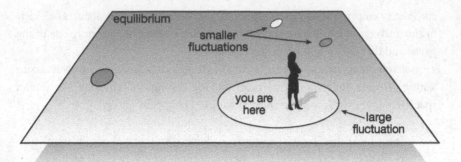

Figure 55: Boltzmann's "multiverse." Space is mostly a collection of particles in equilibrium, but there are occasional local fluctuations to low-entropy states. (Not at all to scale.) We live in the aftermath of one exceptionally large fluctuation.

unobservable. As Boltzmann correctly points out, what matters is that the current universe is in the middle of a transition from a low-entropy state to one of thermal equilibrium. In the midst of such a transition, any living beings who pop up will always label the direction of lower entropy "the past," and the direction of higher entropy "the future."

This is a provocative picture of the universe. On large scales, matter is almost always in a dilute collection of gas at some temperature. But every so often, over the course of billions of years, a series of random accidents conspire to create pockets of anomalously low entropy, which then relax back to equilibrium. You, and I, and all the bustling activity we see around us, are epiphenomena riding the wave of entropy as it bounces back from a random excursion into a wildly improbable state.[182]

So what does a typical downward fluctuation in entropy look like? The answer, of course, is that it looks exactly like the time-reverse of a typical evolution from a low-entropy state back to a high-entropy one. The whole universe wouldn't suddenly zoom from a thin gas of particles into a dense Big-Bang-like state in a matter of minutes; it would, most of the time, experience a series of unlikely accidents spread over billions of years, all of which would decrease the entropy just a little bit. Stars and galaxies would un-form, omelets would turn into eggs, objects in equilibrium would spontaneously develop substantial temperature gradients. All of these would be completely independent events, each individually unlikely, and the combination of all of them is fantastically unlikely. But if you truly have eternity to wait, even the most unlikely things will eventually happen.

SWERVING THROUGH ANTIQUITY

Boltzmann wasn't actually the first to think along these lines, if we allow ourselves a little poetic license. Just as Boltzmann was concerned with understanding the world in terms of atoms, so were his predecessors in ancient Greece and Rome. Democritus (c. 400 B.C.E.) was the most famous atomist, but his teacher Leucippus was probably the first to propose the idea. They were materialists, who hoped to explain the world in terms of objects obeying rules, rather than being driven by an underlying "purpose." In particular, they were interested in rising to the challenge raised by Parmenides, who argued that change was an illusion. The theory of unchanging atoms moving through a void was meant to account for the possibility of motion without imagining that something arises from nothing.

One challenge that the atomists of antiquity faced was to explain the messy complexity of the world around them. The basic tendency of atoms, they believed, was to fall downward in straight lines; that doesn't make for a very interesting universe. It was left to the Greek thinker Epicurus (c. 300 B.C.E.) to propose a solution to this puzzle, in the form of an idea called "the swerve" (*clinamen*).[183] Essentially Epicurus suggested that, in addition to the basic tendency of atoms to move along straight lines, there is a random component to their motion that occasionally kicks them from side to side. It's vaguely reminiscent of modern quantum mechanics, although we shouldn't get carried away. (Epicurus didn't know anything about blackbody radiation, atomic spectra, the photoelectric effect, or any of the other experimental results motivating quantum mechanics.) Part of Epicurus's reason for introducing the swerve was to leave room for free will—basically, to escape the implications of Laplace's Demon, long before that mischievous beast had ever reared his ugly head. But another motivation was to explain how individual atoms could come together to form macroscopic objects, rather than just falling straight down to Earth.

The Roman poet-philosopher Lucretius (c. 50 B.C.E.) was an avid atomist and follower of Epicurus; he was a primary inspiration for Virgil's poetry. His poem "On the Nature of Things (De Rerum Natura)" is a remarkable work, concerned with elucidating Epicurean philosophy and applying it to everything from cosmology to everyday life. He was especially interested in dispelling superstitious beliefs; imagine Carl Sagan writing in Latin hexameter. A famous section of "On the Nature of Things" counsels against the fear of death, which he sees as simply a transitional event in the endless play of atoms.

Lucretius applied atomism, and in particular the idea of the swerve, to the question of the origin of the universe. Here is what he imagines happening:

> For surely the atoms did not hold council, assigning
> Order to each, flexing their keen minds with
> Questions of place and motion and who goes where.
> But shuffled and jumbled in many ways, in the course
> Of endless time they are buffeted, driven along,
> Chancing upon all motions, combinations.
> At last they fall into such an arrangement
> As would create this universe.[184]

The opening lines here should be read in a semi-sarcastic tone of voice. Lucretius is mocking the idea that the atoms somehow planned the cosmos; rather, they just jumbled around chaotically. But through those random motions, if we wait long enough we will witness the creation of our universe.

The resemblance to Boltzmann's scenario is striking. We should always be careful, of course, not to credit ancient philosophers with a modern scientific understanding; they came from a very different perspective, and worked under a different set of presuppositions than we do today. But the parallelism between the creation scenarios suggested by Lucretius and Boltzmann is more than just a coincidence. In both cases, the task was to explain the emergence of the apparent complexity we see around us without appealing to an overall design, but simply by considering the essentially random motions of atoms. It is no surprise that a similar conclusion is reached: the idea that our observable universe is a random fluctuation in an eternal cosmos. It's perfectly fair to call this the "Boltzmann-Lucretius scenario" for the origin of the universe.

Can the real world possibly be like that? Can we live in an eternal universe that spends most of its time in equilibrium, with occasional departures that look like what we see around us? Here we need to rely on the mathematical formalism developed by Boltzmann and his colleagues, to which Lucretius didn't have recourse.

UN-BREAKING AN EGG

The problem with the Boltzmann-Lucretius scenario is not that you can't make a universe that way—in the context of Newtonian spacetime, with everlasting atoms bumping against one another and occasionally giving rise to random downward fluctuations of entropy, it's absolutely going to happen that you create a region of the size and shape of our universe if you wait long enough.

The problem is that the numbers don't work. Sure, you can fluctuate into something that looks like our universe—but you can fluctuate into a lot of other things as well. And the other things win, by a wide margin.

Rather than weighing down our brains with the idea of a huge collection of particles fluctuating into something like the universe we see around us (or even just our galaxy), let's keep things simple by considering one of our favorite examples of entropy in action: a single egg. An unbroken egg is quite orderly and has a very low entropy; if we break the egg the entropy increases, and if we whisk the ingredients together the entropy increases even more. The maximum-entropy state will be a soup of individual molecules; details of the configuration will depend on the temperature, the presence of a gravitational field, and so on, but none of that will matter for our present purposes. The point is that it won't look anything like an unbroken egg.

Imagine we take such an egg and seal it in an absolutely impenetrable box, one that will last literally forever, undisturbed by the rest of the universe. For convenience, we put the egg-in-a-box out in space, far away from any gravity or external forces, and imagine that it floats undisturbed for all eternity. What happens inside that box?

Even if we initially put an unbroken egg inside the box, eventually it would break, just through the random motions of its molecules. It will spend some time as a motionless, broken egg, differentiated into yolk and white and shell. But if we wait long enough, further random motions will gradually cause the yolk and white and even the shell to disintegrate and mix, until we reach a truly high-entropy state of undifferentiated egg molecules. That's equilibrium, and it will last an extraordinarily long time.

But if we continue to wait, the same kind of random motions that caused the egg to break in the first place will stir those molecules into lower-entropy configurations. All of the molecules may end up on one side of the box, for example. And after a very long time indeed, random motions will re-create something that looks like a broken egg (shell, yolk, and white), or even an unbroken egg! That seems preposterous, but it's the straightforward implication of Poincaré's recurrence theorem, or of taking seriously the ramifications of random fluctuations over extraordinarily long timescales.

Most of the time, the process of forming an egg through random fluctuations of the constituent molecules will look just like the time-reverse of the process by which an unbroken egg decays into high-entropy goop. That is, we will first fluctuate into the form of a broken egg, and then the broken pieces will by chance arrange themselves into the form of an unbroken egg. That's just a consequence of time-reversal symmetry; the most common evolutions from high entropy to low entropy look exactly like the most common evolutions from low entropy to high entropy, just played in reverse.

Here is the rub. Let's imagine that we have such an egg sealed in an impenetrable box, and we peek inside after it's been left to its own devices for an absurdly long time—much greater than the recurrence time. It's overwhelmingly likely that what we will see is something very much like equilibrium: a homogeneous mixture of egg molecules. But suppose we get extremely lucky, and we find what looks like a broken egg—a medium-entropy state, with some shards of eggshell and a yolk running into the egg whites. A configuration, in other words, that looks exactly what we would expect if there had recently been a pristine egg, which for some reason had been broken.

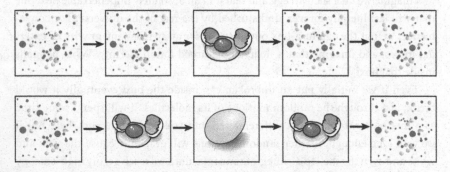

Figure 56: An egg trapped for all eternity in an impenetrable box. Most of the time the box will contain egg molecules in high-entropy equilibrium. Occasionally it will fluctuate into the medium-entropy configuration of a broken egg, as in the top row. Much more rarely, it will fluctuate all the way to the low-entropy form of an unbroken egg, and then back again, as in the bottom row.

Could we actually conclude, from this broken egg, that there had recently been an unbroken egg in the box? Not at all. Remember our discussion at the end of Chapter Eight. Given a medium-entropy configuration, and no other knowledge or assumptions such as a Past Hypothesis (which would clearly be inappropriate in the context of this ancient sealed box), it is overwhelmingly likely to have evolved from a higher-entropy past, just as it is overwhelmingly likely to evolve toward a higher-entropy future. Said conversely, given a broken egg, it is no more likely to have evolved *from* an unbroken egg than it is likely to evolve *to* an unbroken egg. Which is to say, not bloody likely.

BOLTZMANN BRAINS

The egg-in-a-box example illustrates the fundamental problem with the Boltzmann-Lucretius scenario: We can't possibly appeal to a Past Hypothesis that asserts the existence of a low-entropy past state, because the universe (or the egg) simply cycles through every possible configuration it can have, with a predictable frequency. There is no such thing as an "initial condition" in a universe that lasts forever.

The idea that the universe spends most of its time in thermal equilibrium, but we can appeal to the anthropic principle to explain why our local environment isn't in equilibrium, makes a strong prediction—and that prediction is dramatically falsified by the data. The prediction is simply that *we should be as close to equilibrium as possible*, given the requirement that we (under some suitable definition of "we") be allowed to exist. Fluctuations happen, but large fluctuations (such as creating an unbroken egg) are much more rare than smaller fluctuations (such as creating a broken egg). We can see this explicitly back in Figure 54, where the curve exhibits many small fluctuations and only a few larger ones. And the universe we see around us would have to be a large fluctuation indeed.[185]

We can be more specific about what the universe would look like if it were an eternal system fluctuating around equilibrium. Boltzmann invoked the anthropic principle (although he didn't call it that) to explain why we wouldn't find ourselves in one of the very common equilibrium phases: In equilibrium, life cannot exist. Clearly, what we want to do is find the most common conditions within such a universe that are hospitable to life. Or, if we want to be a bit more careful, perhaps we should look for conditions that are not only hospitable to life, but hospitable to the particular kind of intelligent and self-aware life that we like to think we are.

Maybe this is a way out? Maybe, we might reason, in order for an advanced scientific civilization such as ours to arise, we require a "support system" in the form of an entire universe filled with stars and galaxies, originating in some sort of super-low-entropy early condition. Maybe that could explain why we find such a profligate universe around us.

No. Here is how the game should be played: You tell me the particular thing you insist must exist in the universe, for anthropic reasons. A solar system, a planet, a particular ecosystem, a type of complex life, the room you are sitting in now, whatever you like. And then we ask, "Given that requirement, what is the most likely state of the *rest* of the universe in the Boltzmann-Lucretius scenario, in addition to the particular thing we are asking for?"

And the answer is always the same: The most likely state of the rest of the universe is to be in equilibrium. If we ask, "What is the most likely way for an infinite box of gas in equilibrium to fluctuate into a state containing a pumpkin pie?," the answer is "By fluctuating into a state that consists of a pumpkin pie floating by itself in an otherwise homogeneous box of gas." Adding anything else to the picture, either in space or in time—an oven, a baker, a previously existing pumpkin patch—only makes the scenario less likely, because the entropy would have to dip lower to make that happen. By far the easiest way to get a pumpkin pie in this context is for it to gradually fluctuate all by itself out of the surrounding chaos.[186]

Sir Arthur Eddington, in a lecture from 1931, considered a perfectly reasonable anthropic criterion:

> A universe containing mathematical physicists [under these assumptions] will at any assigned date be in the state of maximum disorganization which is not inconsistent with the existence of such creatures.[187]

Eddington presumes that what you really need to make a good universe is a mathematical physicist. Sadly, if the universe is an eternally fluctuating collection of molecules, the most frequently occurring mathematical physicists will be all by themselves, surrounded by randomness.

We can take this logic to its ultimate conclusion. If what we want is a single planet, we certainly don't need a hundred billion galaxies with a hundred billion stars each. And if what we want is a single person, we certainly don't need an entire planet. But if in fact what we want is a single intelligence, able to think about the world, we don't even need an entire person—we just need his or her brain.

So the *reductio ad absurdum* of this scenario is that the overwhelming majority of intelligences in this multiverse will be lonely, disembodied brains, who fluctuate gradually out of the surrounding chaos and then gradually dissolve back into it. Such sad creatures have been dubbed "Boltzmann brains" by Andreas Albrecht and Lorenzo Sorbo.[188] You and I are not Boltzmann brains—we are what one might call "ordinary observers," who did not fluctuate all by ourselves from the surrounding equilibrium, but evolved gradually from an earlier state of very low entropy. So the hypothesis that our universe is a random fluctuation around an equilibrium state in an eternal spacetime seems to be falsified.

You may have been willing to go along with this line of reasoning when only an egg was involved, but draw up short when we start comparing the number of disembodied brains to the number of ordinary observers. But the logic is exactly the same, *if* (and it's a big "if") we are considering an eternal universe full of randomly fluctuating particles. In such a universe, we know what kinds of fluctuations there are, and how often they happened; the more the entropy changes, the less likely that fluctuation will be. No matter how many ordinary observers exist in our universe today, they would be dwarfed by the total number of Boltzmann brains to come. Any given observer is a collection of particles in some particular state, and that state will occur infinitely often, and the number of times it will be surrounded by high-entropy chaos is enormously higher than the number of times it will arise as part of an "ordinary" universe.

Just to be careful, though—are you *really* sure you are not a Boltzmann brain? You might respond that you feel the rest of your body, you see other objects around you, and for that matter you have memories of a lower-entropy past: All things that would appear inconsistent with the idea that you are a disembodied brain recently fluctuated out of the surrounding molecules. The problem is, these purported statements about the outside world are actually just statements about your brain. Your feelings, visions, and memories are all contained within the state of your brain. We could certainly imagine that a brain with exactly these sensations fluctuated out of the surrounding chaos. And, as we have argued, it's much more likely for that brain to fluctuate by itself than to be part of a giant universe. In the Boltzmann-Lucretius scenario, we don't have recourse to the Past Hypothesis, so it is overwhelmingly likely that all of our memories are false.

Nevertheless, we are perfectly justified in dismissing this possibility, as long as we think carefully about what precisely we are claiming. It's not right to say, "I know I am not a Boltzmann brain, so clearly the universe is not a random fluctuation." The right thing to say is, "If I were a Boltzmann brain, there would be a strong prediction: Everything else about the universe should be in equilibrium. But it's not. Therefore the universe is not a random fluctuation." If we insist on being strongly skeptical, we might wonder whether not only our present mental states, but also all of the additional sensory data we are apparently accumulating, represent a random fluctuation rather than an accurate reconstruction of our surroundings. Strictly speaking, that certainly is possible, but it's cognitively unstable in the same sense that we discussed in the last chapter. There is no sensible way to live and think and behave if that is the case, so there is no warrant for believing it. Better to accept the universe around us as it appears (for the most part) to be.

This point was put with characteristic clarity by Richard Feynman, in his famous *Lectures on Physics*:

> [F]rom the hypothesis that the world is a fluctuation, all of the pre-dictions are that if we look at a part of the world we have never seen before, we will find it mixed up, and not like the piece we just looked at. If our order were due to a fluctuation, we would not expect order any-where but where we have just noticed it . . .
>
> We therefore conclude that the universe is *not* a fluctuation, and that the order is a memory of conditions when things started. This is not to say that we understand the logic of it. For some reason, the uni-verse at one time had a very low entropy for its energy content, and since then the entropy has increased. So that is the way toward the future. That is the origin of all irreversibility, that is what makes the processes of growth and decay, that makes us remember the past and not the future, remember the things which are closer to that moment in history of the universe when the order was higher than now, and why we are not able to remember things where the disorder is higher than now, which we call the future.[189]

WHO ARE WE IN THE MULTIVERSE?

There is one final loophole that must be closed before we completely shut the door on the Boltzmann-Lucretius scenario. Let's accept the implications of conven-tional statistical mechanics, that small fluctuations in entropy happen much more frequently than large fluctuations, and that the overwhelming majority of intel-ligent observers in a universe fluctuating eternally around equilibrium will find themselves alone in an otherwise high-entropy environment, not evolving natu-rally from a prior configuration of enormously lower entropy.

One might ask: So what? Why should I be bothered that *most* observers (under any possible definition of "observers") find themselves alone as freak fluctuations in a high-entropy background? All I care about is who I am, not what most observ-ers are like. As long as there is one instance of the universe I see around me some-where in the eternal lifetime of the larger world (which there will be), isn't that all I need to proclaim that this picture is consistent with the data?

In other words, the Boltzmann-brain argument makes an implicit assumption: that we are somehow "typical observers" in the universe, and that therefore we should make predictions by asking what most observers would see.[190] That sounds innocuous, even humble. But upon closer inspection, it leads to conclusions that seem stronger than we can really justify.

Imagine we have two theories of the universe that are identical in every way, except that one predicts that an Earth-like planet orbiting the star Tau Ceti is home to a race of 10 trillion intelligent lizard beings, while the other theory predicts there are no intelligent beings of any kind in the Tau Ceti system. Most of us would say that we don't currently have enough information to decide between these two theories. But if we are truly typical observers in the universe, the first theory strongly predicts that we are more likely to be lizards on the planet orbiting Tau Ceti, not humans here on Earth, just because there are so many more lizards than humans. But that prediction is not right, so we have apparently ruled out the existence of that many observers without collecting any data at all about what's actually going on in the Tau Ceti system.

Assuming we are typical might seem like a self-effacing move on our part, but it actually amounts to an extremely strong claim about what happens throughout the rest of the universe. Not only "we are typical observers," but "typical observers are like us." Put that way, it seems like a stronger assumption than we have any right to make. (In the literature this is known as the "Presumptuous Philosopher Problem.") So perhaps we shouldn't be comparing the numbers of different kinds of observers in the universe at all; we should only ask whether a given theory predicts that observers like us appear *somewhere*, and if they do we should think of the theory as being consistent with the data. If that were the right way to think about it, we wouldn't have any reason to reject the Boltzmann-Lucretius scenario. Even though most observers would be alone in the universe, some would find themselves in regions like ours, so the theory would be judged to be in agreement with our experience.[191]

The difficulty with this minimalist approach is that it offers us too little handle on what is likely to happen in the universe, instead of too much. Statistical mechanics relies on the Principle of Indifference—the assumption that all microstates consistent with our current macrostate are equally likely, at least when it comes to predicting the future. That's essentially an assumption of typicality: Our microstate is likely to be a typical member of our macrostate. If we're not allowed to assume that, all sorts of statistical reasoning suddenly become inaccessible. We can't say that an ice cube is likely to melt in a glass of warm water, because in an eternal universe there will occasionally be times when the opposite happens. We seem to have taken our concerns about typicality too far.

Instead, we should aim for a judicious middle ground. It's too much to ask that we are typical among all observers in the universe, because that's making a strong claim about parts of the universe we've never observed. But we can at least say that we are typical among observers *exactly like us*—that is, observers with the basic physiology and the same set of memories that we have, the same coarse-grained experience of the universe.[192] That assumption doesn't allow us to draw any unwarranted conclusions about the possible existence of other kinds of intelligent beings elsewhere in the universe. But it is more than enough to rule out the Boltzmann-Lucretius scenario. If the universe fluctuates around thermal equilibrium for all eternity, not only will most observers appear all by themselves from the surrounding chaos, but the same is true for the subset of observers with precisely the features that you or I have—complete with our purported memories of the past. Those memories will generally be false, and fluctuating into them is very unlikely, but it's still much more unlikely than fluctuating the entire universe. Even this minimal necessary condition for carrying out statistical reasoning—we should take ourselves to be chosen randomly from the set of observers exactly like us—is more than enough to put the Boltzmann-Lucretius scenario to rest.

The universe we observe is not a fluctuation—at least, to be more careful, a statistical fluctuation in an eternal universe that spends most of its time in equilibrium. So that's what the universe is not; what it *is*, we still have to work out.

ENDINGS

On the evening of September 5, 1906, Ludwig Boltzmann took a piece of cord, tied it to a curtain rod in the hotel room where he was vacationing in Italy with his family, and hanged himself. His body was discovered by his daughter Emma when she returned to their room that evening. He was sixty-two years old.

The reasons for Boltzmann's suicide remain unclear. Some have suggested that he was despondent over the failure of his ideas concerning atomic theory to gain wider acceptance. But, while many German-speaking scientists of the time remained skeptical about atoms, kinetic theory had become standard throughout much of the world, and Boltzmann's status as a major scientist was unquestioned in Austria and Germany. Boltzmann had been suffering from health problems and was prone to fits of depression; it was not the first time he had attempted suicide.

But his depression was intermittent; only months before his death, he had written an engaging and high-spirited account of his previous year's trip to America to lecture at the University of California at Berkeley, and circulated it among his friends. He referred to California as "Eldorado," but found American water undrinkable, and would drink only beer and wine. This was problematic, as the Temperance movement was strong in America at the time, and Berkeley in particular was completely dry; a recurring theme in Boltzmann's account is his attempts to smuggle wine into various forbidden places.[193] We will probably never know what mixture of failing health, depression, and scientific controversy contributed to his ultimate act.

On the question of the existence of atoms and their utility in understanding the properties of macroscopic objects, any lingering doubts that Boltzmann was right were rapidly dissipating when he died. One of Albert Einstein's papers in his "miraculous year" of 1905 was an explanation of Brownian motion (the seemingly random motion of small particles suspended in air) in terms of collisions with individual atoms; most remaining skepticism on the part of physicists was soon swept away.

Questions about the nature of entropy and the Second Law remain with us, of course. When it comes to explaining the low entropy of our early universe, we won't ever be able to say, "Boltzmann was right," because he suggested a number of different possibilities without ever settling on one in particular. But the terms of the debate were set by him, and we're still arguing over the questions that puzzled him more than a century ago.

11

QUANTUM TIME

Sweet is by convention, bitter by convention, hot by convention, cold by convention, color by convention; in truth there are but atoms and the void.

—Democritus[194]

Many people who have sat through introductory physics courses in high school or college might disagree with the claim "Newtonian mechanics makes intuitive sense to us." They may remember the subject as a bewildering merry-go-round of pulleys and vectors and inclined planes, and think that "intuitive sense" is the last thing that Newtonian mechanics should be accused of making.

But while the process of actually calculating something within the framework of Newtonian mechanics—doing a homework problem, or getting astronauts to the moon—can be ferociously complicated, the underlying concepts are pretty straightforward. The world is made of tangible things that we can observe and recognize: billiard balls, planets, pulleys. These things exert forces, or bump into one another, and their motions change in response to those influences. If Laplace's Demon knew all of the positions and momenta of every particle in the universe, it could predict the future and the past with perfect fidelity; we know that this is outside of our capabilities, but we can imagine knowing the positions and momenta of a few billiard balls on a frictionless table, and at least in principle we can imagine doing the math. After that it's just a matter of extrapolation and courage to encompass the entire universe.

Newtonian mechanics is usually referred to as "classical" mechanics by physicists, who want to emphasize that it's not just a set of particular rules laid down by Newton. Classical mechanics is a way of thinking about the deep structure of the world. Different types of things—baseballs, gas molecules, electromagnetic waves—will follow different specific rules, but those rules will share the same pattern. The essence of that pattern is that everything has some kind of "position," and some kind of "momentum," and that information can be used to predict what will happen next.

This structure is repeated in a variety of contexts: Newton's own theory of gravitation, Maxwell's nineteenth-century theory of electricity and magnetism, and Einstein's general relativity all fit into the classical framework. Classical mechanics isn't a particular theory; it's a paradigm, a way of conceptualizing what a physical theory is, and one that has demonstrated an astonishing range of empirical success. After Newton published his 1687 masterwork, *Philosophiæ Naturalis Principia Mathematica*, it became almost impossible to imagine doing physics any other way. The world is made of things, characterized by positions and momenta, pushed about by certain sets of forces; the job of physics was to classify the kinds of things and figure out what the forces were, and we'd be done.

But now we know better: Classical mechanics isn't correct. In the early decades of the twentieth century, physicists trying to understand the behavior of matter on microscopic scales were gradually forced to the conclusion that the rules would have to be overturned and replaced with something else. That something else is *quantum mechanics*, arguably the greatest triumph of human intelligence and imagination in all of history. Quantum mechanics offers an image of the world that is radically different from that of classical mechanics, one that scientists never would have seriously contemplated if the experimental data had left them with any other choice. Today, quantum mechanics enjoys the status that classical mechanics held at the dawn of the twentieth century: It has passed a variety of empirical tests, and most researchers are convinced that the ultimate laws of physics are quantum mechanical in nature.

But despite its triumphs, quantum mechanics remains somewhat mysterious. Physicists are completely confident in how they *use* quantum mechanics—they can build theories, make predictions, and test against experiments, and there is never any ambiguity along the way. Nevertheless, we're not completely sure we know what quantum mechanics really *is*. There is a respectable field of intellectual endeavor, occupying the time of a substantial number of talented scientists and philosophers, that goes under the name "interpretations of quantum mechanics." A century ago, there was no such field as "interpretations of classical mechanics"—classical mechanics is perfectly straightforward to interpret. We're still not sure what is the best way to think and talk about quantum mechanics.

This interpretational anxiety stems from the single basic difference between quantum mechanics and classical mechanics, which is both simple and world shattering in its implications:

According to quantum mechanics, what we can *observe* about the world is only a tiny subset of what actually *exists*.

Attempts at explaining this principle often water it down beyond recognition. "It's like that friend of yours who has such a nice smile, except when you try to take his picture it always disappears." Quantum mechanics is a lot more profound than that. In the classical world, it might be difficult to obtain a precise measurement of some quantity; we need to be very careful not to disturb the system we're looking at. But there is nothing in classical physics that prevents us from being careful. In quantum mechanics, on the other hand, there is an unavoidable obstacle to making complete and nondisruptive observations of a physical system. It simply can't be done, in general. What exactly happens when you try to observe something, and what actually counts as a "measurement"—those are the locus of the mystery. This is what is helpfully known as the "measurement problem," much as having an automobile roll off a cliff and smash into pieces on the rocks hundreds of feet below might be known as "car trouble." Successful physical theories aren't supposed to have ambiguities like this; the very first thing we ask about them is that they be clearly defined. Quantum mechanics, despite all its undeniable successes, isn't there yet.

None of which should be taken to mean that all hell has broken loose, or that the mysteries of quantum mechanics offer an excuse to believe whatever you want. In particular, quantum mechanics doesn't mean you can change reality just by thinking about it, or that modern physics has rediscovered ancient Buddhist wisdom.[195] There are still rules, and we know how the rules operate in the regimes of interest to our everyday lives. But we'd like to understand how the rules operate in every conceivable situation.

Most modern physicists deal with the problems of interpreting quantum mechanics through the age-old strategy of "denial." They know how the rules operate in cases of interest, they can put quantum mechanics to work in specific circumstances and achieve amazing agreement with experiment, and they don't want to be bothered with pesky questions about what it all means or whether the theory is perfectly well-defined. For our purposes in this book, that is often a pretty good strategy. The problem of the arrow of time was there for Boltzmann and his collaborators before quantum mechanics was ever invented; we can go very far talking about entropy and cosmology without worrying about the details of quantum mechanics.

At some point, however, we need to face the music. The arrow of time is, after all, a fundamental puzzle, and it's possible that quantum mechanics will play a crucial role in resolving that puzzle. And there's something else of more direct interest: That process of measurement, where all of the interpretational tangles of quantum mechanics are to be found, has the remarkable property that it is *irreversible*. Alone among all of the well-accepted laws of physics, quantum measurement is a process that defines an arrow of time: Once you do it, you can't undo it. And that's a mystery.

It's very possible that this mysterious irreversibility is of precisely the same character as the mysterious irreversibility in thermodynamics, as codified in the Second Law: It's a consequence of making approximations and throwing away information, even though the deep underlying processes are all individually reversible. I'll be advocating that point of view in this chapter. But the subject remains controversial among the experts. The one sure thing is that we have to confront the measurement problem head-on if we're interested in the arrow of time.

THE QUANTUM CAT

Thanks to the thought-experiment stylings of Erwin Schrödinger, it has become traditional in discussions of quantum mechanics to use cats as examples.[196] Schrödinger's cat was proposed to help illustrate the difficulties involved in the measurement problem, but we're going start with the basic features of the theory before diving into the subtleties. And no animals will be harmed in our thought experiments.

Imagine your cat, Miss Kitty, has two favorite places in your house: on the sofa and under the dining room table. In the real world, there are an infinite number of positions in space that could specify the location of a physical object such as a cat; likewise, there are an infinite number of momenta, even if your cat tends not to move very fast. We're going to be simplifying things dramatically, in order to get at the heart of quantum mechanics. So let's imagine that we can completely specify the state of Miss Kitty—as it would be described in classical mechanics—by saying whether she is on the sofa or under the table. We're throwing out any information about her speed, or any knowledge of exactly what part of the sofa she's on, and we're disregarding any possible positions that are not "sofa" or "table." From the classical point of view, we are simplifying Miss Kitty down to a two-state system. (Two-state systems actually exist in the real world; for example, the spin of an electron or photon can either point up or point down. The quantum state of a two-state system is described by a "qubit.")

Here is the first major difference between quantum mechanics and classical mechanics: In quantum mechanics, there is *no such thing* as "the location of the cat." In classical mechanics, it may happen that we don't know where Miss Kitty is, so we may end up saying things like "I think there's a 75 percent chance that she's under the table." But that's a statement about our ignorance, not about the world; there really is a fact of the matter about where the cat is, whether we know it or not.

In quantum mechanics, there is no fact of the matter about where Miss Kitty (or anything else) is located. The space of states in quantum mechanics just doesn't work that way. Instead, the states are specified by something called a *wave function*. And the wave function doesn't say things like "the cat is on the sofa" or "the cat is under the table." Rather, it says things like "if we were to look, there would be a 75 percent chance that we would find the cat under the table, and a 25 percent chance that we would find the cat on the sofa."

This distinction between "incomplete knowledge" and "intrinsic quantum indeterminacy" is worth dwelling on. If the wave function tells us there is a 75 percent chance of observing the cat under the table and a 25 percent chance of observing her on the sofa, that does not mean there is a 75 percent chance that the cat *is* under the table and a 25 percent chance that she *is* on the sofa. There is no such thing as "where the cat is." Her quantum state is described by a *superposition* of the two distinct possibilities we would have in classical mechanics. It's not even that "they are both true at once"; it's that there is no "true" place where the cat is. The wave function is the best description we have of the reality of the cat.

It's clear why this is hard to accept at first blush. To put it bluntly, the world doesn't look anything like that. We see cats and planets and even electrons in particular positions when we look at them, not in superpositions of different possibilities described by wave functions. But that's the true magic of quantum mechanics: What we see is not what there is. The wave function really exists, but we don't see it when we look; we see things as if they were in particular ordinary classical configurations.

None of which stops classical physics from being more than good enough to play basketball or put satellites in orbit. Quantum mechanics features a "classical limit" in which objects behave just as they would had Newton been right all along, and that limit includes all of our everyday experiences. For objects such as cats that are macroscopic in size, we never find them in superpositions of the form "75 percent here, 25 percent there"; it's always "99.9999999 percent (or more) here, 0.0000001 percent (or much less) there." Classical mechanics is an approximation to how the macroscopic world operates, but a very good one. The real world runs by the rules of quantum mechanics, but classical mechanics is more than good enough to get us through everyday life. It's only when we start to consider atoms and elementary particles that the full consequences of quantum mechanics simply can't be avoided.

HOW WAVE FUNCTIONS WORK

You might wonder how we know any of this is true. What is the difference, after all, between "there is a 75 percent chance of observing the cat under the table" and "there is a 75 percent chance that the cat is under the table"? It seems hard to imagine an experiment that could distinguish between those possibilities—the only way we would know where it is would be to look for it, after all. But there is a crucially important phenomenon that drives home the difference, known as *quantum interference*. To understand what that means, we have to bite the bullet and dig a little more deeply into how wave functions really work.

In classical mechanics, where the state of a particle is a specification of its position and its momentum, we can think of that state as specified by a collection of numbers. For one particle in ordinary three-dimensional space, there are six numbers: the position in each of the three directions, and the momentum in each of the three directions. In quantum mechanics the state is specified by a wave function, which can also be thought of as a collection of numbers. The job of these numbers is to tell us, for any observation or measurement we could imagine doing, what the probability is that we will get a certain result. So you might naturally think that the numbers we need are just the probabilities themselves: the chance Miss Kitty will be observed on the sofa, the chance she will be observed under the table, and so on.

As it turns out, that's not how reality operates. Wave functions really are wave-like: A typical wave function oscillates through space and time, much like a wave on the surface of a pond. This isn't so obvious in our simple example where there are only two possible observational outcomes—"on sofa" or "under table"—but it becomes more clear when we consider observations with continuous possible outcomes, like the position of a real cat in a real room. The wave function is like a wave on a pond, except it's a wave on the space of all possible outcomes of an observation—for example, all possible positions in a room.

When we see a wave on a pond, the level of the water isn't uniformly higher than what it would be if the pond were undisturbed; sometimes the water goes up, and sometimes it goes down. If we were to describe the wave mathematically, to every point on the pond we would associate an *amplitude*—the height by which the water was displaced and that amplitude would sometimes be positive, sometimes be negative. Wave functions in quantum mechanics work the same way. To every possible outcome of an observation, the wave function assigns a number, which we call the amplitude, and which can be positive or negative. The complete wave function consists of a particular amplitude for every possible observational outcome; those are the numbers that specify the state in quantum mechanics, just as the positions and momenta specify the state in classical mechanics. There is an amplitude that Miss Kitty is under the table, and another one that she is on the sofa.

There's only one problem with that setup: What we care about are probabilities, and the probability of something happening is never a negative number. So it can't be true that the amplitude associated with a certain observational outcome is equal to the probability of obtaining that outcome—instead, there must be a way of calculating the probability if we know what the amplitude is. Happily, the calculation is very easy! To get the probability, you take the amplitude and you square it.

$$(\text{probability of observing } X) = (\text{amplitude assigned to } X)^2.$$

So if Miss Kitty's wave function assigns an amplitude of 0.5 to the possibility that we observe her on the sofa, the probability that we see her there is $(0.5)^2 = 0.25$, or 25 percent. But, crucially, the amplitude could also be -0.5, and we would get exactly the same answer: $(-0.5)^2 = 0.25$. This might seem like a pointless piece of redundancy—two different amplitudes corresponding to the same physical situation—but it turns out to play a key role in how states evolve in quantum mechanics.[197]

INTERFERENCE

Now that we know that wave functions can assign negative amplitudes to possible outcomes of observations, we can return to the question of why we ever need to talk about wave functions and superpositions in the first place, rather than just assigning probabilities to different outcomes directly. The reason is interference, and those negative numbers are crucial in understanding the way interference comes about—we can add two (nonvanishing) amplitudes together and get zero, which we couldn't do if amplitudes were never negative.

To see how this works, let's complicate our model of feline dynamics just a bit. Imagine that we see Miss Kitty leave the upstairs bedroom. From our previous observations of her wanderings through the house, we know quite a bit about how this quantum cat operates. We know that, once she settles in downstairs, she will inevitably end up either on the sofa or under the table, nowhere else. (That is, her final state is a wave function describing a superposition of being on the sofa and being under the table.) But let's say we also know that she has two possible routes to take from the upstairs bed to whatever downstairs resting place she chooses: She will either stop by the food dish to eat, or stop by the scratching post to sharpen her claws. In the real world all of these possibilities are adequately described by classical mechanics, but in our idealized thought-experiment world we imagine that quantum effects play an important role.

Now let's see what we actually observe. We'll do the experiment two separate ways. First, when we see Miss Kitty start downstairs, we very quietly sneak behind her to see which route she takes, via either the food bowl or the scratching post. She actually has a wave function describing a superposition of both possibilities, but when we make an observation we always find a definite result. We're as quiet as possible, so we don't disturb her; if you like, we can imagine that we've placed spy cameras or laser sensors. The technology used to figure out whether she goes to the bowl or the scratching post is completely irrelevant; what matters is that we've observed it.

We find that we observe her visiting the bowl exactly half the time and the scratching post exactly half the time. (We assume that she visits one or the other, but never both, just to keep things as simple as we can.) Any one particular observation doesn't reveal the wave function, of course; it can tell us only that we saw her stop at the post or at the bowl that particular time. But imagine that we do this experiment a very large number of times, so that we can get a reliable idea of what the probabilities are.

But we don't stop at that. We next let her continue on to either the sofa or the table, and after she's had time to settle down we look again to see which place she ends up. Again, we do the experiment enough times that we can figure out the probabilities. What we now find is that it didn't matter whether she stopped at the scratching post or at her food bowl; in both cases, we observe her ending up on the sofa exactly half the time, and under the table exactly half the time, completely independently of whether she first visited the bowl or the scratching post. Apparently the intermediate step along the way didn't matter very much; no matter which alternative we observed along the way, the final wave function assigns equal probability to the sofa and the table.

Next comes the fun part. This time, we simply choose not to observe Miss Kitty's intermediate step along her journey; we don't keep track of whether she stops at the scratching post or the food bowl. We just wait until she's settled on the sofa or under the table, and we look at where she is, reconstructing the final probabilities assigned by the wave functions. What do we expect to find?

In a world governed by classical mechanics, we know what we should see. When we did our spying on her, we were careful that our observations shouldn't have affected how Miss Kitty behaved, and half the time we found her on the sofa and half the time on the table no matter what route she took. Clearly, even if we don't observe what she does along the way, it shouldn't matter—in either case we end up with equal probabilities for the final step, so even if we don't observe the intermediate stage we should still end up with equal probabilities.

But we don't. That's not what we see, in this idealized thought-experiment world where our cat is a truly quantum object. What we see when we choose not to observe whether she goes via the food bowl or the scratching post is that she ends up on the sofa 100 percent of the time! We never find her under the table—the final wave function assigns an amplitude of zero to that possible outcome. Apparently, if all this is to be believed, the very presence of our spy cameras changed her wave function in some dramatic way. The possibilities are summarized in the table.

WHAT ROUTE WE SEE MS. KITTY TAKE	FINAL PROBABILITIES
Scratching post	50% sofa, 50% table
Food bowl	50% sofa, 50% table
We don't look	100% sofa, 0% table

This isn't just a thought experiment; it's been done. Not with real cats, who are unmistakably macroscopic and well described by the classical limit, but with individual photons, in what is known as the "double slit experiment." A photon passes through two possible slits, and if we don't watch which slit it goes through, we get one final wave function, but if we do, we get a completely different one, no matter how unobtrusive our measurements were.

Here's how to explain what is going on. Let's imagine that we do observe whether Miss Kitty stops by the bowl or the scratching post, and we see her stop by the scratching post. After she does so, she evolves into a superposition of being on the sofa and being under the table, with equal probability. In particular, due to details in Miss Kitty's initial condition and certain aspects of quantum feline dynamics, the final wave function assigns equal *positive* amplitudes to the sofa possibility and the table possibility. Now let's consider the other intermediate step, that we see her stop by the food bowl. In that case, the final wave function assigns a negative amplitude to the table, and a positive one to the sofa—equal but opposite numbers, so that the probabilities turn out precisely the same.[198]

But if we don't observe her at the scratching-post/food-bowl juncture, then (by the lights of our thought experiment) she is in a superposition of the two possibilities at this intermediate step. In that case, the rules of quantum mechanics instruct

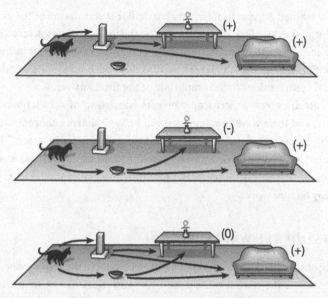

Figure 57: Alternative evolutions for Miss Kitty's wave function. At the top, we observe her stop at the scratching post, after which she could go to either the table or the sofa, both with positive amplitude. In the middle, we observe her go to the food bowl, after which she could go to either the table or the sofa, but this time the table has a negative amplitude (still a positive probability). At the bottom, we don't observe her intermediate journey, so we add the amplitudes from the two possibilities. We end up with zero amplitude for the table (since the positive and negative contributions cancel), and positive amplitude for the sofa.

us to add the two possible contributions to the final wave function—one from the route where she stopped by the scratching post, and one from the food bowl. In either case, the amplitudes for ending up on the sofa were positive numbers, so they reinforce each other. But the amplitudes for ending up under the table were opposite for the two intermediate cases—so when we add them together, they precisely cancel. Individually, Miss Kitty's two possible intermediate paths left us with a nonzero probability that she would end up under the table, but when both paths were allowed (because we didn't observe which one she took), the two amplitudes interfered.

That's why the wave function needs to include negative numbers, and that's how we know that the wave function is "real," not just a bookkeeping device to keep track of probabilities. We have an explicit case where the individual probabilities would have been positive, but the final wave function received contributions from two intermediate steps, which ended up canceling each other.

Let's catch our breath to appreciate how profound this is, from our jaundiced classically trained perspective. For any particular instantiation of the experiment, we are tempted to ask: Did Miss Kitty stop by the food bowl or the scratching post? The only acceptable answer is: no. She didn't do either one. She was in a superposition of both possibilities, which we know because both possibilities ended up giving crucial contributions to the amplitude of the final answer.

Real cats are messy macroscopic objects consisting of very large numbers of molecules, and their wave functions tend to be very sharply concentrated around something resembling our classical notion of a "position in space." But at the microscopic level, all this talk of wave functions and superpositions and interference becomes brazenly demonstrable. Quantum mechanics strikes us as weird, but it's the way Nature works.

COLLAPSE OF THE WAVE FUNCTION

The thing about this discussion that tends to rub people the wrong way—and with good reason—is the crucial role played by observation. When we observed what the cat was doing at the scratching-post/food-bowl juncture, we got one answer for the final state; when we didn't make any such observation, we got a very different answer. That's not how physics is supposed to work; the world is supposed to evolve according to the laws of Nature, whether we are observing it or not. What counts as an "observation," anyway? What if we set up surveillance cameras along the way but then never looked at the tapes? Would that count as an observation? (Yes, it would.) And what precisely happened when we did make our observation?

This is an important collection of questions, and the answers are not completely clear. There is no consensus within the physics community about what really constitutes an observation (or "measurement") in quantum mechanics, nor about what happens when an observation occurs. This is the "measurement problem" and is the primary focus of people who spend their time thinking about interpretations of quantum mechanics. There are many such interpretations on the market, and we're going to discuss two: the more or less standard picture, known as the "Copenhagen interpretation," and a view that seems (to me) to be a bit more respectable and likely to conform to reality, which goes under the forbidding name the "many-worlds interpretation." Let's look at Copenhagen first.[199]

The Copenhagen interpretation is so named because Niels Bohr, who in many ways was the godfather of quantum mechanics, helped to develop it from his institute in Copenhagen in the 1920s. The actual history of this perspective is complicated, and certainly involves a great deal of input from Werner Heisenberg, another quantum pioneer. But the history is less crucial to our present purposes than the status of the Copenhagen view as what is enshrined in textbooks as the standard picture. Every physicist learns this first, and then gets to contemplate other alternatives (or choose not to, as the case may be).

The Copenhagen interpretation of quantum mechanics is as easy to state as it is hard to swallow: when a quantum system is subjected to a measurement, its wave function *collapses*. That is, the wave function goes instantaneously from describing a superposition of various possible observational outcomes to a completely different wave function, one that assigns 100 percent probability to the outcome that was actually measured, and 0 percent to anything else. That kind of wave function, concentrated entirely on a single possible observational outcome, is known as an "eigenstate." Once the system is in that eigenstate, you can keep making the same kind of observation, and you'll keep getting the same answer (unless something kicks the system out of the eigenstate into another superposition). We can't say with certainty which eigenstate the system will fall into when an observation is made; it's an inherently stochastic process, and the best we can do is assign a probability to different outcomes.

We can apply this idea to the story of Miss Kitty. According to the Copenhagen interpretation, our choice to observe whether she stopped by the food bowl or the scratching post had a dramatic effect on her wave function, no matter how sneaky we were about it. When we didn't look, she was in a superposition of the two possibilities, with equal amplitude; when she then moved on to the sofa or the table, we added up the contributions from each of the intermediate steps, and found there was interference. But if we chose to observe her along the way, we collapsed her wave function. If we saw her stop at the scratching post, once that observation was made she was in a state that was no longer a superposition—she was 100 percent scratching post, 0 percent food bowl. Likewise if we saw her stop at the food bowl, with the amplitudes reversed. In either case, there was nothing left to interfere with, and her wave function evolved into a state that gave her equal probabilities to end up on the sofa or under the table.[200]

There is good news and bad news about this story. The good news is that it fits the data. If we imagine that wave functions collapse every time we make an observation—no matter how unobtrusive our observational strategy may be—and that they end up in eigenstates that assign 100 percent probability to the outcome we observed, we successfully account for all of the various quantum phenomena known to physicists.

The bad news is that this story barely makes sense. What counts as an "observation"? Can the cat herself make an observation, or could a nonliving being? Surely we don't want to suggest that the phenomenon of *consciousness* is somehow playing a crucial role in the fundamental laws of physics? (No, we don't.) And does the purported collapse really happen instantaneously, or is it gradual but just very fast?

IRREVERSIBILITY

At heart, the thing that bugs us about the Copenhagen interpretation of quantum mechanics is that it treats "observing" as a completely distinct kind of natural phenomenon, one that requires a separate law of nature. In classical mechanics, everything that happens can be accounted for by systems evolving according to Newton's laws. But if we take the collapse of the wave function at face value, a system described by quantum mechanics evolves according to two completely separate kinds of rules:

1. When we're not looking at it, a wave function evolves smoothly and predictably. The role that Newton's laws play in classical mechanics is replaced by the *Schrödinger equation* in quantum mechanics, which operates in a precisely analogous way. Given the state of the system at any one time, we can use the Schrödinger equation to evolve it reliably into the future and into the past. The evolution conserves information and is completely reversible.

2. When we observe it, a wave function collapses. The collapse is not smooth, or perfectly predictable, and information is not conserved. The amplitude (squared) associated with any particular outcome tells us the probability that the wave function will collapse to a state that is concentrated entirely on that outcome. Two different wave functions can very easily collapse to exactly the same state after an observation is made; therefore, wave function collapse is not reversible.

Madness! But it works. The Copenhagen interpretation takes concepts that would seem to be nothing more than useful approximations to some deeper underlying truth—distinguishing between a "system" that is truly quantum mechanical and an "observer" who is essentially classical—and imagines that these categories play a crucial role in the fundamental architecture of reality. Most physicists, even those who use quantum mechanics every day in their research, get along perfectly well speaking the language of the Copenhagen interpretation, and choosing not to worry about the puzzles it presents. Others, especially those who think carefully about the foundations of quantum mechanics, are convinced that we need to do better. Unfortunately there is no strong consensus at present about what that better understanding might be.

For many people, the breakdown of perfect predictability is a troubling feature of quantum mechanics. (Einstein is among them; that's the origin of his complaint that "God does not play dice with the universe.") If the Copenhagen interpretation is right, there could be no such thing as Laplace's Demon in a quantum world; at least, not if that world contained observers. The act of observing introduces a truly random element into the evolution of the world. Not *completely* random—a wave function may give a very high probability to observing one thing, and a very low probability to observing something else. But *irreducibly* random, in the sense that there is no piece of missing information that would allow us to predict outcomes with certainty, if only we could get our hands on it.[201] Part of the glory of classical mechanics had been its clockwork reliability—even if Laplace's Demon didn't really exist, we knew he could exist in principle. Quantum mechanics destroys that hope. It took a long while for people to get used to the idea that probability enters the laws of physics in some fundamental way, and many are still discomforted by the concept.

One of our questions about the arrow of time is how we can reconcile the irreversibility of macroscopic systems described by statistical mechanics with the apparent reversibility of the microscopic laws of physics. But now, according to quantum mechanics, it seems that the microscopic laws of physics aren't necessarily reversible. The collapse of the wave function is a process that introduces an intrinsic arrow of time into the laws of physics: Wave functions collapse, but they don't un-collapse. If we observe Miss Kitty and see that she is on the sofa, we know that she is an eigenstate (100 percent on the sofa) right after we've done the measurement. But we don't know what state she was in *before* we did the measurement. That information, apparently, has been destroyed. All we know is that the wave function must have had some nonzero amplitude for the cat to be on the sofa—but we don't know how much, or what the amplitude for any other possibilities might have been.

So the collapse of the wave function—if, indeed, that's the right way to think about quantum mechanics—defines an intrinsic arrow of time. Can that be used to somehow explain "the" arrow of time, the thermodynamic arrow that appears in the Second Law and on which we've blamed all the various macroscopic differences between past and future?

Probably not. Although irreversibility is a key feature of the arrow of time, not all irreversibilities are created equal. It's very hard to see how the fact that wave functions collapse could, by itself, account for the Past Hypothesis. Remember, it's not hard to understand why entropy increases; what's hard to understand is why it was ever low to begin with. The collapse of the wave function doesn't seem to offer any direct help with that problem.

On the other hand, quantum mechanics is very likely to play some sort of role in the ultimate explanation, even if the intrinsic irreversibility of wave function collapse doesn't directly solve the problem all by itself. After all, we believe that the laws of physics are fundamentally quantum mechanical at heart. It's quantum mechanics that sets the rules and tells us what is and is not allowed in the world. It's perfectly natural to expect that these rules will come into play when we finally do begin to understand why our universe had such a low entropy near the Big Bang. We don't know exactly where this journey is taking us, but we're savvy enough to anticipate that certain tools will prove useful along the way.

UNCERTAINTY

Our discussion of wave functions has glossed over one crucial property. We've said that wave functions assign an amplitude to any possible outcome of an observation we could imagine doing. In our thought experiment, we restricted ourselves to only one kind of observation—the location of the cat—and only two possible outcomes at a time. A real cat, or an elementary particle or an egg or any other object, has an infinite number of possible positions, and the relevant wave function in each case assigns an amplitude to every single possibility.

More important, however, there are things we can observe other than positions. Remembering our experience with classical mechanics, we might imagine observing the momentum rather than the position of our cat. And that's perfectly possible; the state of the cat is described by a wave function that assigns an amplitude to every possible momentum we could imagine measuring. When we do such a measurement and get an answer, the wave function collapses into an "eigenstate of momentum," where the new state assigns nonzero amplitude only to the particular momentum we actually observed.

But if that's true, you might think, what's to stop us from putting the cat into a state where both the position and momentum are determined exactly, so it becomes just like a classical state? In other words, why can't we take a cat with an arbitrary wave function, observe its position so that it collapses to one definite value, and then observe its momentum so that it also collapses to a definite value? We should be left with something that is completely determined, no uncertainty at all.

The answer is that there are no wave functions that are simultaneously concentrated on a single value of position and also on a single value of momentum.

Indeed, the hope for such a state turns out to be maximally frustrated: If the wave function is concentrated on a single value of position, the amplitudes for different momenta will be spread out as widely as possible over all the possibilities. And vice versa: If the wave function is concentrated on a single momentum, it is spread out widely over all possible positions. So if we observe the position of an object, we lose any knowledge of what its momentum is, and vice versa.[202] (If we measure the position only approximately, rather than exactly, we can retain some knowledge of the momentum; this is what actually happens in real-world macroscopic measurements.)

That's the true meaning of the Heisenberg Uncertainty Principle. In quantum mechanics, it is possible to "know exactly" what the position of a particle is—more precisely, it's possible for the particle to be in a position eigenstate, where there is a 100 percent probability of finding it in a certain position. Likewise, it is possible to "know exactly" what the momentum is. But we can never know precisely the position and momentum at the same time. So when we go to measure the properties that classical mechanics would attribute to a system—both position and momentum—we can never say for certain what the outcomes will be. That's the uncertainty principle.

The uncertainty principle implies that there must be some spread of the wave function over different possible values of either position or momentum, or (usually) both. No matter what kind of system we look at, there is an unavoidable quantum unpredictability when we try to measure its properties. The two observables are complementary: When the wave function is concentrated in position, it's spread out in momentum, and vice versa. Real macroscopic systems that are well described by the classical limit of quantum mechanics find themselves in compromise states, where there is a small amount of uncertainty in both position and momentum. For large enough systems, the uncertainty is relatively small enough that we don't notice at all.

Keep in mind that there really is no such thing as "the position of the object" or "the momentum of the object"—there is only a wave function assigning amplitudes to the possible outcomes of observations. Nevertheless, we often can't resist falling into the language of *quantum fluctuations*—we say that we can't pin the object down to a single position, because the uncertainty principle forces it to fluctuate around just a bit. That's an irresistible linguistic formulation, and we won't be so uptight that we completely refrain from using it, but it doesn't accurately reflect what is really going on. It's not that there is a position and a momentum, each of which keeps fluctuating around; it's that there is a wave function, which can't simultaneously be localized in position and momentum.

In later chapters we will explore applications of quantum mechanics to much grander systems than single particles, or even single cats—quantum field theory, and also quantum gravity. But the basic framework of quantum mechanics remains the same in each case. Quantum field theory is the marriage of quantum mechanics with special relativity, and explains the particles we see around us as the observable features of the deeper underlying structure—quantum fields—that make up the world. The uncertainty principle will forbid us from precisely determining the position and momentum of every particle, or even the exact number of particles. That's the origin of "virtual particles," which pop in and out of existence even in empty space, and ultimately it will lead to the phenomenon of Hawking radiation from black holes.

One thing we *don't* understand is quantum gravity. General relativity provides an extremely successful description of gravity as we see it operate in the world, but the theory is built on a thoroughly classical foundation. Gravity is the curvature of spacetime, and in principle we can measure the spacetime curvature as precisely as we like. Almost everyone believes that this is just an approximation to a more complete theory of quantum gravity, where spacetime itself is described by a wave function that assigns different amplitudes to different amounts of curvature. It might even be the case that entire universes pop in and out of existence, just like virtual particles. But the quest to construct a complete theory of quantum gravity faces formidable hurdles, both technical and philosophical. Overcoming those obstacles is the full-time occupation of a large number of working physicists.

THE WAVE FUNCTION OF THE UNIVERSE

There is one fairly direct way of addressing the conceptual issues associated with wave function collapse: Simply deny that it ever happens, and insist that ordinary smooth evolution of the wave function suffices to explain everything we know about the world. This approach—brutal in its simplicity, and far-reaching in its consequences—goes under the name of the "many-worlds interpretation" of quantum mechanics and is the leading competitor to the Copenhagen interpretation. To understand how it works, we need to take a detour into perhaps the most profound feature of quantum mechanics of all: entanglement.

When we introduced the idea of a wave function we considered a very minimalist physical system, consisting of a single object (a cat). We would obviously like to be able to move beyond that, to consider systems with multiple parts—perhaps a cat and also a dog. In classical mechanics, that's no problem; if the state of one object is described by its position and its momentum, the state of two objects is just the state of both objects individually—two positions and two momenta. So it would be the most natural thing in the world to guess that the correct quantum mechanical description of a cat and a dog would simply be two wave functions, one for the cat and one for the dog.

That's not how it works. In quantum mechanics, no matter how many individual pieces make up the system you are thinking about, there is *only one wave function*. Even if we consider the entire universe and everything inside it, there is still only one wave function, sometimes redundantly known as the "wave function of the universe." People don't always like to talk that way, for fear of sounding excessively grandiose, but at bottom that's simply the way quantum mechanics works. (Other people enjoy the grandiosity for its own sake.)

Let's see how this plays out when our system consists of a cat and a dog, Miss Kitty and Mr. Dog. As before, we imagine that when we look for Miss Kitty, there are only two places we can find her: on the sofa or under the table. Let's also imagine that there are only two places we can ever observe Mr. Dog: in the living room or out in the yard. According to the initial (but wrong) guess that each object has its own wave function, we would describe Miss Kitty's location as a superposition of under the table and on the sofa, and separately describe Mr. Dog's location as a superposition of in the living room or in the yard.

But instead, quantum mechanics instructs us to consider every possible alternative for the entire system—cat plus dog—and assign an amplitude to every distinct possibility. For the combined system, there are four possible answers to the question "What do we see when we look for the cat and the dog?" They can be summarized as follows:

(table, living room)

(table, yard)

(sofa, living room)

(sofa, yard)

Here, the first entry tells us where we see Miss Kitty, and the second where we see Mr. Dog. According to quantum mechanics, the wave function of the universe assigns every one of these four possibilities a distinct amplitude, which we would square to get the probability of observing that alternative.

You may wonder what the difference is between assigning amplitudes to the locations of the cat and dog separately, and assigning them to the combined locations. The answer is *entanglement*—properties of any one subset of the whole can be strongly correlated with properties of other subsets.

ENTANGLEMENT

Let's imagine that the wave function of the cat/dog system assigns zero amplitude to the possibility (table, yard) and also zero amplitude to (sofa, living room). Schematically, that means the state of the system must be of the form

$$(\text{table, living room}) + (\text{sofa, yard}).$$

This means there is a nonzero amplitude that the cat is under the table and the dog is in the living room, and also a nonzero amplitude that the cat is on the sofa and the dog is in the yard. Those are the only two possibilities allowed by this particular state, and let's imagine that they have equal amplitude.

Now let's ask: What do we expect to see if we look for only Miss Kitty? An observation collapses the wave function onto one of the two possibilities, (table, living room) or (sofa, yard), with equal probability, 50 percent each. If we simply don't care about what Mr. Dog is doing, we would say that there is an equal probability for observing Miss Kitty under the table or on the sofa. In that sense, it's fair to say that we have no idea where Miss Kitty is going to be before we look.

Now let's imagine that we instead look for Mr. Dog. Again, there is a 50 percent chance each for the possibilities (table, living room) or (sofa, yard), so if we don't care what Miss Kitty is doing, it's fair to say that we have no idea where Mr. Dog is going to be before we look.

Here is the kicker: Even though we have no idea where Mr. Dog is going to be before we look, if we first choose to look for Miss Kitty, once that observation is complete we know exactly where Mr. Dog is going to be, even without ever looking for him! That's the magic of entanglement. Let's say that we saw Miss Kitty on the sofa. That means that, given the form of the wave function we started with, it must have collapsed onto the possibility (sofa, yard). We therefore know with certainty (assuming we were right about the initial wave function) that Mr. Dog will be in the yard if we look for him. We have collapsed Mr. Dog's wave function without ever observing him. Or, more correctly, we have collapsed the wave function of the universe, which has important consequences for Mr. Dog's whereabouts, without ever interacting with Mr. Dog directly.

This may or may not seem surprising to you. Hopefully, we've been so clear and persuasive in explaining what wave functions are all about that the phenomenon of entanglement seems relatively natural. And it should; it's part and parcel of the machinery of quantum mechanics, and a number of clever experiments have demonstrated its validity in the real world. Nevertheless, entanglement can lead to consequences that—taken at face value—seem inconsistent with the spirit of relativity, if not precisely with the letter of the law. Let's stress: There is no real incompatibility between quantum mechanics and special relativity (general relativity, where gravity comes into the game, is a different story). But there is a tension between them that makes people nervous. In particular, things seem to happen faster than the speed of light. When you dig deeply into what those "things" are, and what it means to "happen," you find that nothing really bad is going on—nothing has actually moved faster than light, and no real information can be conveyed outside anyone's light cone. Still, it rubs people the wrong way.

THE EPR PARADOX

Let's go back to our cat and dog, and imagine that they are in the quantum state described above, a superposition of (table, living room) and (sofa, yard). But now let's imagine that if Mr. Dog is out in the yard, he doesn't just sit there; he runs away. Also, he is very adventurous, and lives in the future, when we have regular rocket flights to a space colony on Mars. Mr. Dog—in the alternative where he starts in the yard, not in the living room—runs away to the spaceport, stows away on a rocket, and flies to Mars, completely unobserved the entire time. It's only when he clambers out of the rocket into the arms of his old friend Billy, who had graduated from high school and joined the Space Corps and been sent on a mission to the Red Planet, that the state of Mr. Dog is actually observed, collapsing the wave function.

What we're imagining, in other words, is that the wave function describing the cat/dog system has evolved smoothly according to the Schrödinger equation from

$$(\text{table, living room}) + (\text{sofa, yard})$$

to

$$(\text{table, living room}) + (\text{sofa, Mars}).$$

There's nothing impossible about that—implausible, maybe, but as long as nobody made any observations during the time it took the evolution to happen, we'll end up with the wave function in this superposition.

But the implications are somewhat surprising. When Billy unexpectedly sees Mr. Dog bounding out of the spaceship on Mars, he makes an observation and collapses the wave function. If he knew what the wave function was to begin with, featuring an entangled state of cat and dog, Billy *immediately* knows that Miss Kitty is on the sofa, not under the table. The wave function has collapsed to the possibility (sofa, Mars). Not only is Miss Kitty's state now known even without anyone interacting with her; it seems to have been determined instantaneously, despite the fact that it takes at least several minutes to travel between Mars and Earth even if you were moving at the speed of light.

This feature of entanglement—the fact that the state of the universe, as described by its quantum wave function, seems to change "instantaneously" throughout space, even though the lesson of special relativity was supposed to be that there's no unique definition of what "instantaneously" means—bugs the heck out of people. It certainly bugged Albert Einstein, who teamed up with Boris Podolsky and Nathan Rosen in 1935 to write a paper pointing out this weird possibility, now known as the "EPR paradox."[203] But it's not really a "paradox" at all; it might fly in the face of our intuition, but not of any experimental or theoretical requirements.

The important feature of the apparently instantaneous collapse of a wave function that is spread across immense distances is that it cannot be used to actually transmit any information faster than light. The thing that bothers us is that, before Billy observed the dog, Miss Kitty back here on Earth was not in any definite location—we had a 50/50 chance to observe her on the sofa or under the table. Once Billy observes Mr. Dog, we now have a 100 percent chance of observing her to be on the sofa. But so what? We don't actually know that Billy did any such observation—for all we know, if we looked for Mr. Dog we would find him in the living room. For Billy's discovery to make any difference to us, he would have to come tell us about it, or send us a radio transmission—one way or another, he would have to communicate with us by conventional slower-than-light means.

Entanglement between two far-apart subsystems seems mysterious to us because it violates our intuitive notions of "locality"—things should only be able to directly affect other nearby things, not things arbitrarily far away. Wave functions just don't work like that; there is one wave function that describes the entire universe all at once, and that's the end of it. The world we observe, meanwhile, still respects a kind of locality—even if wave functions collapse instantaneously all over space, we can't actually take advantage of that feature to send signals faster than light. In other words: As far as things actually bumping into you and affecting your life, it's still true that they have to be right next to you, not far away.

On the other hand, we shouldn't expect that even this weaker notion of locality is truly a sacred principle. In the next chapter we'll talk a little bit about quantum gravity, where the wave function applies to different configurations of spacetime itself. In that context, an idea like "objects can affect each other only when they are nearby" ceases to have any absolute meaning. Spacetime itself is not absolute, but only has different amplitudes for being in different configurations—so the notion of "the distance between two objects" becomes a little fuzzy. These are ideas that have yet to be fully understood, but the final theory of everything is likely to exhibit non-locality in some very dramatic ways.

MANY WORLDS, MANY MINDS

The leading contender for an alternative to the Copenhagen view of quantum mechanics is the so-called *many-worlds interpretation*. "Many worlds" is a scary and misleading name for what is really a very straightforward idea. That idea is this: There is no such thing as "collapse of the wave function." The evolution of states in quantum mechanics works just like it does in classical mechanics; it obeys a deterministic rule—the Schrödinger equation—that allows us to predict the future and past of any specific state with perfect fidelity. And that's all there is to it.

The problem with this claim is that we appear to *see* wave functions collapsing all the time, or at least to observe the effects of the collapse. We can imagine arranging Miss Kitty in a quantum state that has equal amplitudes for finding her on the sofa or under the table; then we look for her, and see her under the table. If we look again immediately thereafter, we're going to see her under the table 100 percent of the time; the original observation (in the usual way of talking about these things) collapsed the wave function to a table-eigenstate. And that way of thinking has empirical consequences, all of which have been successfully tested in real experiments.

The response of the many-worlds advocate is simply that you are thinking about it wrong. In particular, you have misidentified *yourself* in the wave function of the universe. After all, you are part of the physical world, and therefore you are also subject to the rules of quantum mechanics. It's not right to set yourself off as some objective classical observing apparatus; we need to take your own state into account in the wave function.

So, this new story goes, we shouldn't just start with a wave function describing Miss Kitty as a superposition of (sofa) and (table); we should include your own configuration in the description. In particular, the relevant feature of your description is what you have observed about Miss Kitty's position. There are three possible states you could be in: You could have seen her on the sofa, you could have seen her under the table, and you might not have looked yet. To start with, the wave function of the universe (or at least the bit of it we're describing here) gives Miss Kitty equal amplitude to be on the sofa or under the table, while you are uniquely in the state of not having looked yet. This can be schematically portrayed like this:

(sofa, you haven't yet looked) + (table, you haven't yet looked).

Now you observe where she is. In the Copenhagen interpretation, we would say that the wave function collapses. But in the many-worlds interpretation, we say that your own state becomes entangled with that of Miss Kitty, and the combined system evolves into a superposition:

(sofa, you see her on the sofa) + (table, you see her under the table).

There is no collapse; the wave function evolves smoothly, and there is nothing special about the process of "observation." What is more, the entire procedure is reversible—given the final state, we could use the Schrödinger equation to uniquely recover the original state. There is no intrinsically quantum mechanical arrow of time in this interpretation. For many reasons, this is an altogether more elegant and satisfying picture of the world than that provided by the Copenhagen picture.

The problem, meanwhile, should be obvious: The final state has you in a super-position of two different outcomes. The difficulty with that, of course, is that you never *feel* like you're in such a superposition. If you actually did make an obser-vation of a system that was in a quantum superposition, after the observation you would always *believe* that you had observed some specific outcome. The problem with the many-worlds interpretation, in other words, is that it doesn't seem to accord with our experience of the real world.

But let's not be too hasty. Who is this "you" of which we are speaking? It's true: The many-worlds interpretation says that the wave function of the universe evolves into the superposition shown above, with an amplitude for you seeing the cat on the sofa, and another amplitude for you seeing her under the table. Here is the crucial step: The "you" that does the seeing and perceiving and believing is not that super-position. Rather, "you" are either one of those alternatives, or the other. That is, there are now two different "yous," one who saw Ms. Kitty on the sofa and another who saw her under the table, and they both honestly *exist* there in the wave function. They share the same prior memories and experiences—before they observed the cat's loca-tion, they were in all respects the same person—but now they have split off into two different "branches of the wave function," never to interact with each other again.

These are the "many worlds" in question, although it should be clear that the label is somewhat misleading. People sometimes raise the objection to the many-worlds interpretation that it's simply too extravagant to be taken seriously—all those different "parallel realities," infinite in number, just so that we don't have to believe in wave function collapse. That's silly. Before we made an observation, the universe was described by a single wave function, which assigned a particular amplitude to every possible observational outcome; after the observation, the universe is described by a single wave function, which assigns a particular amplitude to every possible observational outcome. Before and after, the wave function of the universe is just a particular point in the space of states describing the universe, and that space of states didn't get any bigger or smaller. No new "worlds" have really been created; the wave function still contains the same amount of information (after all, in this interpretation its evolution is reversible). It has simply evolved in such a way that there are now a greater number of distinct subsets of the wave function describing individual conscious beings such as ourselves. The many-worlds interpretation of quantum mechanics may or may not be right, but to object to it on the grounds that "Gee, that's a lot of worlds," is wrong-headed.

The many-worlds interpretation was not originally formulated by Bohr, Heisenberg, Schrödinger, or any of the other towering figures of the early days of quantum mechanics. It was proposed in 1957 by Hugh Everett III, who was a graduate student working with John Wheeler at Princeton.[204] At the time (and for decades thereafter), the dominant view was the Copenhagen interpretation, so Wheeler did the obvious thing: He sent Everett on a trip to Copenhagen, to discuss his novel perspective with Niels Bohr and others. But the trip was not a success—Bohr was utterly unconvinced, and the rest of the physics community exhibited little interest in Everett's ideas. He left academic physics to work for the Defense Department, and eventually founded his own computer firm. In 1970, theoretical physicist Bryce DeWitt (who, along with Wheeler, was a pioneer in applying quantum mechanics to gravity) took up the cause of the many-worlds interpretation and helped popularize it among physicists. Everett lived to see a resurgence of interest in his ideas within the physics community, but he never returned to active research; he passed away suddenly of a heart attack in 1982, at the age of fifty-one.

DECOHERENCE

Despite its advantages, the many-worlds interpretation of quantum mechanics isn't really a finished product. There remain unanswered questions, from the deep and conceptual—why are conscious observers identified with discrete branches of the wave function, rather than superpositions?—to the dryly technical—how do we justify the rule that "probabilities are equal to amplitudes squared" in this formalism? These are real questions, to which the answers aren't perfectly clear, which is (one reason) why the many-worlds interpretation doesn't enjoy universal acceptance. But a great deal of progress has been made over the last few decades, especially involving an intrinsically quantum mechanical phenomenon known as *decoherence*. There are great hopes—although little consensus—that decoherence can help us understand why wave functions *appear* to collapse, even if the many-worlds interpretation holds that such collapse is only apparent.

Decoherence occurs when the state of some small piece of the universe—your brain, for example—becomes so entangled with parts in the wider environment that it is no longer subject to interference, the phenomenon that truly makes something "quantum." To get a feeling for how this works, let's go back to the example of the entangled state of Miss Kitty and Mr. Dog. There are two alternatives, with equal amplitudes: the cat is under the table and the dog is in the living room, or the cat is on the sofa and the dog is in the yard:

$$(\text{table, living room}) + (\text{sofa, yard}).$$

We saw how, if someone observed the state of Mr. Dog, the wave function would (in the Copenhagen language) collapse, leaving Miss Kitty in some definite state.

But now let's do something different: Imagine that nobody observes the state of Mr. Dog, but we simply ignore him. Effectively, we throw away any information about the entanglement between Miss Kitty and Mr. Dog, and simply ask ourselves: What is the state of Miss Kitty all by herself?

We might think that the answer is a superposition of the form (table)+(sofa), like we had before we had ever introduced the canine complication into the picture. But that's not quite right. The problem is that interference—the phenomenon that convinced us we needed to take quantum amplitudes seriously in the first place—can no longer happen.

In our original example of interference, there were two contributions to the amplitude for Miss Kitty to be under the table: one from the alternative where she passed by her food bowl, and one from where she stopped at her scratching post. But it was crucially important that the two contributions that ultimately canceled were contributions to *exactly the same final alternative* ("Miss Kitty is under the table"). Two contributions to the final wave function are going to interfere only if they involve truly the same alternative for everything in the universe; if they are contributing to different alternatives, they can't possibly interfere, even if the differences involve the rest of the universe, and not Miss Kitty herself.

So when the state of Miss Kitty is entangled with the state of Mr. Dog, interference between alternatives that alter Miss Kitty's state without a corresponding
change in Mr. Dog's becomes impossible. Some contribution to the wave function
can't interfere with the alternative "Miss Kitty is under the table," because that
alternative isn't a complete specification of what can be observed; it could only
interfere with the alternatives "Miss Kitty is under the table and Mr. Dog is in the
living room" that are actually represented in the wave function.²⁰⁵

Therefore, if Miss Kitty is entangled with the outside world but we don't know
the details of that entanglement, it's not right to think of her state as a quantum
superposition. Rather, we should just think of it as an ordinary *classical* distribution of different alternatives. Once we throw away any information about what she
is entangled with, Miss Kitty is no longer in a true superposition; as far as any conceivable experiment is concerned, she is in either one state or the other, even if we
don't know which. Interference is no longer possible.

That's decoherence. In classical mechanics, every object has a definite position,
even if we don't know what the position is and can ascribe probabilities only to the
various alternatives. The miracle of quantum mechanics was that there is no longer
any such thing as "where the object is"; it's in a true simultaneous superposition of
the possible alternatives, which we know must be true via experiments that demonstrate the reality of interference. But if the quantum state describing the object
is entangled with something in the outside world, interference becomes impossible, and we're back to the traditional classical way of looking at things. As far as
we are concerned, the object is in one state or another, even if the best we can do is
assign a probability to the different alternatives—the probabilities are expressing
our ignorance, not the underlying reality. If the quantum state of some particular subset of the universe represents a true superposition that is un-entangled with
the rest of the world, we say it is "coherent"; if the superposition has been ruined by
becoming entangled with something outside, we say that it has become "decoher-
ent." (That's why, in the many-worlds view, setting up surveillance cameras counts
as making an observation; the state of the cat became entangled with the state of
the cameras.)

WAVE FUNCTION COLLAPSE AND THE ARROW OF TIME

In the many-worlds interpretation, decoherence clearly plays a crucial role in the
apparent process of wave function collapse. The point is not that there is something special or unique about "consciousness" or "observers," other than the fact
that they are complicated macroscopic objects. The point is that any complicated

macroscopic object is *inevitably* going to be interacting (and therefore entangled) with the outside world, and it's hopeless to imagine keeping track of the precise form of that entanglement. For a tiny microscopic system such as an individual electron, we can isolate it and put it into a true quantum superposition that is not entangled with the state of any other particles, but for a messy system such as a human being (or a secret surveillance camera, for that matter) that's just not possible.

In that case, our simple picture in which the state of our perceptions becomes entangled with the state of Miss Kitty's location is an oversimplification. A crucial part of the story is played by the entanglement of us with the external world. Let's imagine that Miss Kitty starts out in a true quantum superposition, un-entangled with the rest of the world; but we, complicated creatures that we are, are deeply entangled with the outside world in ways we can't possibly specify. The wave function of the universe assigns distinct amplitudes to all the alternative configurations of the combined system of Miss Kitty, us, and the outside world. After we observe Miss Kitty's location, the wave function evolves into something of the form

(sofa, you see her on the sofa, world$_1$) + (table, you see her under the table, world$_2$),

where the last piece describes the (unknown) configuration of the external world, which will be different in the two cases.

Because we don't know anything about that state, we simply ignore the entanglement with the outside world, and keep the knowledge of Miss Kitty's location and our own mental perceptions. Those are clearly correlated: If she is on the sofa, we believe we have seen her on the sofa, and so forth. But after throwing away the configuration of the outside world, we're no longer in a real quantum superposition. Rather, there are two alternatives that seem for all intents and purposes classical: Miss Kitty is on the sofa and we saw her on the sofa, or she's under the table and we saw her under the table.

That's what we mean when we talk about the branching of the wave function into different "worlds." Some small system in a true quantum superposition is observed by a macroscopic measuring apparatus, but the apparatus is entangled with the outside world; we ignore the state of the outside world and are left with two classical alternative worlds. From the point of view of either classical alternative, the wave function has "collapsed," but from a hypothetical larger point of view where we kept all of the information in the wave function of the universe, there were no sudden changes in the state, just a smooth evolution according to the Schrödinger equation.

This business about throwing away information may make you a little uneasy, but it should also sound somewhat familiar. All we're really doing is coarse-graining, just as we did in (classical) statistical mechanics to define macrostates corresponding to various microstates. The information about our entanglement with the messy external environment is analogous to the information about the position and momentum of every molecule in a box of gas—we don't need it, and in practice can't keep track of it, so we create a phenomenological description based solely on macroscopic variables.

In that sense, the irreversibility that crops up when wave functions collapse appears to be directly analogous to the irreversibility of ordinary thermodynamics. The underlying laws are perfectly reversible, but in the messy real world we throw away a lot of information, and as a result we find apparently irreversible behavior on macroscopic scales. When we observe our cat's location, and our own state becomes entangled with hers, in order to reverse the process we would need to know the precise state of the outside world with which we are also entangled, but we've thrown that information away. It's exactly analogous to what happens when a spoonful of milk mixes into a cup of coffee; in principle we could reverse the process if we had kept track of the position and momentum of every single molecule in the mixture, but in practice we keep track of only the macroscopic variables, so reversibility is lost.

In this discussion of decoherence, a crucial role was played by our ability to take the system to be observed (Miss Kitty, or some elementary particle) and isolate it from the rest of the world in a true quantum superposition. But that's clearly a very special kind of state, much like the low-entropy states we start with by hypothesis when discussing the origin of the Second Law of Thermodynamics. A completely generic state would feature all kinds of entanglements between our small system and the external environment, right from the start.

None of which is intended to give the impression that the application of deco-herence to the many-worlds interpretation manages to swiftly solve all of the interpretive problems of quantum mechanics. But it seems like a step in the right direction, and highlights an important relationship between the macroscopic arrow of time familiar from statistical mechanics and the macroscopic arrow of time exhibited when wave functions collapse. Perhaps best of all, it helps remove ill-defined notions such as "conscious observers" from the vocabulary with which we describe the natural world.

With that in mind, we're going to go back to speaking as if the fundamental laws of physics are all completely reversible on microscopic scales. This isn't an unassailable conclusion, but it has some good arguments behind it—we can keep an open mind while continuing to explore the consequences of this particular point of view. Which leaves us, of course, right where we started: with the task of explaining the apparent lack of reversibility on macroscopic scales in terms of special conditions near the Big Bang. To take that problem seriously, it's time that we start thinking about gravity and the evolution of the universe.

PART FOUR

FROM THE KITCHEN TO
THE MULTIVERSE

12

BLACK HOLES: THE ENDS OF TIME

Time, old gal of mine, will soon dim out.

— Anne Sexton, "For Mr. Death Who Stands with His Door Open"

Stephen Hawking is one of the most willful people on Earth. In 1963, while working toward his doctorate at Cambridge University at the age of twenty-one, he was diagnosed with motor neurone disease. The prognosis was not good, and Hawking was told that he likely did not have long to live. After some soul-searching, he decided to move forward and redouble his commitment to research. We all know the outcome; now well into his seventh decade, Hawking has been the most influential scientist in the field of general relativity since Albert Einstein, and is instantly recognizable worldwide through his efforts to popularize physics.

Among other things, Hawking is a tireless traveler, and he spends some time in California every year. In 1998, I was a postdoctoral research fellow at the Institute for Theoretical Physics at the University of California, Santa Barbara, and Hawking came to visit on his annual sojourn. The institute administrator gave me a simple task: "Pick up Stephen at the airport."

As you might guess, picking up Stephen Hawking at the airport is different than picking up anyone else. For one thing, you're not really "picking him up"; he rents a van that is equipped to carry his wheelchair, for which a special license is required to drive it—a license

Figure 58: Stephen Hawking, who gave us the most important clue we have about the relationship between quantum mechanics, gravity, and entropy.

I certainly didn't have. The actual driving was left to his graduate assistant; my job was merely to meet them at the tiny Santa Barbara airport and show them to the van. By "them," I mean Hawking's entourage: a graduate assistant (usually a physics student who helps with logistics), other graduate students, family members, and a retinue of nurses. But it wasn't a matter of pointing to the van and going on my way. Despite the fact that the graduate assistant was the only person allowed to drive the van, Hawking insisted that the van stay with him at all times, and also wanted to go to a restaurant for dinner before dropping off the assistant at his apartment. Which meant that I tagged along in my car while they all went to dinner, so I could shuttle the assistant back and forth. Hawking was the only one who knew where the restaurant was, but speaking through his voice synthesizer is a slow process; we spent several tense moments stopped in the middle of a busy road while Hawking explained that we had passed the restaurant and would have to turn around.

Stephen Hawking has been able to accomplish remarkable things while working under extraordinary handicaps, and the reason is basically straightforward: He refuses to compromise in any way. He's not going to cut down his travel schedule, or eat at the wrong restaurant, or drink a lesser quality of tea, or curtail his wicked sense of humor, or think less ambitiously about the inner workings of the universe, merely because he is confined to a wheelchair. And that strength of character pushes him scientifically, as well as getting him through life.

In 1973, Hawking was annoyed. Jacob Bekenstein, a young graduate student at Princeton, had written a paper suggesting something crazy: that black holes carried huge amounts of entropy.[206] By this time Hawking was the world's expert on black holes, and (in his own words) he was irritated at Bekenstein, who he thought had misused some of his earlier results.[207] So he set about showing exactly how crazy Bekenstein's idea was—for one thing, if black holes had entropy, you could show that they would have to give off radiation, and everyone knows that black holes are black!

In the end, of course, Hawking ended up surprising everyone, including himself. Black holes do have entropy, and indeed they do give off radiation, once we take into account the subtle consequences of quantum mechanics. No matter how stubborn your personality may be, Nature's rules will not bend to your will, and Hawking was smart enough to accept the radical implications of his discovery. He ended up providing physicists with their single most important clue about the interplay of quantum mechanics and gravity, and teaching us a deep lesson about the nature of entropy.

BLACK HOLES ARE FOR REAL

We have excellent reasons to believe that black holes exist in the real world. Of course we can't see them directly—they're still pretty dark, even if Hawking showed that they're not completely black. But we can see what happens to things around them, and the environment near a black hole is sufficiently unique that we can often be confident that we've located one. Some black holes are formed from the collapse of very massive stars, and frequently these have companion stars orbiting them. Gas from the companion can fall toward the black hole, forming an accretion disk that heats to tremendous temperatures and emits X-ray radiation in copious amounts. Satellite observatories have found a number of X-ray sources that demonstrate all of the qualities you would expect in such an object: In particular, a large amount of high-intensity radiation coming from a very small region of space. Astrophysicists know of no good explanations for these observations other than black holes.

There is also good evidence for supermassive black holes at the centers of galaxies—black holes more than a million times the mass of the Sun. (Still a very tiny fraction of the total mass of a galaxy, which is typically a hundred billion times the mass of the Sun.) In the early stages of galaxy formation, these giant black holes sweep up matter around them in a violent maelstrom, visible to us as quasars. Once the galaxy has settled down a bit, things become calmer, and the quasars "turn off." In our own Milky Way, we nevertheless are pretty sure that there lurks a black hole weighing in at about 4 million solar masses. Even without the blazing radiation of a quasar, observations of stars in the galactic center reveal that they are orbiting in tight ellipses around an invisible object. We can deduce that these stars must be caught in the gravitational field of something that is so dense and massive that it can't be anything *but* a black hole, if general relativity has anything to say about the matter.[208]

BLACK HOLES HAVE NO HAIR

But as much fun as it is to search for real black holes in the universe, it is even more fun to sit and think about them.[209] Black holes are the ultimate thought-experiment laboratory for anyone interested in gravity. And what makes black holes special is their purity.

While observations convince us that black holes exist, they don't give us a lot of detailed information about their properties; we aren't able to get up close to a black hole and poke at it. So when we make confident assertions about this or that feature of black holes, we're always implicitly speaking within some theoretical framework. Unfortunately, scientists don't yet fully understand quantum gravity, the presumed ultimate reconciliation of general relativity with the tenets of quantum mechanics. So we don't have a single correct theory in which to answer our questions once and for all.

Instead, we often investigate questions within one of three different theoretical frameworks:

1. Classical general relativity, as written down by Einstein. This is the best full theory of gravity we currently possess, and it is completely consistent with all of known experimental data. We understand the theory perfectly, in the sense that any well-posed question has a definite answer (even if it might be beyond our calculational abilities to figure it out). Unfortunately it's not right, as it's completely classical rather than quantum mechanical.

2. Quantum mechanics in curved spacetime. This is a framework with a split personality. We treat spacetime, the background through which stuff in the universe moves, as classical, according to the rules of general relativity. But we treat the "stuff" as quantum mechanical, described by wave functions. This is a useful compromise approach for trying to understand a number of real-world problems.

3. Quantum gravity. We don't know the correct theory of quantum gravity, although approaches like string theory are very promising. We're not completely clueless—we know something about how relativity works, and something about how quantum mechanics works. That's often sufficient to make some reasonable guesses about how things should work in an ultimate version of quantum gravity, even if we don't have the full-blown theory.

Classical general relativity is the best understood of these, while quantum gravity is the least well understood; but quantum gravity is the closest to the real world. Quantum mechanics in curved spacetime occupies a judicious middle ground and is the approach Hawking took to investigating black-hole radiation. But it behooves us to understand how black holes work in the relatively safe context of general relativity before moving on to more advanced but speculative ideas.

In classical general relativity, a black hole is just about the purest kind of gravitational field you can have. In the flexible world of thought experiments, we could imagine creating a black hole in any number of ways: from a ball of gas like an ordinary star, or out of a huge planet made of pure gold, or from an enormous sphere of ice cream. But once these things collapse to the point where their gravitational field is so strong that nothing can escape—once they are officially black holes—any indication of what kind of stuff they were made from completely disappears. A black hole made out of a ball of gas the mass of the Sun is indistinguishable from a black hole made from a ball of ice cream the mass of the Sun. The black hole isn't, according to general relativity, just a densely packed version of whatever we started with. It's pure gravitational field—the original "stuff" has disappeared into the singularity, and we're left with a region of strongly curved spacetime.

When we think of the gravitational field of the Earth, we might start by modeling our planet as a perfect sphere of a certain mass and size. But that's clearly just an approximation. If we want to do a little bit better, we'll take into account the fact that the Earth also spins, so it's a little wider near the equator than near the poles. And if we want to be super-careful about it, the exact gravitational field of the Earth changes from point to point in complicated ways; changes in altitude of the surface, as well as changes in density between land and sea or between different kinds of rock, lead to small but measurable variations in the Earth's gravity. All of the local features of the gravitational field of the Earth actually contain quite a bit of information.

Black holes are not like that. Once they form, any bumps and wiggles in the stuff they formed from are erased. There might be a short period right when the formation happens, when the black hole hasn't quite settled down, but it quickly becomes smooth and featureless. Once it has settled, there are three things that we can measure about a black hole: its total mass, how fast it is spinning, and how much electric charge it has. (Real astrophysical black holes usually have close to zero net electric charge, but they are often spinning very rapidly.) And that's it. Two collections of stuff with the same mass, charge, and spin, once they get turned into black holes, will become completely indistinguishable, as far as classical general relativity is concerned. This interesting prediction of general relativity is summed up in a cute motto coined by John Wheeler, the same guy who gave black holes their name: "Black holes have no hair."

This no-hair business should set off some alarm bells. Apparently, if everything we've just said is true, the process of forming a black hole has a dramatic consequence: Information is lost. We can take two very different kinds of initial conditions (one solar mass of hot gas, or one solar mass of ice cream), and they can evolve into precisely the same final condition (a one-solar-mass black hole). But up until now we've been saying that the microscopic laws of physics—of which

Einstein's equation of general relativity is presumably one—have the property that they conserve information. Put another way: Making a black hole seems to be an irreversible process, even though Einstein's equation would appear to be perfectly reversible.

You are right to worry! This is a time puzzle. In classical general relativity, there is a way out: We can say that the information isn't truly *lost*; it's just *lost to you*, as it's hidden behind the event horizon of a black hole. You can decide for yourself whether that seems satisfying or sounds like a cop-out. Either way, we can't just stop there, as Hawking will eventually tell us that black holes evaporate once quantum mechanics is taken into account. Then we have a serious problem, one that has launched a thousand theoretical physics papers.[210]

LAWS OF BLACK-HOLE MECHANICS

You might think that, because nothing can escape from a black hole, it's impossible for its total mass to ever decrease. But that's not quite right, as shown by a very imaginative idea due to Roger Penrose. Penrose knew that black holes could have spin and charge as well as mass, so he asked a reasonable question: Can we use that spin and charge to do useful work? Can we, in other words, extract energy from a black hole by decreasing its spin and charge? (When we think of black holes as single objects at rest, we can use "mass" and "energy" interchangeably, with $E = mc^2$ lurking in the back of our minds.)

The answer is yes, at least at the thought-experiment level we're working at here. Penrose figured out a way we could throw things close to a spinning black hole and have them emerge with *more* energy than they had when they went in, slowing the black hole's rotation in the process of lowering its mass. Essentially, we can convert the spin of the black hole into useful energy. A stupendously advanced civilization, with access to a giant, spinning black hole, would have a tremendous source of energy available for whatever public-works projects they might want to pursue. But not an unlimited source—there is a finite amount of energy we can extract by this process, since the black hole will eventually stop spinning altogether. (In the best-case scenario, we can extract about 29 percent of the total energy of the original rapidly spinning black hole.)

So: Penrose showed that black holes are systems from which we can extract useful work, at least up to a certain point. Once the black hole has no spin, we've used up all the extractable energy, and the hole just sits there. Those words should sound vaguely familiar from our previous discussions of thermodynamics.

Stephen Hawking followed up on Penrose's work to show that, while it's possible to decrease the mass/energy of a spinning black hole, there is a quantity that always either increases or remains constant: the area of the event horizon, which is basically the size of the black hole. The area of the horizon depends on a particular combination of the mass, spin, and charge, and Hawking found that this particular combination never decreases, no matter what we do. If we have two black holes, for example, they can smash into each other and coalesce into a single black hole, oscillating wildly and giving off gravitational radiation.[211] But the area of the new event horizon is always larger than the combined area of the original two horizons—and, as an immediate consequence of Hawking's result, one big black hole can therefore never split into two smaller ones, since the area would go down.[212] For a given amount of mass, we get the maximum-area horizon from a single, uncharged, nonrotating black hole.

So: While we can extract useful work from a black hole up to a point, there is some quantity (the area of the event horizon) that keeps going up during the process and reaches its maximum value when all the useful work has been extracted. Interesting. This really does sound eerily like thermodynamics.

Enough with the suggestive implications; let's make this analogy explicit.[213] Hawking showed that the area of the event horizon of a black hole never decreases; it either increases or stays constant. That's much like the behavior of entropy, according to the Second Law of Thermodynamics. The First Law of Thermodynamics is usually summarized as "energy is conserved," but it actually tells us how different forms of energy combine to make the total energy. There is clearly an analogous rule for black holes: The total mass is given by a formula that includes contributions from the spin and charge.

There is also a Third Law of Thermodynamics: There is a minimum possible temperature, absolute zero, at which the entropy is also a minimum. What, in the case of black holes, is supposed to play the role of "temperature" in this analogy? The answer is the *surface gravity* of a black hole—how strong the gravitational pull of the hole is near the event horizon, as measured by an observer very far away. You might think the surface gravity should be infinite—isn't that the whole point of a black hole? But it turns out that the surface gravity is really a measure of how dramatically spacetime is curved near the event horizon, and it actually gets *weaker* as the black hole gets more and more massive.[214] And there is a minimum value for the surface gravity of a black hole—zero!—which is achieved when all of the black-hole energy comes from charge or spin, none from "mass all by itself."

Finally, there is a Zeroth Law of Thermodynamics: If two systems are in thermal equilibrium with a third system, they are in thermal equilibrium with each other. The analogous statement for black holes is simply "the surface gravity has the same value everywhere on the event horizon of a stationary black hole." And that's true.

So there is a perfect analogy between the laws of thermodynamics, as they were developed over the course of the 1800s, and the "laws of black-hole mechanics," as they were developed in the 1970s. The different elements of the analogy are summarized in the table.

Thermodynamics ⟺ Black holes
Energy ⟺ Mass
Temperature ⟺ Surface gravity
Entropy ⟺ Area

But now we're faced with an important question, the kind that leads to big breakthroughs in science: How seriously should we take this analogy? Is it just an amusing coincidence, or does it reflect some deep underlying truth?

This is a legitimate question, not just a cheap setup for a predictable answer. Coincidences happen sometimes. When scientists stumble across an intriguing connection between two apparently unrelated things, like thermodynamics and black holes, that may be a clue to an important discovery, or it may just be an accident. Different people might have different gut feelings about whether or not such a deep connection is out there to be found. Ultimately, we should be able to attack the problem scientifically and come to a conclusion, but the answer is not obvious ahead of time.

BEKENSTEIN'S ENTROPY CONJECTURE

It was Jacob Bekenstein, then a graduate student working with John Wheeler, who took the analogy between thermodynamics and black-hole mechanics most seriously. Wheeler, when he wasn't busy coining pithy phrases, was enthusiastically pushing forward the field of quantum gravity (and general relativity overall) at a time when the rest of the physics community was more interested in particle physics—the heroic days of the 1960s and 1970s, when the Standard Model was being constructed. Wheeler's influence has been felt not only through his ideas— he and Bryce DeWitt were the first to generalize the Schrödinger equation of quantum mechanics to a theory of gravity—but also through his students. In addition to Bekenstein, Wheeler was the Ph.D. supervisor for an impressive fraction of the scientists who are now leading researchers in gravitational physics, including Kip

Thorne, Charles Misner, Robert Wald, and William Unruh—not to mention Hugh Everett, as well as Wheeler's first student, one Richard Feynman.

So Princeton in the early 1970s was a fruitful environment to be thinking about black holes, and Bekenstein was in the thick of it. In his Ph.D. thesis, he made a simple but dramatic suggestion: The relationship between black-hole mechanics and thermodynamics isn't simply an analogy; it's an identity. In particular, Bekenstein used ideas from information theory to argue that the area of a black-hole event horizon isn't just *like* the entropy; it *is* the entropy of the black hole.[215]

On the face of it, this suggestion seems a little hard to swallow. Boltzmann told us what entropy is: It characterizes the number of microscopic states of a system that are macroscopically indistinguishable. "Black holes have no hair" seems to imply that there are very

Figure 59: Jacob Bekenstein, who first suggested that black holes have entropy.

few states for a large black hole; indeed, for any specified mass, charge, and spin, the black hole is supposed to be unique. But here is Bekenstein, saying that the entropy of an astrophysical-sized black hole is staggeringly large.

The area of an event horizon has to be measured in some kind of units—acres, hectares, square centimeters, what have you. Bekenstein claimed that the entropy of a black hole was approximately equal to the area of its event horizon as measured in units of the *Planck area*. The Planck length, 10^{-33} centimeters, is the very tiny distance at which quantum gravity is supposed to become important; the Planck area is just the Planck length squared. For a black hole with a mass comparable to the Sun, the area of the event horizon is about 10^{77} Planck areas. That's a big number; an entropy of 10^{77} would be larger than the regular entropy in all of the stars, gas, and dust in the entire Milky Way galaxy.

At a superficial level, there is a pretty straightforward route to reconciling the apparent tension between the no-hair idea and Bekenstein's entropy idea: Classical general relativity is just not correct, and we need quantum gravity to understand the enormous number of states implied by the amount of black hole entropy. Or, to put it more charitably, classical general relativity is kind of like thermodynamics, and quantum gravity will be needed to uncover the microscopic "statistical mechanics" understanding of entropy in cases when gravity is important. Bekenstein's proposal seemed to imply that there are really jillions of different ways that spacetime can arrange itself at the microscopic quantum level to make a macroscopic classical black hole. All we have to do is figure out what those ways are. Easier said than done, as it turns out; more than thirty-five years later, we still don't have a firm grasp on the nature of those microstates implied by the black-hole entropy formula. We think that a black hole is like a box of gas, but we don't know what the "atoms" are—although there are some tantalizing clues.

But that's not a deal-breaker. Remember that the actual Second Law was formulated by Carnot and Clausius before Boltzmann ever came along. Maybe we are in a similar stage of progress right now where quantum gravity is concerned. Perhaps the properties of mass, charge, and spin in classical general relativity are simply macroscopic observables that don't specify the full microstate, just as temperature and pressure are in ordinary thermodynamics.

In Bekenstein's view, black holes are not some weird things that stand apart from the rest of physics; they are thermodynamic systems just like a box of gas would be. He proposed a "Generalized Second Law," which is basically the ordinary Second Law with black-hole entropy included. We can take a box of gas with a certain entropy, throw it into a black hole, and calculate what happens to the total entropy before and after. The answer is: It goes up, if we accept Bekenstein's claim that the black-hole entropy is proportional to the area of the event horizon. Clearly such a scenario has some deep implications for the relationship between entropy and spacetime, which are worth exploring more carefully.

HAWKING RADIATION

Along with Wheeler's group at Princeton, the best work in general relativity in the early 1970s was being done in Great Britain. Stephen Hawking and Roger Penrose, in particular, were inventing and applying new mathematical techniques to the study of curved spacetime. Out of these investigations came the celebrated singularity theorems—when gravity becomes sufficiently strong, as in black holes or near the Big Bang, general relativity necessarily predicts the existence of singularities—as well as Hawking's result that the area of black-hole event horizons would never decrease.

So Hawking paid close attention to Bekenstein's work, but he wasn't very happy with it. For one thing, if you're going to take the analogy between area and entropy seriously, you should take the other parts of the thermodynamics/black-hole-mechanics analogy just as seriously. In particular, the surface gravity of a black hole (which is large for small black holes with negligible spin and charge, smaller for large black holes or ones with substantial spin or charge) should be proportional to its *temperature*. But that would seem to be, on the face of it, absurd. When you heat things up to high temperature, they glow, like molten metal or a burning flame. But black holes don't glow; they're black. *So there*, we can imagine Hawking thinking across the Atlantic.

Inveterate traveler that he is, Hawking visited the Soviet Union in 1973 to talk about black holes. Under the leadership of Yakov Zel'dovich, Moscow featured a group of experts in relativity and cosmology that rivaled those in Princeton or Cambridge. Zel'dovich and his colleague Alexander Starobinsky told Hawking about some work they had done to understand the Penrose process—extracting energy from a rotating black hole—in the light of quantum mechanics. According to the Moscow group, quantum mechanics implied that a spinning black hole would *spontaneously* emit radiation and lose energy; there was no need for a super-advanced civilization to throw things at it.

Hawking was intrigued but didn't buy the specific arguments that Zel'dovich and Starobinsky had offered.[216] So he set out to understand the implications of quantum mechanics in the context of black holes by himself. It's not a simple problem. "Quantum mechanics" is a very general idea: The space of states consists of wave functions rather than positions and momenta, and you can't observe the wave function exactly without dramatically altering it. Within that framework, we can think of different types of quantum systems, from individual particles to collections of superstrings. The founders of quantum mechanics focused, sensibly enough, on relatively simple systems, consisting of a small number of atoms moving slowly with respect to one another. That's still what most physics students learn when they first study quantum mechanics.

When particles become very energetic and start moving at speeds close to the speed of light, we can no longer ignore the lessons of relativity. For one thing, the energy of two particles that collide with each other can become so large that they create multiple new particles, through the miracle of $E = mc^2$. Through decades of effort on the part of theoretical physicists, the proper formalism to reconcile quantum mechanics with special relativity was assembled, in the form of "quantum field theory."

The basic idea of quantum field theory is simple: The world is made of fields, and when we observe the wave functions of those fields, we see particles. Unlike a particle, which exists at some certain point, a field exists everywhere in space; the electric field, the magnetic field, and the gravitational field are all familiar examples. At every single point in space, every field that exists has some particular value (although that value might be zero). According to quantum field theory, *everything* is a field—there is an electron field, various kinds of quark fields, and so

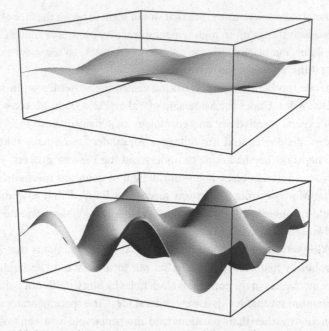

Figure 60: Fields have a value at every point in space. When we observe a quantum field, we don't see the field itself, but a collection of particles. A gently oscillating field, at the top, corresponds to a small number of particles; a wildly oscillating field, at the bottom, corresponds to a large number of particles.

on. But when we look at a field, we see particles. When we look at the electric and magnetic fields, for example, we see photons, the particles of electromagnetism. A weakly vibrating electromagnetic field shows up as a small number of photons; a wildly vibrating electromagnetic field shows up as a large number of photons.[217]

Quantum field theory reconciles quantum mechanics with special relativity. This is very different from "quantum gravity," which would reconcile quantum mechanics with *general* relativity, the theory of gravity and spacetime curvature. In quantum field theory, we imagine that spacetime itself is perfectly classical, whether it's curved or not; the fields are subject to the rules of quantum mechanics, while spacetime simply acts as a fixed background. In full-fledged quantum gravity, by contrast, we imagine that even spacetime has a wave function and is completely quantum mechanical. Hawking's work was in the context of quantum field theory in a fixed curved spacetime.

Field theory was not something Hawking was an expert in; despite being lumped with general relativity under the umbrella of "impressive-sounding theories of modern physics that seem inscrutable to outsiders," the two areas are quite different, and an expert in one might not know much about the other. So he set out to learn. Sir Martin Rees, who is one of the world's leading theoretical astrophysicists and currently Astronomer Royal of Britain, was at the time a young scientist at Cambridge; like Hawking, he had received his Ph.D. a few years before under the supervision of Dennis Sciama. By this time, Hawking was severely crippled by his disease; he would ask for a book on quantum field theory, and Rees would prop it up in front of him. While Hawking stared silently at the book for hours on end, Rees wondered whether the toll of his condition was simply becoming too much for him.[218]

Far from it. In fact, Hawking was applying the formalism of quantum field theory to the question of radiation from black holes. He was hoping to derive a formula that would reproduce Zel'dovich and Starobinsky's result for rotating black holes, but instead he kept finding something unbelievable: Quantum field theory seemed to imply that even nonrotating black holes should radiate. Indeed, they should radiate in exactly the same way as a system in thermal equilibrium at some fixed temperature, with the temperature being proportional to the surface gravity, just as it had been in the analogy between black holes and thermodynamics.

Hawking, much to his own surprise, had proven Bekenstein right. Black holes really do behave precisely as ordinary thermodynamic objects. That means, among other things, that the entropy of a black hole actually is proportional to the area of its event horizon; that connection is not just an amusing coincidence. In fact, Hawking's calculation (unlike Bekenstein's argument) allowed him to pinpoint the precise constant of proportionality: 1/4. That is, if L_P is the Planck length, so L_P^2 is the Planck area, the entropy of a black hole is 1/4 of the area of its horizon as measured in units of the Planck area:

$$S_{BH} = A/(4L_P^2).$$

You are allowed to imagine that the subscript BH stands for "Black Hole" or "Bekenstein-Hawking," as you prefer. This formula is the single most important clue we have about the reconciliation of gravitation with quantum mechanics.[219] And if we want to understand why entropy was small near the Big Bang, we have to understand something about entropy and gravity, so this is a logical starting point.

EVAPORATION

To really understand how Hawking derived the startling result that black holes radiate requires a subtle mathematical analysis of the behavior of quantum fields in curved space. Nevertheless, there is a favorite hand-waving explanation that conveys enough of the essential truth that everyone in the world, including Hawking himself, relies upon it. So why not us?

The primary idea is that quantum field theory implies the existence of "virtual particles" as well as good old-fashioned real particles. We encountered this idea briefly in Chapter Three, when we were talking about vacuum energy. For a quantum field, we might think that the state of lowest energy would be when the field was absolutely constant—just sitting there, not changing from place to place or time to time. If it were a classical field, that would be right, but just as we can't pin down a particle to one particular position in quantum mechanics, we can't pin down a field to one particular configuration in quantum field theory. There will always be some intrinsic uncertainty and fuzziness in the value of the quantum field. We can think of this inherent jitter in the quantum field as particles popping in and out of existence, one particle and one antiparticle at a time, so rapidly that we can't observe them. These virtual particles can never be detected directly; if we see a particle, we know it's a real one, not a virtual one. But virtual particles can interact with real (non-virtual) ones, subtly altering their properties, and those effects have been observed and studied in great detail. They really are there.

What Hawking figured out is that the gravitational field of a black hole can turn virtual particles into real ones. Ordinarily, virtual particles appear in pairs: one particle and one antiparticle.[220] They appear, persist for the briefest moment, and then annihilate, and no one is the wiser. But a black hole changes things, due to the presence of the event horizon. When a virtual particle/antiparticle pair pops into existence very close to the horizon, one of the partners can fall in, and obviously has no choice but to continue on to the singularity. The other partner, meanwhile, is now able to escape to infinity. The event horizon has served to rip apart the virtual pair, gobbling up one of the particles. The one that escapes is part of the Hawking radiation.

At this point a crucial property of virtual particles comes into play: Their energy can be anything at all. The total energy of a virtual particle/antiparticle pair is exactly zero, since they must be able to pop into and out of the vacuum. For real particles, the energy is equal to the mass times the speed of light squared when the particle is at rest, and grows larger if the particle is moving; consequently, it can never be negative. So if the real particle that escapes the black hole has positive energy, and the total energy of the original virtual pair was zero, that means the partner that fell into the black hole must have a *negative* energy. When it falls in, the total mass of the black hole goes down.

Eventually, unless extra energy is added from other sources, a black hole will

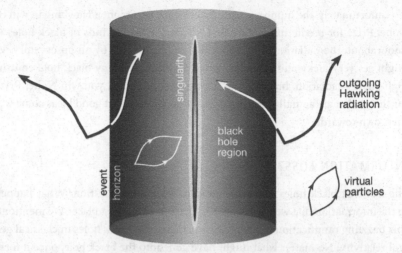

Figure 61: Hawking radiation. In quantum field theory, virtual particles and antiparticles are constantly popping in and out of the vacuum. But in the vicinity of a black hole, one of the particles can fall into the event horizon, while the other escapes to the outside world as Hawking radiation.

evaporate away entirely. Black holes are not, as it turns out, places were time ends once and for all; they are objects that exist for some period of time before they eventually disappear. In a way, Hawking radiation has made black holes a lot more ordinary than they seemed to be in classical general relativity.

An interesting feature of Hawking radiation is that *smaller* black holes are *hotter*. The temperature is proportional to the surface gravity, which is greater for less massive black holes. The kinds of astrophysical black holes we've been talking about, with masses equal to or much greater than that of the Sun, have extremely low Hawking temperatures; in the current universe, they are not evaporating at all, as they are taking in a lot more energy from objects around them than they are losing energy from Hawking radiation. That would be true even if the only external source of energy were the cosmic microwave background, at a temperature of about 3 Kelvin. In order for a black hole to be hotter than the microwave background is today, it would have to be less than about 10^{14} kilograms—about the mass of Mt. Everest, and much smaller than any known black hole.[221] Of course, the microwave background continues to cool down as the universe expands; so if we wait long enough, the black holes will be hotter than the surrounding universe, and begin to lose mass. As they do so, they heat up, and lose mass even faster; it's a runaway process and, once the black hole has been whittled down to a very small size, the end comes quickly in a dramatic explosion.

Unfortunately, the numbers make it very hard for Stephen Hawking to win the Nobel Prize for predicting black hole radiation. For the kinds of black holes we know about, the radiation is far too feeble to be detected by an observatory. We might get very lucky and someday detect an extremely tiny black hole emitting high-energy radiation, but the odds are against it[222]—and you win Nobel Prizes for things that are actually seen, not just for good ideas. But good ideas come with their own rewards.

INFORMATION LOSS?

The fact that black holes evaporate away raises a deep question: What happens to the information that went into making the hole in the first place? We mentioned this puzzling ramification of the no-hair principle for black holes in classical general relativity: No matter what might have gone into the black hole, once it forms the only features it has are its mass, charge, and spin. Previous chapters made a big deal about the fact that the laws of physics preserve the information needed to specify a state as the universe evolves from moment to moment. At first blush, a black hole would seem to destroy that information.

Imagine that, in frustration at the inability of modern physics to provide a compelling explanation for the arrow of time, you throw your copy of this book onto an open fire. Later, you worry that you might have been a bit hasty, and you want to get the book back. Too bad, it's already burnt into ashes. But the laws of physics tell us that all the information contained in the book is still available in principle, no matter how hard it might be to reconstruct in practice. The burning book evolved into a very particular arrangement of ashes and light and heat; if we could exactly capture the complete microstate of the universe after the fire, we could theoretically run the clock backward and figure out whether the book that had burned was this one or, for example, *A Brief History of Time*. (Laplace's Demon would know which book it was.) That's very theoretical, because the entropy increased by a large amount along the way, but in principle it could happen.

If instead of throwing the book into a fire, we had thrown it into a black hole, the story would be different. According to classical general relativity, there is no way to reconstruct the information; the book fell into a black hole, and we can measure the resulting mass, charge, and spin, but nothing more. We might console ourselves that the information is still in there somewhere, but we can't get to it.

Once Hawking radiation is taken into account, this story changes. Now the black hole doesn't last forever; if we're sufficiently patient, it will completely evaporate away. If information is not lost, we should be in the same situation we were in with the fire, where in principle it's possible to reconstruct the contents of the book from properties of the outgoing radiation.

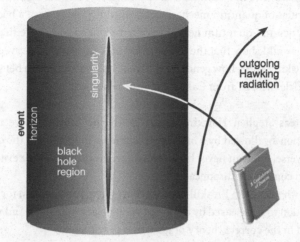

outgoing
Hawking
radiation

singularity

event
horizon

black
hole
region

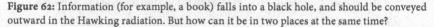

Figure 62: Information (for example, a book) falls into a black hole, and should be conveyed outward in the Hawking radiation. But how can it be in two places at the same time?

The problem with that expectation arises when we think about how Hawking radiation originates from virtual particles near the event horizon of a black hole. Looking at Figure 62 we can imagine a book falling through the horizon, all the way to the singularity (or whatever should replace the singularity in a better theory of quantum gravity), taking the information contained on its pages along with it. Meanwhile, the radiation that purportedly carries away the same information has already left the black hole. How can the information be in two places at once?[223] As far as Hawking's calculation is concerned, the outgoing radiation is the same for every kind of black hole, no matter what went into making it. At face value, it would appear that the information is simply destroyed; it would be as if, in our earlier checkerboard examples, there was a sort of blob that randomly spit out gray and white squares without any consideration for the prior state.

This puzzle is known as the "black hole information-loss paradox." Because direct experimental information about quantum gravity is hard to come by, thinking about ways to resolve this paradox has been a popular pastime among theoretical physicists over the past few decades. It has been a real controversy within the physics community, with different people coming down on different sides of the debate. Very roughly speaking, physicists who come from a background in general relativity (including Stephen Hawking) have tended to believe that information really is lost, and that black hole evaporation represents a breakdown of the conventional rules of quantum mechanics; meanwhile, those from a background in particle physics and quantum field theory have tended to believe that a better understanding would show that the information was somehow preserved.

In 1997, Hawking and fellow general-relativist Kip Thorne made a bet with John Preskill, a particle theorist from Caltech. It read as follows:

> Whereas Stephen Hawking and Kip Thorne firmly believe that information swallowed by a black hole is forever hidden from the outside universe, and can never be revealed even as the black hole evaporates and completely disappears,
>
> And whereas John Preskill firmly believes that a mechanism for the information to be released by the evaporating black hole must and will be found in the correct theory of quantum gravity,
>
> Therefore Preskill offers, and Hawking/Thorne accept, a wager that:
>
> When an initial pure quantum state undergoes gravitational collapse to form a black hole, the final state at the end of black hole evaporation will always be a pure quantum state.
>
> The loser(s) will reward the winner(s) with an encyclopedia of the winner's choice, from which information can be recovered at will.
>
> Stephen W. Hawking, Kip S. Thorne, John P. Preskill
> Pasadena, California, 6 February 1997

In 2004, in a move that made newspaper headlines, Hawking conceded his part of the bet; he admitted that black hole evaporation actually does preserve information. Interestingly, Thorne has not (as of this writing) conceded his own part of the bet; furthermore, Preskill accepted his winnings (*Total Baseball: The Ultimate Baseball Encyclopedia*, 8th edition) only reluctantly, as he believes the matter is still not settled.[224]

What convinced Hawking, after thirty years of arguing that information was lost in black holes, that it was actually preserved? The answer involves some deep ideas about spacetime and entropy, so we have to lay some background.

HOW MANY STATES CAN FIT IN A BOX?

We are delving into such detail about black holes in a book that is supposed to be about the arrow of time for a very good reason: The arrow of time is driven by an increase in entropy, which ultimately originates in the low entropy near the Big Bang, which is a period in the universe's history when gravity is fundamentally important. We therefore need to know how entropy works in the presence of gravity, but we're held back by our incomplete understanding of quantum gravity. The one clue we have is Hawking's formula for the entropy of a black hole, so we would like to follow that clue to see where it leads. And indeed, efforts to understand black-hole entropy and the information-loss paradox have had dramatic consequences for our understanding of spacetime and the space of states in quantum gravity.

Consider the following puzzle: How much entropy can fit in a box? To Boltzmann and his contemporaries, this would have seemed like a silly question—we could fit as much entropy as we liked. If we had a box full of gas molecules, there would be a maximum-entropy state (an equilibrium configuration) for any particular number of molecules: The gas would be evenly distributed through the box at constant temperature. But we could certainly squeeze more entropy into the box if we wanted to; all we would have to do is add more and more molecules. If we were worried that the molecules took up a certain amount of space, so there was some maximum number we could squeeze into the box, we might be clever and consider a box full of photons (light particles) instead of gas molecules. Photons can be piled on top of one another without limit, so we should be able to have as many photons in the box as we wish. From that point of view, the answer seems to be that we can fit an infinite (or at least arbitrarily large) amount of entropy in any given box. There is no upper limit.

That story, however, is missing a crucial ingredient: gravity. As we put more and more stuff into the box, the mass inside keeps growing.[225] Eventually, the stuff we are putting into the box suffers the same fate as a massive star that has exhausted its nuclear fuel: It collapses under its own gravitational pull and forms a black hole. Every time that happens, the entropy increases—the black hole has more entropy than the stuff of which it was made. (Otherwise the Second Law would prevent black holes from forming.)

Unlike boxes full of atoms, we can't make black holes with the same size but different masses. The size of a black hole is characterized by the "Schwarzschild radius," which is precisely proportional to its mass.[226] If you know the mass, you know the size; contrariwise, if you have a box of fixed size, there is a maximum mass black hole you can possibly fit into it. But if the entropy of the black hole is proportional to the area of its event horizon, that means *there is a maximum amount of entropy you can possibly fit into a region of some fixed size, which is achieved by a black hole of that size.*

That's a remarkable fact. It represents a dramatic difference in the behavior of entropy once gravity becomes important. In a hypothetical world in which there was no such thing as gravity, we could squeeze as much entropy as we wanted into any given region, but gravity stops us from doing that.

The importance of this insight comes when we hearken back to Boltzmann's understanding of entropy as (the logarithm of) the number of microstates that are macroscopically indistinguishable. If there is some finite maximum amount of entropy we can fit into a region of fixed size, that means there are only a finite number of possible states within that region. That's a deep feature of quantum gravity, radically different from the behavior of theories without gravity. Let's see where this line of reasoning takes us.

THE HOLOGRAPHIC PRINCIPLE

To appreciate how radical the lesson of black-hole entropy is, we have to first appreciate the cherished principle it apparently overthrows: *locality*. That's the idea that different places in the universe act more or less independently of one another. An object at some particular location can be influenced by its immediate surroundings, but not by things that are far away. Distant things can influence one another indirectly, by sending some signal from place to place, such as a disturbance in the gravitational field or an electromagnetic wave (light). But what happens here doesn't directly influence what happens in some other region of the universe.

Think back to the checkerboards. What happened at one moment in time was influenced by what happened at the previous moment in time. But what happened at one point in "space"—the collection of squares across a single row—was completely unrelated to what happened at any other point in space at the same time. Along any particular row, we were free to imagine any distribution of white and gray squares we chose. There were no rules along the lines of "when there is a gray square here, the square twenty slots to the right has to be white." And when squares did "interact" with one another as time passed, it was always with the squares right next to them. Similarly, in the real world, things bump into one another and influence other things when they are close by, not when they are far apart. That's locality.

Locality has an important consequence for entropy. Consider, as usual, a box of gas, and calculate the entropy of the gas in the box. Now let's mentally divide the box in two, and calculate the entropy in each half. (We don't need to imagine a physical barrier, just consider the left side of the box and the right side separately.) What is the relationship between the total entropy of the box and the separate entropy of the two halves?

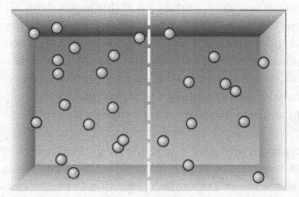

Figure 63: A box of gas, mentally divided into two halves. The total entropy in the box is the sum of the entropies of each half.

The answer is: You get the entropy in the whole box by simply adding the entropy of one half to the entropy of the other half. This would seem to be a direct consequence of Boltzmann's definition of the entropy—indeed, it's the entire reason why that definition has a logarithm in it. We have a certain number of allowed microstates in one half of the box, and a certain number in the other half. The total number of microstates is calculated as follows: For every possible microstate of the left side, we are allowed to choose any of the possible microstates on the right side. So we get the total number of microstates by *multiplying* the number of microstates on the left by the number of microstates on the right. But the entropy is the logarithm of that number, and the logarithm of "X times Y" is "the logarithm of X" *plus* "the logarithm of Y."

So the entropy of the total box is simply the sum of the entropies of the two sub-boxes. Indeed, that would work no matter how we divided up the original box, or how many sub-boxes we divided it into; the total entropy is always the sum of the sub-entropies. This means that the maximum entropy we can have in a box is always going to be proportional to the *volume* of the box—the more space we have, the more entropy we can have, and it scales directly with the addition of more volume.

But notice the sneaky assumption in that argument: We were able to count the number of states in one half of the box, and then multiply by the number in the other half. In other words, what happened in one half of the box was assumed to be totally independent of what happened in the other half. And that is exactly the assumption of locality.

When gravity becomes important, all of this breaks down. Gravity puts an upper limit on the amount of entropy we can squeeze into a box, given by the largest black hole that can fit in the box. But the entropy of a black hole isn't proportional to the volume enclosed—it's proportional to the *area* of the event horizon. Area and volume are very different! If we have a sphere 1 meter across, and we increase it in size until it's 2 meters across, the volume inside goes up by a factor of 8 (2^3), but the area of the boundary only goes up by a factor of 4 (2^2).

The upshot is simple: Quantum gravity doesn't obey the principle of locality. In quantum gravity, what goes on over here is not completely independent from what goes on over there. The number of things that can possibly go on (the number of possible microstates in a region) isn't proportional to the volume of the region; it's proportional to the area of a surface we can draw that encloses the region. The real world, described by quantum gravity, allows for much less information to be squeezed into a region than we would naïvely have imagined if we weren't taking gravity into account.

This insight has been dubbed the *holographic principle*. It was first proposed by Dutch Nobel laureate Gerard 't Hooft and American string theorist Leonard Susskind, and later formalized by German-American physicist Raphael Bousso (formerly a student of Stephen Hawking).[227] Superficially, the holographic principle might sound a bit dry. Okay, the number of possible states in a region is proportional to the size of the region squared, not the size of the region cubed. That's not the kind of line that's going to charm strangers at cocktail parties.

Here is why holography is important: It means that spacetime is not fundamental. When we typically think about what goes on in the universe, we implicitly assume something like locality; we describe what happens at this location, and at that location, and give separate specifications for every possible location in space. Holography says that we can't really do that, in principle—there are subtle correlations between things that happen at different locations, which cut down on our freedom to specify a configuration of stuff through space.

An ordinary hologram displays what appears to be a three-dimensional image by scattering light off of a special two-dimensional surface. The holographic principle says that the universe is like that, on a fundamental level: Everything you think is happening in three-dimensional space is secretly encoded in a two-dimensional surface's worth of information. The three-dimensional space in which we live and breathe could (again, in principle) be reconstructed from a much more compact description. We may or may not have easy access to what that description actually is—usually we don't, but in the next section we'll discuss an explicit example where we do.

Perhaps none of this should be surprising. As we discussed in the previous chapter, there is a type of non-locality already inherent in quantum mechanics, before gravity ever gets involved; the state of the universe describes all particles at once, rather than referring to each particle separately. So when gravity is in the game, it's natural to suppose that the state of the universe would include all of spacetime at once. But still, the type of non-locality implied by the holographic principle is different than that of quantum mechanics alone. In quantum mechanics, we could imagine particular wave functions in which the state of a cat was entangled with the state of a dog, but we could just as easily imagine states in which they were not entangled, or where the entanglement took on some different form. Holography seems to be telling us that there are some things that just can't happen, that the information needed to encode the world is dramatically compressible. The implications of this idea are still being explored, and there are undoubtedly more surprises to come.

HAWKING GIVES IN

The holographic principle is a very general idea; it should be a feature of whatever theory of quantum gravity eventually turns out to be right. But it would be nice to have one very specific example that we could play with to see how the consequences of holography work themselves out. For example, we think that the entropy of a black hole in our ordinary three-dimensional space is proportional to the two-dimensional area of its event horizon; so it should be possible, in principle, to specify all of the possible microstates corresponding to that black hole in terms of different things that could happen on that two-dimensional surface. That's a goal of many theorists working in quantum gravity, but unfortunately we don't yet know how to make it work.

In 1997, Argentine-American theoretical physicist Juan Maldacena revolutionized our understanding of quantum gravity by finding an explicit example of holography in action.[228] He considered a hypothetical universe nothing like our own—for one thing, it has a negative vacuum energy (whereas ours seems to have a positive vacuum energy). Because empty space with a positive vacuum energy is called "de Sitter space," it is convenient to label empty space with a negative vacuum energy "anti–de Sitter space." For another thing, Maldacena considered five dimensions instead of our usual four. And finally, he considered a very particular theory of gravitation and matter—"supergravity," which is the supersymmetric version of general relativity. Supersymmetry is a hypothetical symmetry between bosons (force particles) and fermions (matter particles), which plays a crucial role in many theories of modern particle physics; happily, the details aren't crucial for our present purposes.

Maldacena discovered that this theory—supergravity in five-dimensional anti–de Sitter space—is completely equivalent to an entirely different theory—a *four*-dimensional quantum field theory *without gravity at all.* Holography in action: Everything that can possibly happen in this particular five-dimensional theory with gravity has a precise analogue in a theory without gravity, in one dimension less. We say that the two theories are "dual" to each other, which is a fancy way of saying that they look very different but really have the same content. It's like having two different but equivalent languages, and Maldacena has uncovered the Rosetta stone that allows us to translate between them. There is a one-to-one correspondence between states in a particular theory of gravity in five dimensions and a particular nongravitational theory in four dimensions. Given a state in one, we can translate it into a state in the other, and the equations of motion for each theory will evolve the respective states into new states that correspond to each other according to the same translation dictionary (at least in principle; in practice we can work out simple examples, but more complicated situations become intractable). Obviously the correspondence needs to be nonlocal; you can't match up individual points in a four-dimensional space to points in a five-dimensional space. But you can imagine matching up states in one theory, defined at some time, to states in the other theory.

If that doesn't convince you that spacetime is not fundamental, I can't imagine what would. We have an explicit example of two different versions of precisely the same theory, but they describe spacetimes with different numbers of dimensions! Neither theory is "the right one"; they are completely equivalent to each other.

Maldacena's discovery helped persuade Stephen Hawking to concede his bet with Preskill and Thorne (although Hawking, as is his wont, worked things out his own way before becoming convinced). Remember that the issue in question was whether the process of black hole evaporation, unlike evolution according to ordinary quantum mechanics, destroys information, or whether the information that goes into a black hole somehow is carried out by the Hawking radiation.

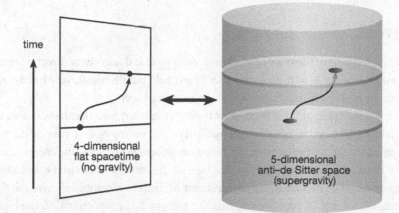

time

4-dimensional
flat spacetime
(no gravity)

5-dimensional
anti–de Sitter space
(supergravity)

Figure 64: The Maldacena correspondence. A theory of gravity in a five-dimensional anti–de Sitter space is equivalent to a theory without gravity in four-dimensional flat spacetime.

If Maldacena is right, we can consider that question in the context of five-dimensional anti–de Sitter space. That's not the real world, but the ways in which it differs from the real world don't seem extremely relevant for the information-loss puzzle—in particular, we can imagine that the negative cosmological constant is very small, and essentially unimportant. So we make a black hole in anti–de Sitter space and then let it evaporate. Is information lost? Well, we can translate the question into an analogous situation in the four-dimensional theory. But that theory doesn't have gravity, and therefore obeys the rules of ordinary quantum mechanics. There is no way for information to be lost in the four-dimensional nongravitational theory, which is supposed to be completely equivalent to the five-dimensional theory with gravity. So, if we haven't missed some crucial subtlety, the information must somehow be preserved in the process of black hole evaporation.

That is the basic reason why Hawking conceded his bet, and now accepts that black holes don't destroy information. But you can see that the argument, while seemingly solid, is nevertheless somewhat indirect. In particular, it doesn't provide us with any concrete physical understanding of how the information actually gets into the Hawking radiation. Apparently it happens, but the explicit mechanism remains unclear. That's why Thorne hasn't yet conceded his part of the bet, and why Preskill accepted his encyclopedia only with some reluctance. Whether or not we accept that information is preserved, there's clearly more work to be done to understand exactly what happens when black holes evaporate.

A STRING THEORY SURPRISE

There is one part of the story of black-hole entropy that doesn't bear directly on the arrow of time but is so provocative that I can't help but discuss it, very briefly. It's about the nature of black-hole microstates in string theory.

The great triumph of Boltzmann's theory of entropy was that he was able to explain an observable macroscopic quantity—the entropy—in terms of microscopic components. In the examples he was most concerned with, the components were the atoms constituting a gas in a box, or the molecules of two liquids mixing together. But we would like to think that his insight is completely general; the formula $S = k \log W$, proclaiming that the entropy S is proportional to the logarithm of the number of ways W that we can rearrange the microstates, should be true for any system. It's just a matter of figuring out what the microstates are, and how many ways we can rearrange them. In other words: What are the "atoms" of this system?

Hawking's formula for the entropy of a black hole seems to be telling us that there are a very large number of microstates corresponding to any particular macroscopic black hole. What are those microstates? They are not apparent in classical general relativity. Ultimately, they must be states of quantum gravity. There's good news and bad news here. The bad news is that we don't understand quantum gravity very well in the real world, so we are unable to simply list all of the different microstates corresponding to a macroscopic black hole. The good news is that we can use Hawking's formula as a *clue*, to test our ideas of how quantum gravity might work. Even though physicists are convinced that there must be some way to reconcile gravity with quantum mechanics, it's very hard to get direct experimental input to the problem, just because gravity is an extremely weak force. So any clue we discover is very precious.

The leading candidate for a consistent quantum theory of gravity is *string theory*. It's a simple idea: Instead of the elementary constituents of matter being point-like particles, imagine that they are one-dimensional pieces of "string." (You're not supposed to ask what the strings are made of; they're not made of anything more fundamental.) You might not think you could get much mileage out of a suggestion like that—okay, we have strings instead of particles, so what?

The fascinating thing about string theory is that it's a very constraining idea. There are lots of different theories we could imagine making from the idea of elementary particles, but it turns out that there are very few consistent quantum mechanical theories of strings—our current best guess is that there is only one. And that one theory necessarily comes along with certain ingredients—extra dimensions of space, and supersymmetry, and higher-dimensional branes (sort of like strings, but two or more dimensions across). And, most important, it comes with gravity. String theory was originally investigated as a theory of nuclear forces, but that didn't turn out very well, for an unusual reason—the theory kept predicting the existence of a force like gravity! So theorists decided to take that particular lemon and make lemonade, and study string theory as a theory of quantum gravity.[229]

If string theory is the correct theory of quantum gravity—we don't know yet whether it is, but there are promising signs—it should be able to provide a microscopic understanding of where the Bekenstein-Hawking entropy comes from. Remarkably, it does, at least for some certain very special kinds of black holes.

The breakthrough was made in 1996 by Andrew Strominger and Cumrun Vafa, building on some earlier work of Leonard Susskind and Ashoke Sen.[230] Like Maldacena, they considered five-dimensional spacetime, but they didn't have a negative vacuum energy and they weren't primarily concerned with holography. Instead, they took advantage of an interesting feature of string theory: the ability to "tune" the strength of gravity. In our everyday world, the strength of the gravitational force is set by Newton's gravitational constant, denoted G. But in string theory the strength of gravity becomes variable—it can change from place to place and time to time. Or, in the flexible and cost-effective world of thought experiments, you can choose to look at a certain configuration of stuff with gravity "turned off" (G set to zero), and then look at the same configuration with gravity "turned on" (G set to a value large enough that gravity is important).

So Strominger and Vafa looked at a configuration of strings and branes in five dimensions, carefully chosen so that the setup could be analyzed with or without gravity. When gravity was turned on, their configuration looked like a black hole, and they knew what the entropy was supposed to be from Hawking's formula. But when gravity was turned off, they basically had the string-theory equivalent of a box of gas. In that case, they could calculate the entropy in relatively conventional ways (albeit with some high-powered math appropriate to the stringy stuff they were considering).

And the answer is: The entropies agree. At least in this particular example, a black hole can be smoothly turned into a relatively ordinary collection of stuff, where we know exactly what the space of microstates looks like, and the entropy from Boltzmann's formula matches that from Hawking's formula, down to the precise numerical factor.

We don't have a fully general understanding of the space of states in quantum gravity, so there are still many mysteries as far as entropy is concerned. But in the particular case examined by Strominger and Vafa (and various similar situations examined subsequently), the space of microstates predicted by string theory seems to exactly match the expectation from Hawking's calculation using quantum field theory in curved spacetime.[231] It gives us hope that further investigations along the same lines will help us understand other puzzling features of quantum gravity—including what happened at the Big Bang.

13

THE LIFE OF THE UNIVERSE

Time is a great teacher, but unfortunately it kills all its pupils.

—Hector Berlioz

What *should* the universe look like?

This might not be a sensible question. The universe is a unique entity; it's different in kind from the things we typically think about, all of which exist *in* the universe. Objects within the universe belong to larger collections of objects, all of which share common properties. By observing these properties we can get a feel for what to expect from that kind of thing. We expect that cats usually have four legs, ice cream is usually sweet, and supermassive black holes lurk at the centers of spiral galaxies. None of these expectations is absolute; we're talking about tendencies, not laws of nature. But our experience teaches us to expect that certain kinds of things usually have certain properties, and in those unusual circumstances where our expectations are not met, we might naturally be moved to look for some sort of explanation. When we see a three-legged cat, we wonder what happened to its other leg.

The universe is different. It's all by itself, not a member of a larger class. (Other universes might exist, at least for certain definitions of "universe"; but we certainly haven't observed any.) So we can't use the same kind of inductive, empirical reasoning—looking at many examples of something, and identifying common features—to justify any expectations for what the universe should be like.[232]

Nevertheless, scientists declare that certain properties of the universe are "natural" all the time. In particular, I'm going to suggest that the low entropy of the early universe is surprising, and argue that there is likely to be an underlying explanation. When we notice that an unbroken egg is in a low-entropy configuration compared to an omelet, we have recourse to a straightforward explanation: The egg is not a closed system. It came out of a chicken, which in turn is part of an ecosystem here on Earth, which in turn is embedded in a universe that has a low-entropy past. But the universe, at least at first glance, does seem to be a closed system—it was not hatched out of a Universal Chicken or anything along those lines. A truly closed physical system with a very low entropy is surprising and suggests that something bigger is going on.[233]

The right attitude toward any apparently surprising feature of the observed universe, such as the low early entropy or the small vacuum energy, is to treat it as a potential clue to a deeper understanding. Observations like this aren't anywhere near as definitive as a straightforward experimental disagreement with a favored theory; they are merely suggestive. In the backs of our minds, we're thinking that if the configuration of the universe were chosen randomly from all possible configurations, it would be in a very high-entropy state. It's not, so therefore the state of the universe isn't just chosen randomly. Then how is it chosen? Is there some process, some dynamical chain of events, that leads inevitably to the seemingly non-random configuration of our universe?

OUR HOT, SMOOTH EARLY DAYS

If we think of the universe as a physical system in a randomly chosen configuration, the question "What should the universe look like?" is answered by "It should be in a high-entropy state." We therefore need to understand what a high-entropy state of the universe would look like.

Even this formulation of the question is not quite right. We don't actually care about the particular state of the universe right this moment; after all, yesterday it was different, and tomorrow it will be different again. What we really care about is the *history* of the universe, its evolution through time. But understanding what would constitute a natural history presupposes that we understand something about the space of states, including what high-entropy states look like.

Cosmologists have traditionally done a very sloppy job of addressing this issue. There are a couple of reasons for this. One is that the expansion of the universe from a hot, dense early state is such an undeniable brute *fact* that, once you've become accustomed to the idea, it seems hard to imagine any alternative. You begin to see your task as a theoretical cosmologist as one of explaining why our universe began in the particular hot, dense early state that it did, rather than some different hot, dense early state. This is temporal chauvinism at its most dangerous—unthinkingly trading in the question "Why does the universe evolve in the way it does?" for "Why were the initial conditions of the universe set up the way they were?"

The other thing standing in the way of more productive work on the entropy of the universe is the inescapable role of gravity. By "gravity" we mean everything having to do with general relativity and curved spacetime—everyday stuff like apples falling and planets orbiting stars, but also black holes and the expansion of the universe. In the last chapter, we focused on the one example where we think we know the entropy of an object with a strong gravitational field: a black hole. That example does not seem immediately helpful when thinking about the whole universe, which is not a black hole; it bears a superficial resemblance to a *white* hole (since there is a singularity in the past), but even that is of little help, since we are inside it rather than outside. Gravity is certainly important to the universe, and that's especially true at early times when space was expanding very rapidly. But appreciating that it's important doesn't help us address the problem, so most people simply put it aside.

There is one other strategy, which appears innocent at first but really hides a potentially crucial mistake. That's to simply separate out gravity from everything else, and calculate the entropy of the matter and radiation within spacetime while forgetting that of spacetime itself. Of course, it's hard to be a cosmologist and ignore the fact that space is expanding; however, we can take the expansion of space as a given, and simply consider the state of the "stuff" (particles of ordinary matter, dark matter, radiation) within such a background. The expanding universe acts to dilute away the matter and cool off the radiation, just as if the particles were all contained in a piston that was gradually being pulled out to create more room for them to breathe. It's possible to calculate the entropy of the stuff in that particular background, exactly as it's possible to calculate the entropy of a collection of molecules inside an expanding piston.

At any one time in the early universe, we have a gas of particles at a nearly constant temperature and nearly constant density from place to place. In other words, a configuration that looks pretty much like thermal equilibrium. It's not exactly thermal equilibrium, because in equilibrium nothing changes, and in the expanding universe things are cooling off and diluting away. But compared to the rate at which particles are bumping into one another, the expansion of space is relatively slow, so the cooling off is quite gradual. If we just consider matter and radiation in the early universe, and neglect any effects of gravity other than the overall expansion, what we find is a sequence of configurations that are very close to thermal equilibrium at a gradually declining density and temperature.[234]

But that's a woefully incomplete story, of course. The Second Law of Thermodynamics says, "The entropy of a closed system either increases or remains constant"; it doesn't say, "The entropy of a closed system, ignoring gravity, either increases or remains constant." There's nothing in the laws of physics that allows us to neglect gravity in situations where it's important—and in cosmology it's of paramount importance.

By ignoring the effects of gravity on the entropy, and just considering the matter and radiation, we are led to nonsensical conclusions. The matter and radiation in the early universe was close to thermal equilibrium, which means (neglecting gravity) that it was in its *maximum entropy* state. But today, in the late universe, we're clearly not in thermal equilibrium (if we were, we'd be surrounded by nothing but gas at constant temperature), so we are clearly *not* in a configuration of maximum entropy. But the entropy didn't go down—that would violate the Second Law. So what is going on?

What's going on is that it's not okay to ignore gravity. Unfortunately, including gravity is not so easy, as there is still a lot we don't understand about how entropy works when gravity is included. But as we'll see, we know enough to make a great deal of progress.

WHAT WE MEAN BY OUR UNIVERSE

For the most part, up until now I have stuck to well-established ground: either reviewing things that all good working physicists agree are correct, or explaining things that are certainly true that all good working physicists *should* agree are correct. In the few genuinely controversial exceptions (such as the interpretation of quantum mechanics), I tried to label them clearly as unsettled. But at this point in the book, we start becoming more speculative and heterodox—I have my own favorite point of view, but there is no settled wisdom on these questions. I'll try to continue distinguishing between certainly true things and more provisional ideas, but it's important to be as careful as possible in making the case.

First, we have to be precise about what we mean by "our universe." We don't see all of the universe; light travels at a finite speed, and there is a barrier past which we can't see—in principle given by the Big Bang, in practice given by the moment when the universe became transparent about 380,000 years after the Big Bang. Within the part that we do see, the universe is homogenous on large scales; it looks pretty much the same everywhere. There is a corresponding strong temptation to take what we see and extrapolate it shamelessly to the parts we can't see, and imagine that the entire universe is homogenous throughout its extent—either through a volume of finite size, if the universe is "closed," or an infinitely big volume, if the universe is "open."

But there's really no good reason to believe that the universe we don't see matches so precisely with the universe we do see. It might be a simple starting assumption, but it's nothing more than that. We should be open to the possibility that the universe eventually looks completely different somewhere beyond the part we can see (even if it keeps looking uniform for quite a while before we get to the different parts).

So let's forget about the rest of the universe, and concentrate on the part we can see—what we've been calling "the observable universe." It stretches about 40 billion light-years around us.[235] But since the universe is expanding, the stuff within what we now call the observable universe was packed into a smaller region in the past. What we do is erect a kind of imaginary fence around the stuff within our currently observable universe, and keep track of what's inside the fence, allowing the fence itself to expand along with the universe (and be smaller in the past). This is known as our *comoving patch* of space, and it's what we have in mind when we say "our observable universe."

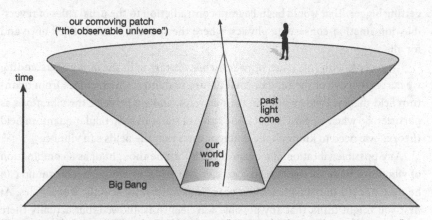

Figure 65: What we call "the observable universe" is a patch of space that is "comoving"—it expands along with the universe. We trace back along our light cones to the Big Bang to define the part of the universe we can observe, and allow that region to grow as the universe expands.

Our comoving patch of space is certainly not, strictly speaking, a closed system. If an observer were located at the imaginary fence, they would notice various particles passing in and out of our patch. But on average, the same number and kind of particles would come in and go out, and in the aggregate they would be basically indistinguishable. (The smoothness of the cosmic microwave background convinces us that the uniformity of our universe extends out to the boundary of our patch, even if we're not sure how far it continues beyond.) So for all practical purposes, it's okay to think of our comoving patch as a closed system. It's not really closed, but it evolves just as if it were—there aren't any important influences from the outside that are affecting what goes on inside.

CONSERVATION OF INFORMATION IN
AN EXPANDING SPACETIME

If our comoving patch defines an approximately closed system, the next step is to think about its space of states. General relativity tells us that space itself, the stage on which particles and matter move and interact, evolves over time. Because of this, the definition of the space of states becomes more subtle than it would have been if spacetime were absolute. Most physicists would agree that information is conserved as the universe evolves, but the way that works is quite unclear in a cosmological context. The essential problem is that more and more things can fit into the universe as it expands, so—naïvely, anyway—it looks as if the space of states is getting bigger. That would be in flagrant contradiction to the usual rules of reversible, information-conserving physics, where the space of states is fixed once and for all.

To grapple with this issue, it makes sense to start with the best understanding we currently have of the fundamental nature of matter, which comes from quantum field theory. Fields vibrate in various ways, and we perceive the vibrations as particles. So when we ask, "What is the space of states in a particular quantum field theory?" we need to know all the different ways that the fields can vibrate.

Any possible vibration of a quantum field can be thought of as a combination of vibrations with different specific wavelengths—just as any particular sound can be decomposed into a combination of various notes with specific frequencies. At first you might think that any possible wavelength is allowed, but actually there are restrictions. The Planck length—the tiny distance of 10^{-33} centimeters at which quantum gravity becomes important—provides a *lower limit* on what wavelengths are allowed. At smaller distances than that, spacetime itself loses its conventional meaning, and the energy of the wave (which is larger when the wavelength is shorter) becomes so large that it would just collapse to a black hole.

Likewise, there is an *upper limit* on what wavelengths are allowed, given by the size of our comoving patch. It's not that vibrations with longer wavelengths can't exist; it's that they just don't matter. Wavelengths larger than the size of our patch are effectively constant throughout the observable universe.

So it's tempting to take "the space of states of the observable universe" as consisting of "vibrations in all the various quantum fields, with wavelengths larger than the Planck length and smaller than the size of our comoving patch." The problem is, that's a space of states that changes as the universe expands. Our patch grows with time, while the Planck length remains fixed. At extremely early times, the universe was very young and expanding very rapidly, and our patch was relatively small. (Exactly how small depends on details of the evolution of the early universe that we don't know.) There weren't that many vibrations you could squeeze into the universe at that time. Today, the Hubble length is enormously larger—about 10^{60} times larger than the Planck length—and there are a huge number of allowed vibrations. Under this way of thinking, it's not so surprising that the entropy of the early universe was small, because the maximum allowed entropy of the universe at that time was small—the maximum allowed entropy increases as the universe expands and the space of states grows.

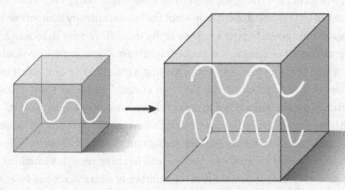

Figure 66: As the universe expands, it can accommodate more kinds of waves. More things can happen, so the space of states would appear to be growing.

But if a space of states changes with time, the evolution clearly can't be information conserving and reversible. If there are more possible states today than there were yesterday, and two distinct initial states always evolve into two distinct final states, there must be some states today that didn't come from anywhere. That means the evolution can't be reversed, in general. All of the conventional reversible laws of physics we are used to dealing with feature spaces of states that are fixed once and for all, not changing with time. The configuration within that space will evolve, but the space of states itself never changes.

We seem to have something of a dilemma. The rules of thumb of quantum field theory in curved spacetime would seem to imply that the space of states grows as the universe expands, but the ideas on which all this is based—quantum mechanics and general relativity—conform proudly to the principle of information conservation. Clearly, something has to give.

The situation is reminiscent of the puzzle of information loss in black holes. There, we (or Stephen Hawking, more accurately) used quantum field theory in curved spacetime to derive a result—the evaporation of black holes into Hawking radiation—that seemed to destroy information, or at least scramble it. Now in the case of cosmology, the rules of quantum field theory in an expanding universe seem to imply fundamentally irreversible evolution.

I am going to imagine that this puzzle will someday be resolved in favor of information conservation, just as Hawking (although not everyone) now believes is true for black holes. The early universe and the late universe are simply two different configurations of the same physical system, evolving according to reversible fundamental laws within precisely the same space of possible states. The right thing to do, when characterizing the entropy of a system as "large" or "small," is to compare it to the largest possible entropy—not the largest entropy compatible with some properties the system happens to have at the time. If we were to look at a box of gas and find that all of the gas was packed tightly into one corner, we wouldn't say, "That's a high-entropy configuration, as long as we restrict our attention to configurations that are packed tightly into that corner." We would say, "That's a very low-entropy configuration, and there's probably some explanation for it."

All of this confusion arises because we don't have a complete theory of quantum gravity, and have to make reasonable guesses on the basis of the theories we think we do understand. When those guesses lead to crazy results, something has to give. We gave a sensible argument that the number of states *described by vibrating quantum fields* changes with time as the universe expands. If the total space of states remains fixed, it must be the case that many of the possible states of the early universe have an irreducibly quantum-gravitational character, and simply can't be described in terms of quantum fields on a smooth background. Presumably, a better theory of quantum gravity would help us understand what those states might be, but even without that understanding, the basic principle of information conservation assures us that they must be there, so it seems logical to accept that and try to explain why the early universe had such an apparently low-entropy configuration.

Not everyone agrees.[236] A certain perfectly respectable school of thought goes something like this: "Sure, information might be conserved at a fundamental level, and there might be some fixed space of states for the whole universe. But who cares? We don't know what that space of states is, and we live in a universe that started out small and relatively smooth. Our best strategy is to use the rules suggested by quantum field theory, allowing only a very small set of configurations at very early times, and a much larger set at later times." That may be right. Until we have the final answers, the best we can do is follow our intuitions and try to come up with testable predictions that we can compare against data. When it comes to the origin of the universe, we're not there yet, so it pays to keep an open mind.

LUMPINESS

Because we don't have quantum gravity all figured out, it's hard to make definitive statements about the entropy of the universe. But we do have some basic tools at our disposal—the idea that entropy has been increasing since the Big Bang, the principle of information conservation, the predictions of classical general relativity, the Bekenstein-Hawking formula for black-hole entropy—that we can use to draw some reliable conclusions.

One obvious question is: What does a high-entropy state look like when gravity is important? If gravity is not important, high-entropy states are states of thermal equilibrium—stuff tends to be distributed uniformly at a constant temperature. (Details may vary in particular systems, such as oil and vinegar.) There is a general impression that high-entropy states are *smooth*, while lower-entropy states can be *lumpy*. Clearly that's just a shorthand way of thinking about a subtle phenomenon, but it's a useful guide in many circumstances.[237] Notice that the early universe is, indeed, smooth, in accordance with the let's-ignore-gravity philosophy we just examined.

But in the later universe, when stars and galaxies and clusters form, it becomes simply impossible to ignore the effects of gravity. And then we see something interesting: The casual association of "high-entropy" with "smooth" begins to fail, rather spectacularly.

For many years now, Sir Roger Penrose has been going around trying to

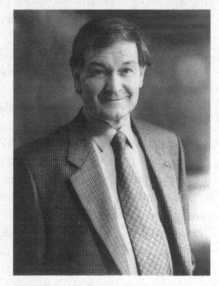

Figure 67: Roger Penrose, who has done more than anyone to emphasize the puzzle of the low-entropy early universe.

convince people that this feature of gravity—things get lumpier as entropy increases in the late universe—is crucially important and should play a prominent role in discussions of cosmology. Penrose became famous in the late 1960s and early 1970s through his work with Hawking to understand black holes and singularities in general relativity, and he is an accomplished mathematician as well as physicist. He is also a bit of a gadfly, and takes great joy in exploring positions that run decidedly contrary to the conventional wisdom in various fields, from quantum mechanics to the study of consciousness.

One of the fields in which Penrose likes to poke holes in cherished beliefs is theoretical cosmology. When I was a graduate student in the late 1980s, theoretical particle physicists and cosmologists mostly took it for granted that some version of inflationary cosmology (discussed in the next chapter) must be true; astronomers tended to be more cautious. Today, this belief is even more common, as evidence from the cosmic microwave background has shown that the small variations in density from place to place in the early universe match very well with what inflation would predict. But Penrose has been consistently skeptical, primarily on the basis of inflation's failure to explain the low entropy of the early universe. I remember reading one of his papers as a student, and appreciating that Penrose was saying something important but feeling that he had missed the point. It took two decades of thinking about entropy before I became convinced that he has mostly been right all along.

We don't have a full picture of the space of microstates in quantum gravity, and correspondingly lack a rigorous understanding of entropy. But there is a simple strategy for dealing with this obstacle: We consider what actually happens in the universe. Most of us believe that the evolution of the observable universe has always been compatible with the Second Law, and entropy has been increasing since the Big Bang, even if we're hazy on the details. If entropy tends to go up, and if there is some process that happens all the time in the universe, but its time-reverse never happens, it probably represents an increase in entropy.

An example of this is "gravitational instability" in the late universe. We've been tossing around the notion of "when gravity is important" and "when gravity is not important," but what is the criterion to decide whether gravity is important? Generally, given some collection of particles, their mutual gravitational forces will always act to pull the particles together—the gravitational force between particles is universally attractive. (In contrast, for example, with electricity and magnetism, which can be either attractive or repulsive depending on what kinds of electric charges you are dealing with.[238]) But there are other forces, generally collected together under the rubric of "pressure," that prevent everything from collapsing to a black hole. The Earth or the Sun or an egg doesn't collapse under its own gravitational pull, because each is supported by the pressure of the material inside it. As a rule of thumb, "gravity is important" means "the gravitational pull of a collection of particles overwhelms the pressure that tries to keep them from collapsing."

In the very early universe, the temperature is high and the pressure is enormous.[239] The local gravity between nearby particles is too weak to bring them together, and the initial smoothness of the matter and radiation is preserved. But as the universe expands and cools, the pressure drops, and gravity begins to take over. This is the era of "structure formation," where the initially smooth distribution of matter gradually begins to condense into stars, galaxies, and larger groups of galaxies. The initial distribution was not perfectly featureless; there were small deviations in density from place to place. In the denser regions, gravity pulled particles even closer together, while the less dense regions lost particles to their denser neighbors and became even emptier. Through gravity's persistent efforts, what was a highly uniform distribution of matter becomes increasingly lumpy.

Penrose's point is this: As structure forms in the universe, entropy increases. He puts it this way:

> Gravitation is somewhat confusing, in relation to entropy, because of its universally attractive nature. We are used to thinking about entropy in terms of an ordinary gas, where having the gas concentrated in small regions represents *low* entropy . . . and where in the *high*-entropy state of thermal equilibrium, the gas is spread uniformly. But with gravity, things tend to be the other way about. A uniformly spread system of gravitating bodies would represent relatively *low* entropy (unless the velocities of the bodies are enormously high and/or the bodies are very small and/or greatly spread out, so that the gravitational contributions become insignificant), whereas *high* entropy is achieved when the gravitating bodies clump together.[240]

All of that is completely correct, and represents an important insight. Under certain conditions, such as those that pertain in the universe on large scales today, even though we don't have a cut-and-dried formula for the entropy of a system including gravity, we can say with confidence that the entropy increases as structure forms and the universe becomes lumpier.

There is another way of reaching a similar conclusion, through the magic of thought experiments. Take the current macrostate of the universe—some collection of galaxies, dark matter, and so forth, distributed in a certain way through space. But now let's make a single change: Imagine that the universe is *contracting* rather than expanding. What should happen?

It should be clear that what *won't* happen is a simple time-reversal of the actual history of the universe from its smooth initial state to its lumpy present—at least, not for the overwhelming majority of microstates within our present macrostate. (If we precisely time-reversed the specific microstate of our present universe, then

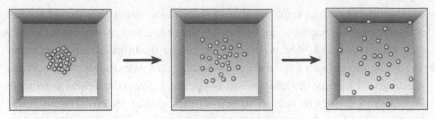

increasing entropy when gravity is unimportant

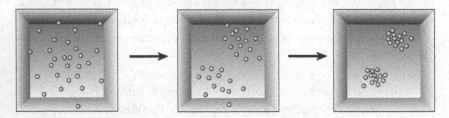

increasing entropy when gravity is important

Figure 68: When gravity is unimportant, increasing entropy tends to smooth things out; when gravity does become important, matter tends to clump together as entropy increases.

of course that is exactly what would happen.) If the distribution of matter in our present universe were to start contracting together, individual stars and galaxies would not begin to disperse and smooth out. Instead, the gravitational force between heavy objects would draw them together, and the amount of lumpy structure would actually increase, even as the universe contracted. Black holes would form, and coalesce together to create bigger black holes. There would ultimately be a sort of Big Crunch, but (as Penrose emphasizes) it wouldn't look anything like the smooth Big Bang from which our universe came. Places where the density was high and black holes formed would crash into a future singularity relatively quickly, while places that were emptier would survive for longer.

This story fits in well with the idea that the space of states within our comoving patch remains fixed, but when the universe is small most of the states cannot be described as vibrating quantum fields in an otherwise smooth space. Such a picture would be completely inadequate to describe the chaotic black-hole-filled mess that we would generically expect from a collapsing universe. But such a messy configuration is just as much an allowed state of the universe as the relatively smooth backgrounds we traditionally deal with in cosmology. Indeed, such a configuration has a higher entropy than a smooth universe (which we know because a collapsing universe would generically evolve into something messy), which means that there are many more microstates of that form than of the form where every-thing is relatively smooth. Why our actual universe is so atypical is, of course, at the heart of our mystery.

THE EVOLUTION OF ENTROPY

We've now assembled enough background knowledge to follow Penrose and take a stab at quantifying how the entropy of our universe changes from early times to today. We know the basic story of how our comoving patch evolves—at early times it was small, and full of hot, dense gas that was very close to perfectly smooth, and at later times it is larger and cooler and more dilute, and contains a distribution of stars and galaxies that is quite lumpy on small scales, although it's still basically smooth over very large distances. So what is its entropy?

At early times, when things were smooth, we can calculate the entropy by simply ignoring the influence of gravity. This might seem to go against the phi-losophy I've thus far been advocating, but we're not saying that gravity is irrele-vant in principle—we're simply taking advantage of the fact that, in practice, our early universe was in a configuration where the gravitational forces between particles didn't play a significant role in the dynamics. Basically, it was just a box of hot gas. And a box of hot gas is something whose entropy we know how to calculate.

The entropy of our comoving patch of space when it was young and smooth is:

$$S_{early} \approx 10^{88}.$$

The "$\approx$" sign means "approximately equal to," as we want to emphasize that this is a rough estimate, not a rigorous calculation. This number comes from simply treating the contents of the universe as a conventional gas in thermal equilibrium, and plugging in the formulas worked out by thermodynamicists in the nineteenth century, with one additional feature: Most of the particles in the early universe are photons and neutrinos, moving at or close to the speed of light, so relativity is important. Up to some numerical factors that don't change the answer very much, the entropy of a hot gas of relativistic particles is simply equal to the total number of such particles. There are about 10^{88} particles within our comoving patch of uni-verse, so that's what the entropy was at early times. (It increases a bit along the way, but not by much, so treating the entropy as approximately constant at early times is a good approximation.)

Today, gravity has become important. It is not very accurate to think of the matter in the universe as a gas in thermal equilibrium with negligible gravity; ordinary matter and dark matter have condensed into galaxies and other structures, and the entropy has increased considerably. Unfortunately, we don't have a reliable formula that tracks the change in entropy during the formation of a galaxy.

What we do have is a formula for the circumstance in which gravity is at its most important: in a black hole. As far as we know, very little of the total mass of the universe is contained in the form of black holes.[241] In a galaxy like the Milky Way, there are a number of stellar-sized black holes (with maybe 10 times the mass of the Sun each), but the majority of the total black hole mass is in the form of a single supermassive black hole at the galactic center. While supermassive black holes are certainly big—often over a million times the mass of the Sun—that's nothing compared to an entire galaxy, where the total mass might be 100 billion times the mass of the Sun.

But while only a tiny fraction of the mass of the universe appears to be in black holes, they contain a huge amount of entropy. A single supermassive black hole, a million times the mass of the Sun, has an entropy according to the Bekenstein-Hawking formula of 10^{90}. That's a hundred times larger than all of the nongravitational entropy in all the matter and radiation in the observable universe.[242]

Even though we don't have a good understanding of the space of states of gravitating matter, it's safe to say that the total entropy of the universe today is mostly in the form of these supermassive black holes. Since there are about 100 billion (10^{11}) galaxies in the universe, it's reasonable to approximate the total entropy by assuming 100 billion such black holes. (They might be missing from some galaxies, but other galaxies will have larger black holes, so this is not a bad approximation.) With an entropy of 10^{90} per million-solar-mass black hole, that gives us a total entropy within our comoving patch today of

$$S_{today} \approx 10^{101}.$$

Mathematician Edward Kasner coined the term *googol* to stand for 10^{100}, a number he used to convey the idea of an unimaginably big quantity. The entropy of the universe today is about ten googols. (The folks from Google used this number as an inspiration for the name of their search engine; now it's impossible to refer to a googol without being misunderstood.)

When we write the current entropy of our comoving patch as 10^{101}, it doesn't seem all that much larger than its value in the early universe, 10^{88}. But that's just the miracle of compact notation. In fact, 10^{101} is ten trillion (10^{13}) times bigger than 10^{88}.

The entropy of the universe has increased by an enormous amount since the days when everything was smooth and featureless.

Still, it's not as big as it could be. What is the maximum value the entropy of the observable universe could have? Again, we don't know enough to say for sure what the right answer is. But we can say that the maximum entropy must be at least a certain number, simply by imagining that all of the matter in the universe were rearranged into one giant black hole. That's an allowed configuration for the physical system corresponding to our comoving patch of universe, so it's certainly possible that the entropy could be that large. Using what we know about the total mass contained in the universe, and plugging once again into the Bekenstein-Hawking entropy formula for black holes, we find that the maximum entropy of the observable universe is at least

$$S_{max} \approx 10^{120}.$$

That's a fantastically big number. A hundred quintillion googols! The maximum entropy the observable universe could have is at least that large.

These numbers drive home the puzzle of entropy that modern cosmology presents to us. If Boltzmann is right, and entropy characterizes the number of possible microstates of a system that are macroscopically indistinguishable, it's clear that the early universe was in an extremely special state. Remember that the entropy is the logarithm of the number of equivalent states, so a state with entropy S is one of 10^S indistinguishable states. So the early universe was in one of

$$10^{10^{88}}$$

different states. But it could have been in any of the

$$10^{10^{120}}$$

possible states that are accessible to the universe. Again, the miracle of typography makes these numbers look superficially similar, but in fact the latter number is enormously, inconceivably larger than the former. If the state of the early universe were simply "chosen randomly" from among all possible states, the chance that it would have looked like it actually did are ridiculously tiny.

The conclusion is perfectly clear: The state of the early universe was *not* chosen randomly among all possible states. Everyone in the world who has thought about the problem agrees with that. What they don't agree on is *why* the early universe was so special—what is the mechanism that put it in that state? And, since we shouldn't be temporal chauvinists about it, why doesn't the same mechanism put the *late* universe in a similar state? That's what we're here to figure out.

MAXIMIZING ENTROPY

We've established that the early universe was in a very unusual state, which we think is something that demands explanation. What about the question we started this chapter with: What should the universe look like? What is the maximum-entropy state into which we can arrange our comoving patch?

Roger Penrose thinks the answer is a black hole.

> What about the maximum-entropy state? Whereas with a gas, the maximum entropy of thermal equilibrium has the gas uniformly spread throughout the region in question, with large *gravitating* bodies, maximum entropy is achieved when all the mass is concentrated in one place—in the form of an entity known as a *black hole*.[243]

You can see why this is a tempting answer. As we've seen, in the presence of gravity, entropy increases when things cluster together, rather than smoothing out. A black hole is certainly as densely packed as things can possibly get. As we discussed in the last chapter, a black hole represents the most entropy we can squeeze into a region of spacetime with any fixed size; that was the inspiration behind the holographic principle. And the resulting entropy is undoubtedly a big number, as we've seen in the case of a supermassive black hole.

But in the final analysis, that's not the best way to think about it.[244] A black hole doesn't maximize the total entropy a system can have—it only maximizes the entropy that can be packed into a region of fixed size. Just as the Second Law doesn't say "entropy tends to increase, not including gravity," it also doesn't say "entropy per volume tends to increase." It just says "entropy tends to increase," and if that requires a big region of space, then so be it. One of the wonders of general relativity—and a crucial distinction with the absolute spacetime of Newtonian mechanics—is that sizes are never fixed. Even without a complete understanding of entropy, we can get a handle on the answer by following in Penrose's footsteps, and simply examining the natural evolution of systems toward higher-entropy states.

Consider a simple example: a collection of matter gathered in one region of an otherwise empty universe, without even any vacuum energy. In other words, a spacetime that is empty almost everywhere, except for some particular place where some matter particles are congregated. Because most of space has no energy in it at all, the universe won't be expanding or contracting, so nothing really happens outside the region where the matter is located. The particles will contract together under their own gravitational force.

Let's imagine that they collapse all the way to a black hole. Along the way, there's no question that the entropy increases. However, the black hole doesn't just sit there for all of eternity—it gives off Hawking radiation, gradually shrinking as it loses energy, and eventually evaporating away completely.

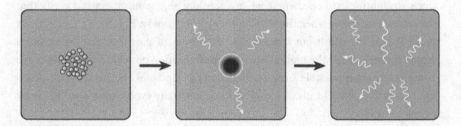

increasing entropy via black-hole formation and evaporation

Figure 69: A black hole has a lot of entropy, but it evaporates into radiation that has even more entropy.

The natural behavior of black holes in an otherwise empty universe is to radiate away into a dilute gas of particles. Because such behavior is natural, we expect it to represent an increase in entropy—and it does. We can explicitly compare the entropy of the black hole to the entropy of the radiation into which it evaporates— and the entropy of the radiation is higher. By about 33 percent, to be specific.[245]

Now, the *density* of entropy has clearly gone down—when we had a black hole, all that entropy was packed into a small volume, whereas the Hawking radiation is emitted gradually and gets spread out over a huge region of space. But again, the density of entropy isn't what we care about; it's just the total amount.

EMPTY SPACE

The lesson of this thought experiment is that the rule of thumb "when we take gravity into consideration, higher-entropy states are lumpy rather than smooth" is not an absolute law; it's merely valid under certain circumstances. The black hole is lumpier (higher contrast) than the initial collection of particles, but the eventual dispersing radiation isn't lumpy at all. In fact, as the radiation scurries off to the ends of the universe, we approach a configuration that grows ever smoother, as the density everywhere approaches zero.

So the answer to the question "What does a high-entropy state look like, when we take gravity into account?" isn't "a lumpy, chaotic maelstrom of black holes," nor is it even "one single giant black hole." The highest-entropy states look like *empty space*, with at most a few particles here and there, gradually diluting away.

That's a counterintuitive claim, worth examining from different angles.[246] The case of a collection of matter that all falls together to form a black hole is a relatively straightforward one, where we can actually plug in numbers and verify that the entropy increases when the black hole evaporates away. But that's a far cry from proving that the result (an increasingly dilute gas of particles moving through empty space) is really the highest-entropy configuration possible. We should try to contemplate other possible answers. The guiding principles are that we want a configuration that other kinds of configurations naturally evolve into, and that itself persists forever.

What if, for example, we had an array of many black holes? We might imagine that black holes filled the universe, so that the radiation from one black hole eventually fell into another one, which kept them from evaporating away. However, general relativity tells us that such a configuration can't last. By sprinkling objects throughout the universe, we've created a situation where space has to either expand or contract. If it expands, the distance between the black holes will continually grow, and eventually they will simply evaporate away. As before, the long-term future of such a universe simply looks like empty space.

If space is contracting, we have a different story. When the entire universe is shrinking, the future is likely to end in a Big Crunch singularity. That's a unique case; on the one hand, the singularity doesn't really last forever (since time ends there, at least as far as we know), but it doesn't evolve into something else, either. We can't rule out the possibility that the future evolution of some hypothetical universe ends in a Big Crunch, but our lack of understanding of singularities in quantum gravity makes it difficult to say very much useful about that case. (And our real world doesn't seem to be behaving that way.)

One clue is provided by considering a collapsing collection of matter (black holes or otherwise) that looks exactly like a contracting universe, but one where the matter only fills a finite region of space, rather than extending throughout it. If the rest of the universe is empty, this local region is exactly like the situation we considered before, where a group of particles collapsed to make a black hole. So what looks from the inside like a universe collapsing to a Big Crunch looks from the outside like the formation of a giant black hole. In that case, we know what the far future will bring: It might take a while, but that black hole will radiate away into nothing. The ultimate state is, once again, empty space.

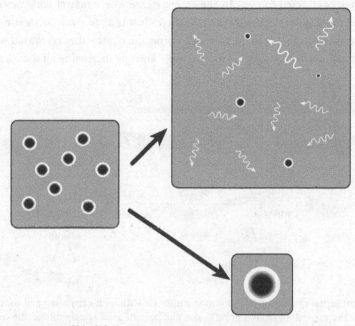

Figure 70: An array of black holes cannot remain static. It will either expand and allow the black holes to evaporate away, approaching empty space (top right), or collapse to make a Big Crunch or a single larger black hole (bottom right).

We can be a little more systematic about this. Cosmologists are used to thinking of universes that are doing the same thing all throughout space, because the observable part of our own universe seems to be like that. But let's not take that for granted; let's ask what could be going on throughout the universe, in perfect generality.

The notion that the space is "expanding" or "contracting" doesn't have to be an absolute property of the entire universe. If the matter in some particular region of space is moving apart and diluting away, it will look locally like an expanding universe, and likewise for contraction when matter moves together. So if we imagine sprinkling particles throughout an infinitely big space, most of the time we will find that some regions are expanding and diluting, while other regions are contracting and growing more dense.

But if that's true, a remarkable thing happens: Despite the apparent symmetry between "expanding" and "contracting," pretty soon the expanding regions begin to win. And the reason is simple: The expanding regions are growing in volume, while the contracting ones are shrinking. Furthermore, the contracting regions don't stay contracted forever. In the extreme case where matter collapses all the way to a black hole, eventually the black holes just radiate away. So starting from initial conditions that contain both expanding and contracting regions, if we wait long enough we'll end up with empty space—entropy increasing all the while.[247]

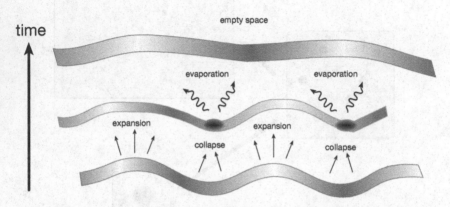

Figure 71: Initial conditions (at bottom) in a universe with both expanding and contracting regions. The expanding regions grow in size and become increasingly dilute. The contracting regions grow denser at first, but at some point will begin to evaporate into the surrounding emptiness.

In each of these examples, the crucial underlying feature is the dynamical nature of spacetime in general relativity. In a fixed, absolute spacetime (such as Boltzmann would have presumed), it makes sense to imagine a universe filled with gas at a uniform temperature and density—thermal equilibrium everywhere. That's a high-entropy state, and a natural guess for what the universe "should" look like. It's no surprise that Boltzmann suggested that our observable universe could be a statistical fluctuation within such a configuration.

But general relativity changes everything. A gas of uniform density in a static spacetime is not a solution to Einstein's equation—the universe would have to be either expanding or contracting. Before Einstein came along it made sense to start your thought experiments by fixing the average density of matter, or the total volume of the region under consideration. But in general relativity, these aren't things you're allowed to simply keep fixed; they will tend to change with time. One way of thinking about it is to realize that general relativity always gives you a way to increase the entropy of any particular configuration: Make the universe bigger, and let the stuff expand to fill the new volume. The ultimate end of such a process is, of course, empty space. That's what counts as a "high-entropy" state once we take gravity into account.

None of these arguments is airtight, of course. They are suggestive of a result that seems to hang together and make sense, once you think it through, but that's far short of a definitive demonstration of anything. The claim that the entropy of some system within the universe can increase by scattering its elements across a vast expanse of space seems pretty safe. But the conclusion that empty space is therefore the highest-entropy state is more tentative. Gravity is tricky, and there's a lot we don't understand about it, so it's not a good idea to become too emotionally invested in any particular speculative scenario.

THE REAL WORLD

Let's apply these ideas to the real world. If high-entropy states are those that look like empty space, presumably our actual observable universe should be evolving toward such a state. (It is.)

We have casually been assuming that when things collapse under the force of gravity, they end up as a black hole before ultimately evaporating away. It's far from obvious that this holds true in the real world, where we see lots of objects held together by gravity, but that are very far from being black holes—planets, stars, and even galaxies.

But the reality is, all of these things will eventually "evaporate" if we wait long enough. We can see this most clearly in the case of a galaxy, which can be thought of as a collection of stars orbiting under their mutual gravitational pull. As individual stars pass by other stars, they interact much like molecules in a box of gas, except that the interaction is solely through gravity (apart from those very rare instances when the stars smack right into one another). These interactions can exchange energy from one star to the other.[248] Over the course of many such encounters, stars will occasionally pick up so much energy that they reach escape velocity and fly away from the galaxy altogether. The rest of the galaxy has now lost some of its energy, and as a consequence it shrinks, so that its stars are clustered more tightly together. Eventually, the remaining stars are going to be packed so closely that they all fall into a black hole at the center. From that point, we return to our previous story.

Similar logic works for any other object in the universe, even if the details might differ. The basic point is that, given some rock or star or planet or what have you, that particular physical system *wants* to be in the highest-entropy arrangement of the constituents from which it is made. That's a little poetic, as inanimate objects don't really have desires, but it reflects the reality that an unfettered evolution of the system would naturally bring it into a higher-entropy configuration.

You might think that the evolution is, in fact, fettered: A planet, for example, might have a higher entropy if its entire mass collapsed into a black hole, but the pressure inside keeps it stable. Here's where the miracle of quantum mechanics comes in. Remember that a planet isn't really a collection of classical particles; it's described by a wave function, just like everything else. That wave function characterizes the probability that we will find the constituents of the planet in any of their possible configurations. One of those possible configurations, inevitably, will be a black hole. In other words, from the point of view of someone observing the planet (or anything else), there is a tiny chance they will find that it has spontaneously collapsed into a black hole. That's the process known as "quantum tunneling."

Do not be alarmed. Yes, it's true, just about everything in the universe—the Earth, the Sun, you, your cat—has a chance of quantum-tunneling into the form of a black hole at any moment. But the chance is very small. It would be many, many times the age of the universe before there were a decent chance of it happening. But in a universe that lasts for all eternity, that means the chances are quite good that it will eventually happen—it's inevitable, in fact. No collection of particles can simply sit undisturbed in the universe forever. The lesson is that matter will find a way to transform into a higher-entropy configuration, if one exists; it might be via tunneling into the form of a black hole, or through much more mundane channels. No matter what kind of lump of matter you have in the universe, it can increase in entropy by evaporating into a thin gruel of particles moving away into empty space.

VACUUM ENERGY

As we discussed back in Chapter Three, there's more than matter and radiation in the universe—there's also dark energy, responsible for making the universe accelerate. We don't know for sure what the dark energy is, but the leading candidate is "vacuum energy," also known as the cosmological constant. Vacuum energy is simply a constant amount of energy inherent in every cubic centimeter of space, one that remains fixed throughout space and time.

The existence of dark energy both simplifies and complicates our ideas about high-entropy states in the presence of gravity. I've been suggesting that the natural behavior of matter is to disperse away into empty space, which is therefore the best candidate for a maximum-entropy state. In a universe like ours, with a vacuum energy that is small but greater than zero, this conclusion becomes even more robust. A positive vacuum energy imparts a perpetual impulse to the expansion of the universe, which helps the general tendency of matter and radiation to dilute away. If, within the next few years, human beings perfect an immortality machine and/or drug, cosmologists who live forever will have to content themselves with observing an increasingly empty universe. Stars will die out, black holes will evaporate, and everything will be pushed away by the accelerating effects of vacuum energy.

In particular, if the dark energy is really a cosmological constant (rather than something that will ultimately fade away), we can be sure that the universe will never re-collapse into a Big Crunch of any sort. After all, the universe is not only expanding but also accelerating, and that acceleration will continue forever. This scenario—which, let's not forget, is the most popular prognosis for the real world according to contemporary cosmologists—vividly highlights the bizarre nature of our low-entropy beginnings. We're contemplating a universe that has existed for a finite time in the past but will persist forever into the future. The first few tens of billions of years of its existence are a hot, busy, complex and interesting mess, which will be followed by an infinite stretch of cold, empty quietness. (Apart from the occasional statistical fluctuation; see next section.) Although it's not much more than a gut feeling, it just seems like a waste to face the prospect of an endless duration of dark loneliness after a relatively exciting few years in our universe's past.

The existence of a positive cosmological constant allows us to actually prove a somewhat rigorous result, rather than just spinning through a collection of thought experiments. The *cosmic no-hair theorem* states that, under the familiar set of "reasonable assumptions," a universe with a positive vacuum energy plus some matter fields will, if it lasts long enough for the vacuum energy to take over, eventually evolve into empty universe with nothing but vacuum energy. The cosmological constant always wins, in other words.[249]

The resulting universe—empty space with a positive vacuum energy—is known as *de Sitter space*, after Dutch physicist Willem de Sitter, one of the first after Einstein to study cosmology within the framework of general relativity. As we mentioned back in Chapter Three, empty space with zero vacuum energy is known as Minkowski space, while empty space with a negative vacuum energy is anti-de Sitter space. Even though spacetime is empty in de Sitter space, it's still curved, because of the positive vacuum energy. The vacuum energy, as we know, imparts a perpetual impulse to the expansion of space. If we consider two particles initially at rest in de Sitter space, they will gradually be pulled apart by the expansion. Likewise, if we trace their motion into the past, they would have been coming toward each other, but ever more slowly as the space between them was pushed apart. Anti–de Sitter space is the reverse; particles are pulled toward each other.

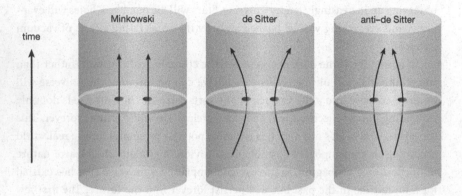

Figure 72: Three different versions of "empty space," with different values of the vacuum energy: Minkowski space when the vacuum energy vanishes, de Sitter when it is positive, and anti–de Sitter when it is negative. In Minkowski space, two particles initially at rest will stay motionless with respect to each other; in de Sitter space they are pushed apart, while in anti–de Sitter space they are pulled together. The larger the magnitude of the vacuum energy, the stronger the pushing or pulling.

Everything we've been arguing points to the idea that de Sitter space is the ultimate endpoint of cosmological evolution when the vacuum energy is positive, and hence the highest-entropy state we can think of in the presence of gravity. That's not a definitive statement—the state of the art isn't sufficiently advanced to allow for definitive statements along these lines—but it's suggestive.

You might wonder how empty space can have a large entropy—entropy is supposed to count the number of ways we can rearrange microstates, and what is there to rearrange if all we have is empty space? But this is just the same puzzle that faced us with black holes. The answer must be that there are a large number of microstates describing the quantum states of space itself, even when it's empty. Indeed, if we believe in the holographic principle, we can assign a definite value to the entropy contained within any observable patch of de Sitter space. The answer is a huge number, and the entropy is *larger* when the vacuum energy is *smaller*.[250] Our own universe is evolving toward de Sitter space, and the entropy of each observable patch will be about 10^{120}. (The fact that this is the same as the entropy we would get by collapsing all the matter in the observable universe into a black hole is a coincidence—it's the same coincidence as the fact that the matter density and vacuum energy are approximately equal at the present time, even though the matter dominated in the past and the vacuum energy will dominate in the future.)

While de Sitter space provides a sensible candidate for a high-entropy state, the idea of vacuum energy complicates our attempts to understand entropy in the context of quantum gravity. The basic problem is that the effective vacuum energy—what you would actually measure as the energy of the vacuum at any particular event in spacetime—can certainly change, at least temporarily. Cosmologists talk about the "true vacuum," in which the vacuum energy takes on its lowest possible value, but also various possible "false vacua," in which the effective vacuum energy is higher. Indeed, it's possible that we might be in a false vacuum right now. The idea that "high entropy" means "empty space" becomes a lot more complicated when empty space can take on different forms, corresponding to different values of the vacuum energy.

That's a *good* thing—we don't want empty space to be the highest-entropy state possible, because we don't live there. In the next couple of chapters we're going to see whether we can't take advantage of different possible values of vacuum energy to somehow make sense of the universe. But first we need to assure ourselves that, without some strategy along those lines, it really would be extremely surprising that we don't live in a universe that is otherwise empty. And that calls for another visit with some of the giants on whose shoulders we are standing, Boltzmann and Lucretius.

WHY DON'T WE LIVE IN EMPTY SPACE?

We began this chapter by asking what the universe should look like. It's not obvious that this is even a sensible question to ask, but if it is, a logical answer would be "it should look like it's in a high-entropy state," because there are a lot more high-entropy states than low-entropy ones. Then we argued that truly high-entropy states basically look like empty space; in a world with a positive cosmological constant, that means de Sitter space, a universe with vacuum energy and nothing else.

So the major question facing modern cosmology is: "Why don't we live in de Sitter space?" Why do we live in a universe that is alive with stars and galaxies? Why do we live in the aftermath of our Big Bang, an enormous conflagration of matter and energy with an extraordinarily low entropy? Why is there so much *stuff* in the universe, and why was it packed so smoothly at early times?

One possible answer would be to appeal to the anthropic principle. We can't live in empty space because, well, it's empty. There's nothing there to do the living. That sounds like a perfectly reasonable assumption, although it doesn't exactly answer the question. Even if we couldn't live precisely in empty de Sitter space, that doesn't explain why our early universe is nowhere near to being empty. Our actual universe seems to be an enormously more dramatic departure from emptiness than would be required by any anthropic criterion.

You might find these thoughts reminiscent of our discussion of the Boltzmann-Lucretius scenario from Chapter Ten. There, we imagined a static universe with an infinite number of atoms, so that there was an average density of atoms throughout space. Statistical fluctuations in the arrangements of those atoms, it was supposed, could lead to temporary low-entropy configurations that might resemble our universe. But there was a problem: That scenario makes a strong prediction, namely, that we (under any possible definition of *we*) should be the smallest possible fluctuation away from thermal equilibrium consistent with our existence. In the most extreme version, we should be disembodied Boltzmann brains, surrounded by a gas with uniform temperature and density. But we're not, and further experiments continue to reveal more evidence that the rest of the universe is not anywhere near equilibrium, so this scenario seems to be ruled out experimentally.

The straightforward scenario Boltzmann had in mind would doubtless be dramatically altered by general relativity. The most important new ingredient is that it's impossible to have a static universe filled with gas molecules. According to Einstein, space filled with matter isn't going to just sit there; it's going to either expand or contract. And if the matter is sprinkled uniformly throughout the universe, and is made of normal particles (which don't have negative energy or pressure), there will inevitably be a singularity in the direction of time where things are getting more dense—a past Big Bang if the universe is expanding, or a future Big Crunch if the universe is contracting. (Or both, if the universe expands for a while and then starts to re-contract.) So this carefree Newtonian picture where molecules persist forever in a happy static equilibrium is not going to be relevant once general relativity comes into the game.

Instead, we should contemplate life in de Sitter space, which replaces a gas of thermal particles as the highest-entropy state. If all you knew about was classical physics, de Sitter space would be truly empty. (Vacuum energy is a feature of space-time itself; there are no particles associated with it.) But classical physics isn't the whole story; the real world is quantum mechanical. And quantum field theory says that particles can be created "out of nothing" in an appropriate curved space-time background. Hawking radiation is the most obvious example.

It turns out, following very similar reasoning to that used by Hawking to investigate black holes, that purportedly empty de Sitter space is actually alive with particles popping into existence. Not a lot of them, we should emphasize—we're talking about an extremely subtle effect. (There are a lot of virtual particles in empty space, but not many real, detectable ones.) Let's imagine that you were sitting in de Sitter space with an exquisitely sensitive experimental apparatus, capable of detecting any particles that happened to be passing your way. What you would discover is that you were actually surrounded by a gas of particles at a constant temperature, just as if you were in a box in thermal equilibrium. And the temperature wouldn't go away as the universe continued to expand—it is a feature of de Sitter space that persists for all eternity.[251]

Admittedly, you wouldn't detect very many particles; the temperature is quite low. If someone asked you what the "temperature of the universe" is right now, you might say 2.7 Kelvin, the temperature of the cosmic microwave background radiation. That's pretty cold; 0 Kelvin is the lowest possible temperature, room temperature is about 300 Kelvin, and the lowest temperature ever achieved in a laboratory on Earth is about 10^{-10} Kelvin. If we allow the universe to expand until all of the matter and cosmic background radiation has diluted away, leaving only those particles that are produced out of de Sitter space by quantum effects, the temperature will be about 10^{-29} Kelvin. Cold by anyone's standards.

Still, a temperature is a temperature, and any temperature above zero allows for fluctuations. When we take quantum effects in de Sitter space into account, the universe acts like a box of gas at a fixed temperature, and that situation will last forever. Even if we have a past that features a dramatic Big Bang, the future is an eternity of ultra-cold temperature that never drops to zero. Hence, we should expect an endless future of thermal fluctuations—including Boltzmann brains and any other sort of thermodynamically unlikely configuration we might have worried about in an eternal box of gas.

And that would seem to imply that *all of the troublesome aspects of the Boltzmann-Lucretius scenario are troublesome aspects of the real world*. If we wait long enough, our universe will empty out until it looks like de Sitter space with a tiny temperature, and stay that way forever. There will be random fluctuations in the thermal radiation that lead to all sorts of unlikely events—including the spontaneous generation of galaxies, planets, and Boltzmann brains. The chance that any one such thing happens at any particular time is small, but we have an eternity to wait, so every allowed thing will happen. In that universe—*our* universe, as far as we can tell—the overwhelming majority of mathematical physicists (or any other kind of conscious observer) will pop out of the surrounding chaos and find themselves drifting alone through space.[252]

The acceleration of the universe was discovered in 1998. Theorists chewed over this surprising result for a while before the problem with Boltzmann brains became clear. It was first broached in a 2002 paper by Lisa Dyson, Matthew Kleban, and Leonard Susskind, with the ominous title "Disturbing Implications of a Cosmological Constant," and amplified in a follow-up paper by Andreas Albrecht and Lorenzo Sorbo in 2004.[253] The resolution to the puzzle is still far from clear. The simplest way out is to imagine that the dark energy is not a cosmological constant that lasts forever, but an ephemeral source of energy that will fade away long before we hit the Poincaré recurrence time. But it's not clear exactly how this would work, and compelling models of decaying dark energy turn out to be difficult to construct.

So the Boltzmann-brain problem—"Why do we find ourselves in a universe evolving gradually from a state of incredibly low entropy, rather than being isolated creatures that recently fluctuated from the surrounding chaos?"—does not yet have a clear solution. And it's worth emphasizing that this puzzle makes the arrow-of-time problem enormously more pressing. Before this issue was appreciated, we had something of a fine-tuning problem: Why did the early universe have such a low entropy? But we were at least allowed to shrug our shoulders and say, "Well, maybe it just did, and there is no deeper explanation." But now that's no longer good enough. In de Sitter space, we can reliably predict the number of times in the history of the universe (including the infinite future) that observers will appear surrounded by cold and forbidding emptiness, and compare them to the observers who will find themselves in comfortable surroundings full of stars and galaxies, and the cold and forbidding emptiness is overwhelmingly likely. This is more than just an uncomfortable fine-tuning; it's a direct disagreement between theory and observation, and a sign that we have to do better.

14

INFLATION AND THE MULTIVERSE

Those who think of metaphysics as the most unconstrained or speculative of disciplines are misinformed; compared with cosmology, metaphysics is pedestrian and unimaginative.

—Stephen Toulmin[254]

On a cool Palo Alto morning in December 1979, Alan Guth pedaled his bike as fast as he could to his office in the theoretical physics group at SLAC, the Stanford Linear Accelerator Center. Upon reaching his desk, he opened his notebook to a new page and wrote:

> SPECTACULAR REALIZATION: this kind of supercooling can explain why the universe today is so incredibly flat—and therefore resolve the fine-tuning paradox pointed out by Bob Dicke in his Einstein Day lectures.

He carefully drew a rectangular box around the words. Then he drew another one.[255]

As a scientist, you live for the day when you hit upon a result—a theoretical insight, or an experimental discovery—so marvelous that it deserves a box around it. The rare double-box-worthy results tend to change one's life, as well as the course of science; as Guth notes, he doesn't have any other double-boxed results anywhere in his notebooks. The one from his days at SLAC is now on display at the Adler Planetarium in Chicago, open to the page with the words above.

Guth had hit on the scenario now known as *inflation*—the idea that the

Figure 73: Alan Guth, whose inflationary universe scenario may help explain why our observed universe is close to smooth and flat.

early universe was suffused with a temporary form of dark energy at an ultra-high density, which caused space to accelerate at an incredible rate (the "supercooling" mentioned above). That simple suggestion can explain more or less everything there is to explain about the conditions we observe in our early universe, from the geometry of space to the pattern of density perturbations observed in the cosmic microwave background. Although we do not yet have definitive proof that inflation occurred, it has been arguably the most influential idea in cosmology over the last several decades.

Which doesn't mean inflation is right, of course. If the early universe was temporarily dominated by dark energy at an ultra-high scale, then we can understand why the universe would evolve into just the state it was apparently in at early times. But there is a danger of begging the important question—why was it ever dominated by dark energy in that way? Inflation doesn't provide any sort of answer by itself to the riddle of why entropy was low in the early universe, other than to assume that it started even lower (which is arguably a bit of a cheat).

Nevertheless, inflation is an extraordinarily compelling idea, which really does seem to match well with the observed features of our early universe. And it leads to some surprising consequences that Guth himself never foresaw when he first suggested the scenario—including, as we'll see, a way to make the idea of a "multiverse" become realistic. It seems likely, in the judgment of most working cosmologists, that some version of inflation is correct—the question is, why did inflation ever happen?

THE CURVATURE OF SPACE

Imagine you take a pencil and balance it vertically on its tip. Obviously its natural tendency will be to fall over. But you could imagine that if you had an extremely stable surface, and you were a real expert at balancing, you could arrange things so that the pencil remained vertical for a very long time. Like, more than 14 billion years.

The universe is somewhat like that, where the pencil represents the *curvature of space*. This can be a more confusing concept than it really should be, because cosmologists sometimes speak about the "curvature of spacetime," and other times about the "curvature of space," and those things are different; you're supposed to figure out from context which one is meant. Just as spacetime can have curvature, space all by itself can as well—and the question of whether space is curved is completely independent of whether spacetime is curved.[256]

One potential problem in discussing the curvature of space by itself is that general relativity gives us the freedom to slice spacetime into three-dimensional copies of space evolving through time in a multitude of ways; the definition of "space" is not unique. Fortunately, in our observed universe there is a natural way to do the slicing: We define "time" such that the density of matter is approximately constant through space on large scales, but diminishing as the universe expands. The distribution of matter, in other words, defines a natural rest frame for the universe. This doesn't violate the precepts of relativity in any way, because it's a feature of a particular configuration of matter, not of the underlying laws of physics.

In general, space could curve in arbitrary ways from place to place, and the discipline of differential geometry was developed to handle the mathematics of curvature. But in cosmology we're lucky in that space is uniform over large distances, and looks the same in every direction. In that case, all you have to do is specify a single number—the "curvature of space"—to tell me everything there is to know about the geometry of three-dimensional space.

The curvature of space can be a positive number, or a negative number, or zero. If the curvature is zero, we naturally say that space is "flat," and it has all the characteristics of geometry as we usually understand it. These characteristics were first set out by Euclid, and include properties like "initially parallel lines stay parallel forever," and "the angles inside a triangle add up to precisely 180 degrees." If the curvature is positive, space is like the surface of a sphere—except that it's three-dimensional. Initially parallel lines do eventually intersect, and angles inside a triangle add up to greater than 180 degrees. If the curvature is negative, space is like the surface of a saddle, or of a potato chip. Initially parallel lines grow apart, and angles inside a triangle—well, you can probably guess.[257]

According to the rules of general relativity, if the universe starts flat, it stays flat. If it starts curved, the curvature gradually diminishes away as the universe expands. However, as we know, the density of matter and radiation also dilutes away. (For right now, forget you've ever heard about dark energy, which changes everything.) When you plug in the equations, the density of matter or radiation decreases faster than the amount of curvature. Relative to matter and radiation, curvature becomes more relevant to the evolution of the universe as space expands.

Therefore: If there is any noticeable amount of curvature whatsoever in the early universe, the universe today should be very obviously curved. A flat universe is like a pencil balanced exactly on its tip; if there were any deviation to the left or right, the pencil would tend to fall pretty quickly onto its side. Similarly, any tiny deviation from perfect flatness at early times should have become progressively more noticeable as time went on. But as a matter of observational fact, the universe looks very flat. As far as anyone can tell, there is no measurable curvature in the universe today at all.[258]

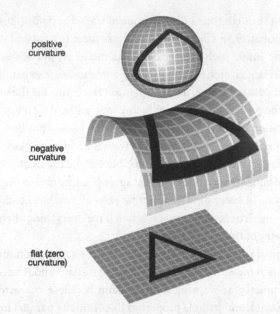

positive
curvature

negative
curvature

flat (zero
curvature)

Figure 74: Ways that space can have a uniform curvature. From top to bottom: positive curvature, as on a sphere; negative curvature, as on a saddle; zero curvature, as on a flat plane.

This state of affairs is known as the *flatness problem*. Because the universe is so flat today, it had to be incredibly flat in the past. But why?

The flatness problem bears a family resemblance to the entropy problem we discussed in the last chapter. In both cases, it's not that there is some blatant disagreement between theory and observation—all we have to do is posit that the early universe had some particular form, and everything follows nicely from there. The problem is that the "particular form" seems to be incredibly unnatural and finely tuned, for no obvious reason. We could say that both the entropy and the spatial curvature of the early universe just were small, and there's no explanation beyond that. But these apparently unnatural features of the universe might be a clue to something important, so it behooves us to take them seriously.

MAGNETIC MONOPOLES

Alan Guth wasn't trying to solve the flatness problem when he hit upon the idea of inflation. He was interested in a very different puzzle, known as the *monopole problem*.

Guth, for that matter, wasn't especially interested in cosmology. In 1979, he was in his ninth year of being a postdoctoral researcher—the phase of a scientist's career in between graduate school and becoming a faculty member, when they concentrate on research without having to worry about teaching duties or other academic responsibilities. (And without the benefit of any job security whatsoever; most postdocs never succeed in getting a faculty job, and eventually leave the field.) Nine years is past the time when a talented postdoc would normally have moved on to become an assistant professor somewhere, but Guth's publication record at this point in his career didn't really reflect the ability that others saw in him. He had labored for a while on a theory of quarks that had fallen out of favor, and was now trying to understand an obscure prediction of the newly popular "Grand Unified Theories": the prediction of magnetic monopoles.

Grand Unified Theories, or GUTs for short, attempt to provide a unified account of all the forces of nature other than gravity. They became very popular in the 1970s, both for their inherent simplicity, and because they made an intriguing prediction: that the proton, the stalwart elementary particle that (along with the electron and the neutron) forms the basis for all the matter around us, would ultimately decay into lighter particles. Giant laboratories were built to search for proton decay, but it hasn't yet been discovered. That doesn't mean that GUTs aren't right; they are still quite popular, but the failure to detect proton decay has left physicists at a loss over how these theories should be tested.

GUTs also predicted the existence of a new kind of particle, the magnetic monopole. Ordinary charged particles are electric monopoles—that is, they have either a positive charge or a negative charge, and that's all there is to it. No one has ever discovered an isolated "magnetic charge" in Nature. Magnets as we know them are always dipoles—they come with a north pole and a south pole. Cut a magnet in half between the poles, and two new poles pop into existence where you made the cut. As far as experimenters can tell, looking for an isolated magnetic pole—a monopole—is a lot like looking for a piece of string with only one end.

But according to GUTs, monopoles should be able to exist. In fact, in the late 1970s people realized that you could sit down and calculate the number of monopoles that should be created in the aftermath of the Big Bang. And the answer is: way too many. The total amount of mass in monopoles, according to these calculations, should be much higher than the total mass in ordinary protons, neutrons, and electrons. Magnetic monopoles should be passing through your body all the time.

There is an easy way out of this, of course: GUTs might not be right. And that still might be the correct solution. But Guth, while thinking about the problem, hit on a more interesting one: inflation.

INFLATION

Dark energy—a source of energy density that is approximately (or exactly) constant throughout space and time, not diluting away as the universe expands—makes the universe accelerate, by imparting a perpetual impulse to the expansion. We believe that most of the energy in the universe, between 70 percent and 75 percent of the total, is currently in the form of dark energy. But in the past, when matter and radiation were denser, dark energy presumably had about the same density it has today, so it would have been relatively unimportant.

Now imagine that, at some other time in the very early universe, there was dark energy with an extraordinarily larger energy density—call it "dark super-energy."[259] It dominated the universe and caused space to accelerate at a terrific rate. Then—for reasons to be specified later—this dark super-energy suddenly decayed into matter and radiation, which formed the hot plasma making up the early universe we usually think about. The decay was almost complete, but not quite, leaving behind the relatively minuscule amount of dark energy that has just recently become important to the dynamics of the universe.

That's the scenario of inflation. Basically, inflation takes a tiny region of space and blows it up to an enormous size. You might wonder what the big deal is—who cares about a temporary phase of dark super-energy, if it just decays into matter and radiation? The reason why inflation is so popular is because it's like confession—it wipes away prior sins.

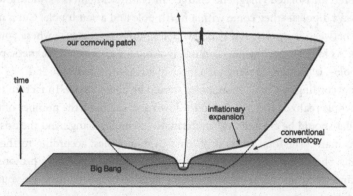

Figure 75: Inflation takes a tiny patch of space and expands it rapidly to a tremendous size. This figure is not at all to scale; inflation occurs in a tiny fraction of a second, and expands space by more than a factor of 10^{26}.

Consider the monopole problem. Monopoles are (if GUTs are correct) produced in copious amounts in the extremely early universe. So imagine that inflation happens pretty early, but later than the production of monopoles. In that case, as long as inflation lasts long enough, space expands by such a tremendous amount that all the monopoles are diluted away practically to nothing. As long as the decay of the dark super-energy into matter and radiation doesn't make any more monopoles (which it won't, if it's not too energetic), voilà—no more monopole problem.

Likewise with spatial curvature. The problem there was that curvature dilutes away more gradually than matter or radiation, so if there were any curvature at all early on it should be extremely noticeable today. But dark energy dilutes away even more gradually than curvature—indeed, it hardly dilutes away at all. So again, if inflation goes on long enough, curvature can get diluted to practically nothing, before matter and radiation are re-created in the decay of the dark super-energy. No more flatness problem.

You can see why Guth was excited about the idea of inflation. He had been thinking about the monopole problem, but from the other side—not trying to solve it, but using it as an argument against GUTs. In his original work on the problem, with Cornell physicist Henry Tye, they had ignored the possible role of dark energy and established that the monopole problem was very hard to solve. But once Guth sat down to study the effects that an early period of dark energy could have, a solution to the monopole problem dropped right into his lap—that's worth at least a single box, right there.

The double-box-worthiness came when Guth understood that his idea would also solve the flatness problem, which he hadn't even been thinking about. Completely coincidentally, Guth had gone to a lecture some time earlier by Princeton physicist Robert Dicke, one of the first people to study the cosmic microwave background. Dicke's lecture, held at a Cornell event called "Einstein Day," pointed out several loose ends in the conventional cosmological model. One of them was the flatness problem, which stuck with Guth, even though his research at the time wasn't especially oriented toward cosmology.

So when he realized that inflation solved not only the monopole problem but also the flatness problem, Guth knew he was onto something big. And indeed he was; almost overnight, he went from being a struggling postdoc to being a hot property on the faculty job market. He chose to return to MIT, where he had been a graduate student, and he's still teaching there today.

THE HORIZON PROBLEM

In working out the consequences of inflation, Guth realized that the scenario offered a solution to yet another cosmological fine-tuning puzzle: the *horizon problem*. Indeed, the horizon problem is arguably the most insistent and perplexing issue in standard Big Bang cosmology.

The problem arises from the simple fact that the early universe looks more or less the same at widely separated points. In the last chapter, we noted that a "typical" state of the early universe, even if we insisted that it be highly dense and rapidly expanding, would tend to be wildly fluctuating and inhomogeneous—it should resemble the time-reverse of a collapsing universe. So the fact that the universe was so smooth is a feature that seems to warrant an explanation. Indeed, it's fair to say that the horizon problem is really a reflection of the entropy problem as we've presented it, although it's usually justified in a different way.

We think of horizons in the context of black holes—the horizon is the place past which, once we get there, we can never return to the outside world. Or, more precisely, we would have to be able to travel faster than light to escape. But in the standard Big Bang model, there's a completely separate notion of "horizon," stemming from the fact that the Big Bang happened a finite time ago. This is a "cosmological horizon," as opposed to the "event horizon" around a black hole. If we draw a light cone from our present location in spacetime into the past, it will intersect the beginning of the universe. And if we now consider the world line of a particle that emerges from the Big Bang outside our light cone, no signal from that world line can ever reach our current event (without going faster than light). We therefore say that such a particle is outside our cosmological horizon, as shown in Figure 76.

That's all well and good, but things start to get interesting when we realize that, unlike an event horizon of a static black hole, our cosmological horizon grows with time as we age along our world line. As we get older, our past light cones encompass more and more of spacetime, and other particle world lines that used to be outside now enter our horizon. (The world lines haven't moved—our horizon has expanded to include them.)

Therefore, events that are far in the past have cosmological horizons that are correspondingly smaller; they are closer (in time) to the Big Bang, so fewer events lie in their past. Consider different points that we observe when we look at the cosmic microwave background on opposite sides of the sky, as shown in Figure 77. The microwave background shows us an image of the moment when the universe became transparent, when the temperature cooled off sufficiently that electrons

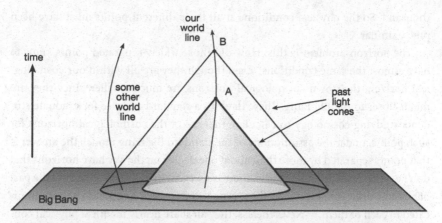

Figure 76: The cosmological horizon is defined by the place where our past light cone meets the Big Bang. As we move forward in time, our horizon grows. A world line that was outside our horizon at moment A comes inside the horizon by the time we get to B.

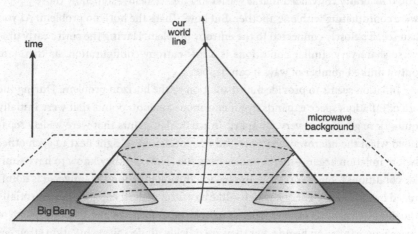

Figure 77: The horizon problem. We look at widely separated points on the cosmic microwave background and see that they are at nearly the same temperature. But those points are far outside each other's horizons; no signal could have ever passed between them. How do they know to be at the same temperature?

and protons got together to form atoms—about 380,000 years after the Big Bang. Depending on the local conditions at these points—the density, expansion rate, and so on—they could appear very different to us here today. But they don't. From our perspective, all the points on the microwave background sky have very similar temperatures; they differ from place to place by only about one part in a hundred thousand. So the physical conditions at all these different points must have been pretty similar.

The horizon problem is this: How did those widely separated points *know* to have almost the same conditions? Even though they are all within our cosmological horizon, their own cosmological horizons are much smaller, since they are much closer to the Big Bang. These days it's a standard exercise for graduate students studying cosmology to calculate the size of the cosmological horizons for such points, under the assumptions of the standard Big Bang model; the answer is that points separated by more than about one degree on the sky have horizons that don't overlap at all. In other words, there is no event in spacetime that is in the past of all these different points, and there is no way that any signal could be communicated to each of them.[260] Nevertheless, they all share nearly identical physical conditions. How did they know?

It's as if you asked several thousand different people to pick a random number between 1 and a million, and they all picked numbers between 836,820 and 836,830. You'd be pretty convinced that it wasn't just an accident—somehow those people were coordinating with one another. But how? That's the horizon problem. As you can see, it's closely connected to the entropy problem. Having the entire early universe share very similar conditions is a low-entropy configuration, as there are only a limited number of ways it can happen.

Inflation seems to provide a neat solution to the horizon problem. During the era of inflation, space expands by an enormous amount; points that were initially quite close get pushed very far apart. In particular, points that were widely separated when the microwave background was formed were right next to each other before inflation began—thereby answering the "How did they know to have similar conditions?" question. More important, during inflation the universe is dominated by dark super-energy, which—like any form of dark energy—has essentially the same density everywhere. There might be other forms of energy in the patch of space where inflation begins, but they are quickly diluted away; inflation stretches space flat, like pulling at the edges of a wrinkled bedsheet. The natural outcome of inflation is a universe that is very uniform on large scales.

TRUE AND FALSE VACUA

Inflation is a simple mechanism to explain the features we observe in the early universe: It stretches a small patch of space to make it flat and wrinkle free, solving the flatness and horizon problems, and dilutes away unwanted relics such as magnetic monopoles. So how does it actually work?

Clearly, the trick to inflation is to have a temporary form of dark super-energy, which drives the expansion of the universe for a while and then suddenly goes away. That might seem difficult, as the defining feature of dark energy is that it is nearly constant through space and time. For the most part that's true, but there can be sudden changes in the density—"phase transitions" where the dark energy abruptly goes down in value, like a bubble bursting. A phase transition of that form is the secret to inflation.

You may wonder what it is that actually creates this dark super-energy that drives inflation. The answer is a quantum field, just like the fields whose vibrations show up as the particles around us. Unfortunately, none of the fields we know—the neutrino field, the electromagnetic field, and so on—are right for the job. So cosmologists simply propose that there is a brand new field, imaginatively dubbed the "inflaton," whose task it is to drive inflation. Inventing new fields out of whole cloth like this is not quite as disreputable as it sounds; the truth is, inflation supposedly takes place at energies far higher than we can directly re-create in laboratories here on Earth. There are undoubtedly any number of new fields that become relevant at such energies, even if we don't know what they are; the question is whether any of them have the right properties to be the inflaton (i.e., give rise to a temporary phase of dark super-energy that expands the universe by a tremendous amount before decaying away).

In our discussions of quantum fields up to this point, we've emphasized that vibrations in such fields give rise to particles. If a field is constant everywhere, so there are no vibrations, you don't see any particles. If all we cared about were particles, the background value of the field—the average value it takes when we imagine smoothing out all the vibrations—wouldn't matter, since it's not directly observable. But the background value of a field can be *indirectly* observable—in particular, it can carry energy, and therefore affect the curvature of spacetime.

The energy associated with a field can arise in different ways. Usually it comes about because the field is changing from point to point in spacetime; there is energy in the stretching associated with the changing field values, much like there is energy in the twists or vibrations in a sheet of rubber. But in addition to that, fields can carry energy just by sitting there with some fixed value. That kind of energy, associated with the value of the field itself rather than changes in the field from place to place or time to time, is known as "potential energy." A rubber sheet that is perfectly flat will have more energy if it's sitting at a high elevation than it will if it's down on the ground; we know that, because we can extract that energy by picking up the sheet and tossing it down. Potential energy can be converted into other sorts of energy.

With a rubber sheet (or any other object sitting in the Earth's gravitational field), the way the potential energy behaves is pretty straightforward: The higher the elevation, the more potential energy it has. With fields, things can become much more complicated. If you're inventing a new theory of particle physics, you have to specify the particular way in which the potential energy depends on the value of each field. There aren't many underlying rules to guide you; every field simply is associated with a potential energy for each possible value, and that's part of the specification of the theory. Figure 78 shows an example of the potential energy for some hypothetical field, as a function of different field values.

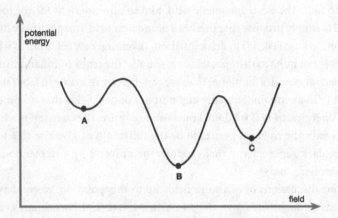

Figure 78: A plot of how the potential energy changes depending on the background value of some hypothetical field such as the inflaton. Fields tend to roll to a low point in the energy curve; in this plot, points A, B, and C represent different phases the vacuum could be in. Phase B has the lowest energy, so it's the "true vacuum," while A and C are "false vacua."

A field that has potential energy, but nothing else (no vibrations, motions, or twists) just sits there, unchanging. The potential energy per cubic centimeter is therefore constant, even as the universe expands. We know what that means: It's vacuum energy. (More carefully, it's one of many possible contributions to the vacuum energy.) You can think of the field like a ball rolling down a hill; it will tend to settle at the bottom of a valley, where the energy is lowest—at least, lower than any other nearby value. There might be other values of the field for which the energy is even lower, but these deeper "valleys" are separated by "hills." In Figure 78, the field could happily live at any of the values A, B, or C, but only B truly has the lowest energy. The values A and C are known as "false vacua," as they appear to be states of lowest energy if you only look at nearby values, while B is known as the "true vacuum," where the energy is truly the lowest. (To a physicist, a "vacuum" is not a machine that cleans your floors, nor does it even necessarily mean "empty space." It's simply "the lowest-energy state of a theory." Looking at the potential energy curve for some field, the bottom of every valley defines a distinct vacuum state.)

Guth put these ideas together to construct the inflationary universe scenario. Imagine that a hypothetical inflaton field found itself at point A, one of the false vacua. The field would be contributing a substantial amount to the vacuum energy, which would cause the universe to rapidly accelerate. Then all we have to do is explain how the field moved from the false vacuum A to the true vacuum B, where we now live—the phase transition that turns the energy locked up in the field into ordinary matter and radiation. Guth's original suggestion was that it occurred when bubbles of true vacuum would appear amidst the false vacuum, and then grow and collide with other bubbles to fill up all of space. This possibility, now known as "old inflation," turns out not to work; either the transition happens too quickly, and you don't get enough inflation, or it happens too slowly, and the universe never stops inflating.

Fortunately, soon after Guth's original paper, an alternative suggestion was made: Rather than the inflaton being stuck in a false vacuum "valley," imagine that it starts out on an elevated plateau—a long stretch that is nearly flat. The field then slowly rolls down the plateau, keeping the energy almost constant but not quite, before ultimately falling off a cliff (the phase transition). This is called "new inflation" and is the most popular implementation of the inflationary universe idea among cosmologists today.[261]

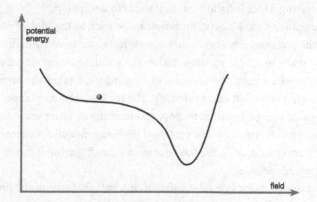

Figure 79: A potential energy curve appropriate for "new inflation." The field is never stuck in a valley, but rolls very slowly down an elevated plateau, before ultimately crashing down to the minimum. The energy density during that phase is not precisely constant, but is nearly so.

But that's not all. Besides offering a solution to the horizon, flatness, and mono-pole problems, inflation comes with a completely unanticipated bonus: It can explain the origin of the small fluctuations in the density of the early universe, which later grew into stars and galaxies.

The mechanism is simple, and inevitable: quantum fluctuations. Inflation does its best to make the universe as smooth as possible, but there is a fundamental limit imposed by quantum mechanics. Things can't become *too* smooth, or we would violate the Heisenberg Uncertainty Principle by pinpointing the state of the universe too precisely. The inevitable quantum fuzziness in the energy density from place to place during inflation gets imprinted on the amount of matter and radiation the inflaton converts into, and that translates into a very specific pre-diction for what kinds of perturbations in density we should see in the early uni-verse. It's those primordial perturbations that imprint temperature fluctuations in the cosmic microwave background, and eventually grow into stars, galaxies, and clusters. So far, the kinds of perturbations predicted by inflation match the obser-vations very well.[262] It's breathtaking to look into the sky at the distribution of gal-axies through space, and imagine that they originated in quantum fluctuations when the universe was a fraction of a second old.

ETERNAL INFLATION

After inflation was originally proposed, cosmologists eagerly started investigat-ing its properties in a variety of different models. In the course of these studies, Russian-American physicists Alexander Vilenkin and Andrei Linde noticed some-thing interesting: Once inflation starts, it tends to never stop.[263]

To understand this, it's actually easiest to go back to the idea of old inflation, although the phenomenon also occurs in new inflation. In old inflation, the infla-ton field is stuck in a false vacuum, rather than rolling slowly down a hill; since space is otherwise empty, the universe during inflation takes the form of de Sit-ter space with a very high energy density. The trick is, how do you get out of that phase—how do you get inflation to stop, and have the de Sitter space turn into the hot expanding universe of the conventional Big Bang model? Somehow we have to convert the energy stored in the false-vacuum state of the inflaton field into ordi-nary matter and radiation.

When a field is stuck in a false vacuum, it wants to decay to the lower-energy true vacuum. But it doesn't do so all at once; the false vacuum decays via the for-mation of bubbles, just like liquid water boils when it turns into water vapor. At random intervals, small bubbles of true vacuum pop into existence within the false vacuum, through the process of quantum fluctuations. Each bubble grows, and the space inside expands. But the space outside the bubble expands even faster, since it's still dominated by the high-energy false vacuum.

So there is a competition: Bubbles of true vacuum appear and grow, but the space in between them is also growing, pushing the bubbles apart. Which one wins? That depends on how quickly the bubbles are created. If they form fast enough, all the bubbles collide, and the energy in the false vacuum gets converted into matter and radiation. But we don't want bubbles to form *too* fast—otherwise we don't get enough inflation to address the cosmological puzzles we want to solve.

Unfortunately for the old-inflation scenario, there is no happy compromise. If we insist that we get enough inflation to solve our cosmological puzzles, it turns out that bubbles must form so infrequently that they never fill up the whole space. Individual bubbles might collide, just by chance; but the total set of bubbles doesn't expand and run into each other fast enough to convert all of the false vacuum into true vacuum. There is always some space in between the bubbles, stuck in the false vacuum, expanding at a terrific rate. Even though bubbles continue to form, the total amount of false vacuum just keeps getting bigger, since space is expanding faster than bubbles are created.

What we're left with is a mess—a chaotic, fractal distribution of bubbles of true vacuum surrounded by regions of false vacuum expanding at a terrific rate. That doesn't seem to look like the smooth, dense early universe with which we are familiar, so old inflation was set aside once new inflation came along.

But there is a loophole: What if our observable universe is contained inside a *single* bubble? Then it wouldn't matter that the space outside was wildly inhomogeneous, with patches of false vacuum and patches of true vacuum—within our single bubble, everything would appear smooth, and we're not able to observe what goes on outside, simply because the early universe is opaque.

There's a good reason why this possibility wasn't considered by Guth when he originally invented old inflation. If you start with the simplest examples of a bubble of true vacuum appearing inside a false vacuum, the interior of that bubble isn't full of matter and radiation—it's completely empty. So you don't go from de Sitter space with a high vacuum energy to a conventional Big Bang cosmology; you go right to empty space, in the form of de Sitter space with a lower value of the vacuum energy (if the energy of the true vacuum is positive). That's not the universe in which we live.

It wasn't until much later that cosmologists realized that this argument was a bit too quick. In fact, there is a way to "reheat" the interior of the true-vacuum bubble, to create the conditions of the Big Bang model: an episode of new inflation

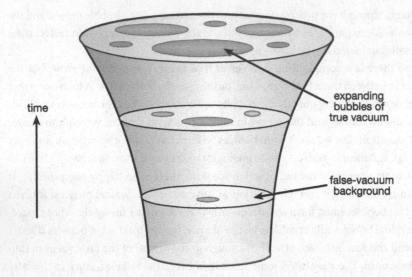

Figure 80: The decay of false-vacuum de Sitter space into bubbles of true vacuum in old inflation. The bubbles never completely collide, and the amount of space in the false-vacuum phase
grows forever; inflation never really ends.

inside the bubble. We imagine that the inflaton field inside the bubble doesn't land
directly at the bottom of its potential, corresponding to the true vacuum; instead,
it lands on an intermediate plateau, from which the field slowly rolls toward that
minimum. In this way, there can be a phase of new inflation within each bubble;
the energy density from the inflaton potential while it's on the plateau can later
be converted into matter and radiation, and we end up with a perfectly plausible
universe.[264]

So old inflation, once it starts, never ends. You can make bubbles of true vacuum that look like our universe, but the region of false vacuum outside always
keeps growing. More bubbles keep appearing, and the process never terminates.
That's the idea of "eternal inflation." It doesn't happen in every model of inflation;
whether or not it occurs depends on details of the inflaton and its potential.[265] But
you don't have to delicately tune the theory too badly to allow for eternal inflation;
it happens in a healthy fraction of inflationary models.

THE MULTIVERSE

There is a lot to say about eternal inflation, but let's just focus on one consequence:
While the universe we see looks very smooth on large scales, on even larger

(unobservable) scales the universe would be very far from smooth. The large-scale uniformity of our observed universe sometimes tempts cosmologists into assuming that it must keep going like that infinitely far in every direction. But that was always an assumption that made our lives easier, not a conclusion from any rigorous chain of reasoning. The scenario of eternal inflation predicts that the universe does *not* continue on smoothly as far as it goes; far beyond our observable horizon, things eventually begin to look very different. Indeed, somewhere out there, inflation is still going on. This scenario is obviously very speculative at this point, but it's important to keep in mind that the universe on ultra-large scales is, if anything, likely to be very different than the tiny patch of universe to which we have direct access.

This situation has led to the introduction of some new vocabulary and the abuse of some old vocabulary. Each bubble of true vacuum, if we set things up correctly, resembles our observable universe in rough outline: The energy that used to be in the false vacuum gets converted into ordinary matter and radiation, and we find a hot, dense, smooth, expanding space. Someone living inside one bubble wouldn't be able to see any of the other bubbles (unless they collided)—they would just see the Big-Bang-like conditions at the beginning of their bubble. This picture actually represents the simplest example of a *multiverse*—each bubble, evolving separately from all the rest, evolves as a universe unto itself.

Obviously we're taking some liberties with the word *universe* here. If we were being more careful, it might be better to use the word *universe* to refer to the totality of everything there is, whether we could see it or not. (And sometimes we do use it that way, just to add to the confusion.) But most cosmologists have been abusing the nomenclature for some time now, and if we want to communicate with other scientists it will be useful to speak the same language. We have heard sentences like "our universe is 14 billion years old" so often that we don't want to go back and correct them all by adding "at least, the observable part of our universe." So instead people often attach the word *universe* to a region of spacetime that resembles our observable universe, starting from a hot, dense state and expanding from there. Alan Guth has suggested the phrase *pocket universes*, which conveys the idea a bit more precisely.

The multiverse, therefore, is just this collection of pocket universes—regions of true vacuum, expanding and cooling after a dramatic beginning—and the background inflating spacetime in which they are embedded. When you think about it, this is a rather mundane conception of the idea of a "multiverse." It's really just a collection of different regions of space, all of which evolve in similar ways to the universe we observe.

An interesting feature of this kind of multiverse has attracted a great deal of attention recently: Local laws of physics can be very different in each of those pocket universes. When we drew the potential energy plot for the inflaton in Figure 78, we illustrated three different vacuum states (A, B, C). But there is nothing to stop there from being many more than that. As we alluded to briefly in Chapter Twelve, string theory seems to predict a huge number of vacuum states—as many as 10^{500}, if not more. Each such state is a different phase in which spacetime can find itself. That means different kinds of particles, with different masses and interactions— basically, completely new laws of physics in each universe. Again, that's a bit of an abuse of language, because the underlying laws (string theory, or whatever) are still the same; but they manifest themselves in different ways, just like water can be solid, liquid, or gas. String theorists these days refer to the "landscape" of possible vacuum states.[266]

But it's one thing for your theory to *permit* many different vacuum states, each with its own laws of physics; it's something else to claim that all the different states actually *exist* somewhere out there in the multiverse. That's where eternal inflation comes in. We told a story in which inflation occurs in a false vacuum state, and ends (within each pocket universe) by evolving into a true vacuum, either by bubble formation or by slowly rolling. But if inflation continues forever, there's nothing to stop it from evolving into different vacuum states in different pocket universes; indeed, that's just what you would expect it should do. So eternal inflation offers a way to take all those possible universes and make them real.

That scenario—if it's right—comes with profound consequences. Most obviously, if you had entertained some hope of uniquely predicting features of physics we observe (the mass of the neutrino, the charge of the electron, and so forth) on the basis of a Theory of Everything, those hopes are now out the window. The local manifestations of the laws of physics will vary from universe to universe. You might hope to make some statistical predictions, on the basis of the anthropic principle; "sixty-three percent of observers in the multiverse will find three families of fermions," or something to that effect. And many people are trying hard to do just that. But it's not clear whether it's even possible, especially since the number of observers experiencing certain features will often end up being infinitely big, in a universe that keeps inflating forever.

For the purposes of this book, we are very interested in the multiverse, but not so much in the details of the landscape of many different vacua, or attempts to wrestle the anthropic principle into a useful set of predictions. Our problem—the small entropy of the observable universe at early times—is so very blatant and dramatic that there's no hope of addressing it via recourse to the anthropic principle;

life could certainly exist in a universe with a much higher entropy. We need to do better, but the idea of a multiverse might very well be a step in the right direction. At the very least, it suggests that what we see might not be nearly all there is, as far as the universe is concerned.

WHAT GOOD IS INFLATION?

Let's put it all together. The story that cosmologists like to tell themselves about inflation[267] goes something like this:

> We don't know what conditions in the extremely early universe were really like. Let's assume it was dense and crowded, but not necessarily smooth; there may have been wild fluctuations from place to place. These may have included black holes, oscillating fields, and even somewhat empty patches. Now imagine that at least one small region of space within this mess is relatively quiet, with its energy density consisting mostly of dark super-energy from the inflaton field. While the rest of space goes on its chaotic way, this particular patch begins to inflate; its volume increases by an enormous amount, while any preexisting perturbations get wiped clean by the inflationary stretching. At the end of the day, that particular patch evolves into a region of space that looks like our universe as described by the standard Big Bang model, regardless of what happens to the rest of the initially fluctuating primordial soup. Therefore, it doesn't require any delicate, unnatural fine-tuning of initial conditions to get a universe that is spatially flat and uniform over large distances; it arises robustly from generic, randomly fluctuating initial conditions.

Note that the goal is to explain why a universe like the one we find ourselves in today would arise *naturally* as the result of dynamical processes in the early universe. Inflation is concerned solely with providing an explanation for some apparently finely tuned features of our universe at early times; if you choose to take the attitude that the early universe is what it is, and it makes no sense to "explain" it, then inflation has nothing to offer to you.

Does it work? Does inflation really explain why our seemingly unnatural initial conditions are actually quite likely? I want to argue that inflation *by itself* doesn't answer these questions at all; it might be part of the final story, but it needs to be supplemented by some ideas about what happened *before* inflation if the idea is to have any force whatsoever. This puts us (that is to say, me) squarely in the minority of contemporary cosmologists, although not completely alone[268]; most workers in the field are confident that inflation operates as advertised to remove the fine-tuning problems that plague the standard Big Bang model. You should be able to make your own judgments, keeping in mind that ultimately it's Nature who decides.

In the last chapter, in order to discuss the evolution of the entropy within our universe, we introduced the idea of our "comoving patch"—the part of the universe that is currently observable to us, considered as a physical system evolving through time. It's reasonable to approximate our patch as a closed system—even though it is not strictly isolated, we don't think that the rest of the universe is influencing what goes on within our patch in any important way. That remains true in the inflationary scenario. Our patch finds itself in a configuration where it is very small, and dominated by dark super-energy; other parts of the universe might look dramatically different, but who cares?

We previously presented the puzzle of the early universe in terms of entropy: Our comoving patch has an entropy today of about 10^{101}, but at earlier times it was approximately 10^{88}, and it could be as large as 10^{120}. So the early universe had a much, much smaller entropy than the current universe. Why? If the state of the universe were chosen randomly among all possible states, it would be extraordinarily unlikely to be in such a low-entropy configuration, so clearly there is more to the story.

Inflation purports to provide the rest of the story. From wildly oscillating initial conditions—which, implicitly or explicitly, are sometimes misleadingly described as "high entropy"—a small patch can naturally evolve into a region with an entropy of 10^{88} that looks like our universe. Having gone through this book, we all know that a truly high-entropy configuration is *not* a wildly oscillating high-energy mess; it's exactly the opposite, a vast and quiet empty space. The conditions necessary for inflation to start are, just like the early universe in the conventional Big Bang story, not at all what we would get if we were picking states randomly from a hat.

In fact it's worse than that. Let's focus in on the tiny patch of space, dominated by dark super-energy, in which inflation starts. What is its entropy? That's a hard question to answer, for the standard reason that we don't know enough about entropy in the presence of gravity, especially not in the high-energy regime relevant for inflation. But we can make a reasonable guess. In the last chapter we discussed how there are only so many possible states that can "fit" into a given region of an expanding universe, at least if they are described by the ordinary assumptions of quantum field theory (which inflation assumes). The states look like vibrating quantum fields, and the vibrations must have wavelengths smaller than the size of the region we are considering, and larger than the Planck length. This means there is a maximum number of possible states that can look like the small patch that is ready to inflate.

The numerical answer will depend on the particular way in which inflation happens, and in particular on the vacuum energy during inflation. But the differences between one model and another aren't that significant, so it suffices to pick an example and stick to it. Let's say that the energy scale during inflation is 1 percent of the Planck scale; pretty high, but low enough that we're safely avoiding complications from quantum gravity. In that case, the estimated entropy of our comoving patch at the beginning of inflation is:

$$S_{inflation} \approx 10^{12}.$$

That's an incredibly small value, compared either to the 10^{120} it could have been or even the 10^{88} it would soon become. It reflects the fact that every single degree of freedom that goes into describing our current universe must have been delicately packed into an extremely smooth, small patch of space, in order for inflation to get going.

The secret of inflation is thereby revealed: It explains why our observable universe was in such an apparently low-entropy, finely tuned early state by assuming that it started in an *even lower-entropy* state before that. That's hardly surprising, if we believe the Second Law and expect entropy to grow with time, but it doesn't seem to address the real issue. Taken at face value, it would seem very surprising indeed that we would find our comoving patch of universe in the kind of low-entropy configuration necessary to start inflation. You can't solve a fine-tuning problem by appealing to an even greater fine-tuning.

OUR COMOVING PATCH REVISITED

Let's think this through, because we're deviating from orthodoxy here and it behooves us to be careful.

We have been making two crucial assumptions about the evolution of the observable universe—our comoving patch of space and all of the stuff within it. First, we're assuming that the observable universe is essentially *autonomous*—that is, it evolves as a closed system, free from outside influences. Inflation does not violate this assumption; once inflation begins, our comoving patch rapidly turns into a smooth configuration, and that configuration evolves independently of the rest of the universe. This assumption can obviously break down before inflation, and play a role in setting up the initial conditions; but inflation itself does not take advantage of any hypothetical external influences in attempting to explain what we currently see.

The other assumption is that the dynamics of our observable universe are *reversible*—they conserve information. This seemingly innocuous point implies a great deal. There is a space of states that is fixed once and for all—in particular, it is the same at early times as at late times—and the evolution within that space takes different starting states to different ending states (in the same amount of time). The early universe looks very different from the late universe—it's smaller, denser, expanding more rapidly, and so on. But (under our assumptions of reversible dynamics) that doesn't mean the space of states has changed, only that the particular kind of state the universe is in has changed.

The early universe, to belabor the point, is the same physical system as the late universe, just in a very different configuration. And the entropy of any given microstate of that system reflects how many other microstates look similar from a macroscopic point of view. If we were to randomly choose a configuration of the physical system we call the observable universe, it would be overwhelmingly likely to be a state of very high entropy—that is, close to empty space.[269]

To be honest, however, people tend not to think that way, even among professional cosmologists. We tend to reason that the early universe is a small, dense place, so that when we imagine what states it might be in, we can restrict our attention to small, dense configurations that are sufficiently smooth and well behaved so that the rules of quantum field theory apply. But there is absolutely no justification for doing so, at least within the dynamics itself. When we imagine what possible states the early universe could have been in, we need to include unknown states that are outside the realm of validity of quantum field theory. For that matter, we should include all of the possible states of the *current* universe, as they are simply different configurations of the same system.

The size of the universe is not conserved—it evolves into something else. When we consider statistical mechanics of gas molecules in a box, it's okay to keep the number of molecules fixed, because that reflects the reality of the underlying dynamics. But in a theory with gravity, the "size of the universe" isn't fixed. So it makes no sense—again, just based on the known laws of physics, without recourse to some new principles outside those laws—to assume from the start that the early universe must be small and dense. That's something we need to explain.

All of which is somewhat problematic for the conventional justification that we put forward for the inflationary universe scenario. According to the previous story, we admit that we don't know what the early universe was like, but we imagine that it was characterized by wild fluctuations. (The current universe, of course, is not characterized by such fluctuations, so already there is something to be explained.) Among those fluctuations, every once in a while a region will come into existence that is dominated by dark super-energy, and the conventional inflation story follows. After all, how hard can it be to randomly fluctuate into the right conditions to start inflation?

The answer is that it can be incredibly hard. If we truly randomly chose a configuration for the degrees of freedom within that region, we would be overwhelmingly likely to get a high-entropy state: a large, empty universe.[270] Indeed, simply by comparing entropies, we'd be much more likely to get our current universe, with a hundred billion galaxies and all the rest, than we would be to get a patch ready to inflate. And if we're not randomly choosing configurations of those degrees of freedom—well, then, what are we doing? That's beyond the scope of the conventional inflationary story.

These problems are not specific to the idea of inflation. They would plague any possible scenario that claimed to provide a dynamical explanation for the apparent fine-tuning of our early universe, while remaining consistent with our two assumptions (our comoving patch is a closed system, and its dynamics are reversible). The problem is that the early universe has a low entropy, which means that there are a relatively small number of ways for the universe to look like that. And, while information is conserved, there is no possible dynamical mechanism that can take a very large number of states and evolve them into a smaller number of states. If there were, it would be easy to violate the Second Law.

SETTING THE STAGE

This discussion has intentionally emphasized the hidden skeletons in the closet of the inflationary universe scenario—there are plenty of other books that will emphasize the arguments in its favor.[271] But let's be clear: The problem isn't really with inflation; it's with how the theory is usually marketed. We often hear that inflation removes the pressing need for a theory of initial conditions, as inflation will begin under fairly generic circumstances, and once it begins all our problems are solved.

The truth is almost the converse: Inflation has a lot going for it, but it makes the need for a theory of initial conditions much more pressing. Hopefully I've made the case that neither inflation nor any other mechanism can, by itself, explain our low-entropy early universe under the assumptions of reversibility and autonomous evolution. It's possible, of course, that reversibility should be the thing to go; perhaps the fundamental laws of physics violate reversibility at a fundamental level. Even though that's intellectually conceivable, I'll argue that it's hard to make such an idea match what we actually see in the world.

A less drastic strategy would be to move beyond the assumption of autonomous evolution. We knew all along that treating our comoving patch as a closed system was, at best, an approximation. It seems like an extremely good approximation right now, or even at any time in the history of the universe about which we actually have empirical data. But surely it breaks down at the very beginning. Inflation could play a crucial role in explaining the universe we see, but only if we can discard the idea that "we just randomly fluctuated into it," and come up with a particular reason why the conditions necessary for inflation ever came to pass.

In other words, it seems like the most straightforward way out of our conundrum is to abandon the goal of explaining the unnatural early universe purely in terms of the autonomous evolution of our comoving patch, and instead try to embed our observable universe into a bigger picture. This brings us back to the idea of the multiverse—a larger structure in which the universe we observe is just a tiny part. If something like that is true, we are at least able to contemplate the idea that the evolution of the multiverse naturally gives rise to conditions under which inflation can begin; after that, the story proceeds as above.

So we want to ask, not what the physical system making up our observable universe should look like, but what a multiverse should look like, and whether it would naturally give rise to regions that look like the universe we see. Ideally, we'd want it to happen without putting in time asymmetry by hand at any step along the way. We want to explain not only how we can get the right conditions to start inflation, but why it might be natural to have a large swath of spacetime (our observable universe) that features those conditions at one end of time and empty space at the other. This is a program that is far from complete, although we have some ideas. We're deep into speculative territory by now, but if we keep our wits about us we should be able to take a safe journey without being devoured by dragons.

15

THE PAST THROUGH TOMORROW

The eternal silence of these infinite spaces fills me with dread.

—Blaise Pascal, *Pensées*[272]

Over the course of this book, we've explored the meaning of the arrow of time, as embodied in the Second Law of Thermodynamics, and its relationship with cosmology and the origin of the universe. Finally we have enough background to put it all together and address the question once and for all: Why was the entropy of our observable universe low at early times? (Or even better, so as not to succumb to asymmetric language right from the start: Why do we live in the temporal vicinity of an extremely low-entropy state?)

We'll address the question, but we don't know the answer. There are ideas, and some ideas seem more promising than others, but all of them are somewhat vague, and we certainly haven't yet put the final pieces together. That's science for you. In fact, that's the exciting part of science—when you have some clues assembled, and some promising ideas, but are still in the process of nailing down the ultimate answers. Hopefully the prospects sketched in this chapter will serve as a useful guide to wherever cosmologists go next in their attempts to address these deep issues.[273]

At the risk of being repetitive, let's review the conundrum one last time, so that we can establish what would count as an acceptable solution to the problem.

All of the macroscopic manifestations of the arrow of time—our ability to turn eggs into omelets but not vice versa, the tendency of milk to mix into coffee but never spontaneously unmix, the fact that we can remember the past but not the future—can be traced to the tendency of entropy to increase, in accordance with the Second Law of Thermodynamics. In the 1870s, Boltzmann explained the microscopic underpinnings of the Second Law: Entropy counts the number of microstates corresponding to each macrostate, so if we start (for whatever reason) in a relatively low-entropy state, it's overwhelmingly likely that the entropy will increase toward the future. However, due to the fundamental reversibility of the laws of physics, if the only thing we have to go on is the fact that the current state is low entropy, we would with equal legitimacy expect the entropy to have been larger in the past. The real world doesn't seem to work that way, so we need something else to go on. That something else is the Past Hypothesis: the assumption that the very early universe found itself in an extremely low-entropy state, and we are currently witnessing its relaxation to a state of high entropy. The question of why the Past Hypothesis is true belongs to the realm of cosmology. The anthropic principle is woefully inadequate for the task, since we could easily find ourselves constituted as random fluctuations (Boltzmann brains) in an otherwise empty de Sitter space. Likewise, inflation by itself doesn't address the question, as it requires an even lower-entropy starting point than the conventional Big Bang cosmology. So the question remains: Why does the Past Hypothesis hold within our observable patch of the universe?

Let's see if we can't make some headway on this.

EVOLVING THE SPACE OF STATES

Start with the most obvious hypothesis: Deep down, the fundamental laws of physics simply aren't reversible. I've tried to be careful to allude to the existence of this possibility all along, but have always spoken as if it's a long shot, or not worthy of our serious attention. There are good reasons for that, though they are not airtight.

A reversible system is one that has a space of states, fixed once and for all, and a rule for evolving those states forward in time that conserves information. Two different states, beginning at some initial time, will evolve predictably into two different states at some specific later time—never into the same future state. That way we can reverse the evolution, since every state the system could currently be in has a unique predecessor at every moment in time.

One way to violate reversibility would be to let the space of states itself actually evolve with time. Perhaps there were simply fewer possible states the universe could have been in at early times, so the small entropy is not so surprising. In that case, many possible microstates that live in the same macrostate as the current universe simply have no possible past state from which they could have come.

Indeed, this is how many cosmologists implicitly speak about what happens in an expanding universe. If we restrict ourselves to "states that look like gentle vibrations of quantum fields around a smooth background," it's certainly true that this particular part of the space of states grows with time, as space itself (in the old-fashioned three-dimensional sense of "space") becomes larger. But that's very different from imagining that the *entire* space of states is actually changing with time. Almost nobody would claim to support such a position, if they sat down and thought through what it really meant. I explicitly rejected this possibility when I argued that the early universe was finely tuned—among all the states it could have been in, we included states that look like the universe today, as well as various choices with even higher entropy.

The weirdest thing about the idea that the space of states changes with time is that it requires an *external* time parameter—a concept of "time" that lives outside the actual universe, and through which the universe evolves. Ordinarily, we think of time as part of the universe—a coordinate on spacetime, measured by various sorts of predictably repetitive clocks. The question "What time is it?" is answered by reference to things going on within the universe—that is to say, to features of the state the universe is currently in. ("The little hand is on the three, and the big hand is on the twelve.") But if the space of states truly changes with time, that conception becomes insufficient. At any one moment, the universe is actually in one specific state. It makes no sense to say, "The space of states is smaller when the universe is in state X than when it is in state Y." The space of states, by definition, includes all of the states the universe could hypothetically be in.

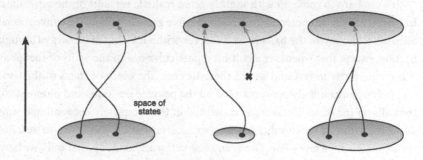

Figure 81: On the left we have reversible laws of physics: The system evolves within a fixed space of states, such that different initial states evolve uniquely to different final states. The middle example is irreversible, because the space of states grows with respect to some external time parameter; some states at later times have no predecessors from which they could have come. On the right we have another form of irreversibility, where the space of states remains fixed, but different initial states evolve into the same final state.

So for the space of states to change with time, we would have to posit a notion of time that is not merely measured by features of the state of the universe, but exists outside the universe as we conventionally understand it. Then it would make sense to say, "When this external time parameter was equal to a certain value, the space of states of the universe was relatively small, and when it had progressed to some other value the space of states had grown larger."

There's not much to say about this idea. It's possible, but very few people advocate it as an approach to the arrow-of-time problem.[274] It would require a dramatic rethinking of the way we currently understand the laws of physics; nothing about our current framework suggests the existence of a time parameter lurking outside the universe itself. So for now, we can't rule it out, but it doesn't give us a warm and fuzzy feeling.

IRREVERSIBLE MOTIONS

The other way to invent laws of physics that are intrinsically irreversible is to stick once and for all with some space of states, but posit dynamical laws that don't conserve information. That's what we saw with checkerboard D back in Chapter Seven; when diagonal lines of gray squares bumped into the vertical line, they simply ceased to exist. There was no way of knowing, from the state at one particular moment in time, precisely what state it came from in the past, since there was no way of reconstructing what diagonals had been lurking around before running afoul of a vertical column.

It's not hard to come up with slightly more realistic versions of the same idea. In Chapter Eight we contemplated an irreversible game of billiards: a conventional billiards table, where the balls moved forever without losing any energy through friction, except that whenever a ball hit a particular one of the walls of the table, it came perfectly to rest and stayed there forever. The space of states of this system never changes; it always consists of all the possible positions and momenta of the balls on the table. The entropy is defined in the completely conventional way, as the logarithm of the number of states with certain macroscopic properties. But the dynamics are irreversible: Given any one ball stuck to the special wall, we have no way of knowing how long it's been there. And the entropy of this system flouts the Second Law with impunity; gradually, as more balls get stuck, the system takes up a smaller and smaller portion of the space of states, and the entropy decreases without any intervention from the outside world.

The laws of physics as we know them—putting aside the important question of wave function collapses in quantum mechanics—seem to be reversible. But we don't know the final laws of physics; all we have are very good approximations. Is it conceivable that the real laws of physics are fundamentally irreversible, and that explains the arrow of time?

First let's untangle a potential misconception about what that would really mean. To "explain" the arrow of time means to come up with a set of laws of physics, and an "initial" state of the universe, so that we naturally (without fine-tuning) witness a change in entropy over time of the sort we observe around us. In particular, if we simply assume that the initial conditions have low entropy, there is nothing to be explained—the entropy will tend to go up, in accordance with Boltzmann, and we're done. In that case there's simply no need to posit irreversible laws of physics; the reversible ones are up to the task. But the problem is that such a low-entropy boundary condition seems unnatural.

So if we wish to explain the arrow of time in a natural way by invoking irreversible fundamental laws, the idea would be to postulate a *high*-entropy condition—a "generic" state of the universe—and imagine that the laws of physics, when acting on that state, naturally work to *decrease* its entropy. That would count as a real explanation of the arrow. It might seem that this setup gets it backward—it predicts that entropy goes down, rather than up. But the essence of the arrow of time is simply that entropy changes in a consistent direction. As long as that is true, observers who lived in such a world would always "remember" the direction of time that had a lower entropy; likewise, relationships of cause and effect would always put the causes on the lower-entropy side of things, as that is the direction with fewer allowed choices. In other words, such observers would *call* the high-entropy end of time the "future," and the low-entropy end "the past," even though the fundamental laws of physics in this world would only precisely reconstruct the past from the future, and not vice versa.

Such a universe is certainly conceivable. The problem is, it seems like it would be dramatically different from *our* universe.

Think carefully about what would have to happen for this scenario to work. The universe, for whatever reason, finds itself in a randomly chosen high-entropy state, which looks like empty de Sitter space. Now our postulated irreversible laws of physics act on that state to decrease the entropy. The result—if all this is to have any chance of working out—should be the history of our actual universe, just reversed in time compared to how we traditionally think about it. In other words: Out of the initial emptiness, some photons miraculously focus on a point in space to create a white hole. That white hole gradually grows in mass through the accretion of additional photons (Hawking radiation in reverse). Gradually a collection of additional white holes come into view from far away, arrayed almost uniformly through space. All of these white holes start belching out gas into the universe, which implodes to make stars, which spiral gently away from the white holes to form galaxies. These stars absorb more radiation from the outside world, and use the energy to break down heavy elements into lighter ones. As the galaxies continue to move toward one another in an increasingly rapid contraction of space, the stars disperse into a uniform distribution of gas. Ultimately the universe collapses to a Big Crunch, as matter and radiation form an extremely smooth and uniform distribution near the end of time.

This is the real history of our observable universe, just played backward in time. It's a perfectly good solution to the laws of physics as we currently understand them; all we have to do is start with the state near the Big Bang, evolve it forward in time to whatever high-entropy microstate it eventually becomes, and then time-reverse that state. But the hypothesis we're currently considering is very different: It says that an evolution of this form would happen for *almost any* high-entropy state of empty de Sitter space. That's a lot to ask of some laws of physics. It's one thing to imagine entropy going down as a result of irreversible laws, but it's another thing entirely to imagine it going down in precisely the right way to produce a time-reversed history of our universe.

We can be more specific about where our discomfort with this scenario comes from. We don't need to think about the whole universe to experience the arrow of time: It's right here in our kitchen. We drop an ice cube into a glass of warm water, and the ice melts as the water cools off, eventually reaching a uniform temperature. The fundamental-irreversibility hypothesis claims that this can be explained by the deep laws of physics, *starting* with the uniformly cool glass of liquid water. In other words, the laws of physics purportedly act on the water to separate out different molecules into the form of an ice cube sitting in a glass of warm water, in precisely the way we would expect had we started with the ice cube and water, only backward in time.

But that's crazy. For one thing: How does it know? Some glasses of cool water were, five minutes ago, glasses of warm water with ice cubes; but others were just glasses of cool water even five minutes ago. Even though there are relatively few microstates corresponding to each low-entropy macrostate, there are a lot more individual low-entropy macrostates than there are high-entropy ones. (More formally, each low-entropy state contains more information than a high-entropy one.)

The problem is closely tied to the issue of complexity I talked about at the end of Chapter Nine. In the real world, as the universe evolves from a low-entropy Big Bang to a high-entropy future, it creates delicate complex structures along the way. The initially uniform gas doesn't simply disperse as the universe expands; it first collapses into stars and planets, which increase the entropy locally, and sustain intricate ecosystems and information-processing subsystems along the way.

It's extremely hard, bordering on impossible, to imagine all of that arising from an initially high-entropy state that gets evolved according to some irreversible laws of physics. This is not an airtight argument, but it seems likely that we will have to look somewhere else for an explanation of the arrow of time in the real world.

A SPECIAL BEGINNING

From here on, we'll be operating under the hypothesis that the fundamental laws of physics are truly reversible: The space of allowed states remains fixed, and the dynamical rules of time evolution conserve the information contained in each state. So how can we possibly hope to account for the low-entropy condition in our observable universe?

For Boltzmann, thinking in the context of an absolute Newtonian space and time, this was quite a puzzle. But general relativity and the Big Bang model offer a new possibility, namely: There was a beginning to the universe, including to time itself, and that the beginning state was one with very low entropy. And you're not allowed to ask why.

Sometimes, the condition "you're not allowed to ask why" is rephrased as follows: "We posit a *new law of nature*, which holds that the initial state of the universe had a very low entropy." It's not clear why these two formulations are really any different. In our usual understanding of the laws of physics, two ingredients are required to completely specify the evolution of a physical system: a set of dynamical laws that can be used to evolve the system from one state to another through time, and a boundary condition that fixes which state the system is in at some particular moment. But, even though both the laws and the boundary condition are necessary, they seem like very different things; it's not clear what is to be gained by thinking of the boundary condition as a "law." A dynamical law demonstrates its validity over and over again; at every moment, the law takes the current state and evolves it into the next state. But the boundary condition is just imposed once and for all; its nature is more like an empirical fact about the universe than an additional law of physics. There isn't any substantive distinction between the statements "the early universe had a low entropy" and "it is a law of physics that the early universe had a low entropy" (unless we imagine that there are many universes, all with the same boundary condition).[275]

Be that as it may, it's undoubtedly possible that this is the most we'll ever be able to say: The low entropy of the early universe is not to be explained via a better understanding of the dynamical laws of physics, but is simply a brute fact, or (if you prefer) an independent law of nature. An example of this approach has been explicitly advocated by Roger Penrose, who has suggested what he calls the "Weyl curvature hypothesis"—a new law of nature that distinguishes explicitly between spacetime singularities that are in the past and those that are in the future. The basic idea is that past singularities have to be smooth and featureless, while future singularities can be arbitrarily messy and complicated.[276] This is an explicit violation of time-reversal symmetry, which would ensure that the Big Bang had a low entropy.

The real problem with a proposal like this is its essentially ad hoc nature.[277] Asserting that past singularities had to be very smooth doesn't help you understand anything else about the universe. It "explains" time asymmetry by putting it in by hand. Nevertheless, one could think of it as a placeholder for a more fundamental understanding: If some deeper principles were uncovered that led to a fundamental distinction between initial and final singularities, such that the curvature of the former were constrained but not the latter, we would certainly have made substantial progress toward understanding the origin of the arrow of time. But even this formulation suggests that the real agenda is to keep looking for something deeper.

A SYMMETRIC UNIVERSE

If the fundamental laws of physics are reversible, and we don't allow ourselves to simply impose time-asymmetric boundary conditions, the remaining possibility seems to be that the evolution of the universe actually is time-symmetric itself, despite appearances to the contrary. It's not hard to imagine how that might happen, if we are open to the possibility that the universe will eventually stop expanding and re-collapse. Before the discovery of dark energy, many cosmologists found a re-collapsing universe philosophically attractive: Einstein and Wheeler, among others, were drawn to the notion of a universe that was finite in both space and time. A future Big Crunch would provide a pleasing symmetry to the history of a universe that began with a Big Bang.

In the conventional picture, however, any such symmetry would be dramatically marred by the Second Law. Everything we know about the evolution of the entropy of the universe is readily explained by assuming that the entropy was very low near the beginning; from there, it naturally increases with time. If the universe were to re-collapse, there is nothing in the known laws of physics that would prevent the entropy from continuing to increase. The Big Crunch would be a messy, high-entropy place, in stark contrast to the pristine smoothness of the Big Bang.

In an attempt to restore the overall symmetry of the history of the universe, people have occasionally contemplated the need for an additional law of physics: a boundary condition in the *future* (a "Future Hypothesis," in addition to the Past Hypothesis), which would guarantee that entropy was low near the Crunch as well as the Bang. This idea, suggested by Thomas Gold (better known as a pioneer of the Steady State model) and others, would imply that the arrow of time would reverse at the moment the universe hit its maximum size, and it would always be true that entropy increased in the direction of time toward which the universe was expanding.[278]

The Gold universe never really caught on among cosmologists, for a simple reason: There's no good reason for there to be a future boundary condition of any particular sort. Sure, it restores the overall symmetry of time, but nothing we have experienced in the universe demands such a condition, nor does it follow from any other underlying principles.

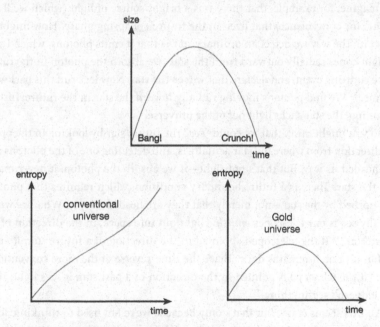

Figure 82: At the top, the size of a re-collapsing universe through time; at bottom, two possible scenarios for the evolution of entropy. By conventional lights, we would expect the entropy to increase even as the universe collapsed, as shown at bottom left. In the Gold universe, the entropy is constrained to decrease by a low-entropy future boundary condition.

On the other hand—there's no good reason for there to be a *past* boundary condition, either, except for the stubborn fact that we need to invoke one to explain the universe we actually see.[279] Huw Price has championed the Gold universe as something that cosmologists should take seriously—at least at the level of a thought experiment, if not as a model for the real world—for just this reason.[280] We don't know why entropy was low near the Big Bang, but it was; therefore, the fact that we don't know why the entropy should be low near a Big Crunch is not a sufficient reason to discard the possibility. Indeed, without introducing time asymmetry by hand, it stands to reason that whatever unknown principle of physics enforces the low entropy at the Bang could also do so at the Crunch.

It's interesting to approach this scenario like real scientists, and ask whether there could be any testable consequences of a low-entropy future condition. Even if such a condition existed, it would be easy enough to avoid any prospective consequences, just by putting the Big Crunch very far in the future. But if it were relatively near in time (a trillion years from now, say, rather than a googol years), we might be able to *see* the effects of the future decrease in entropy.[281]

Imagine, for example, that there was a bright source of light (which we'll call a "star" for convenience) that lived in the future collapsing phase. How might we detect it? The way we detect an ordinary star is that it emits photons, which travel on light cones radially outward from the star; we absorb the photon in the future of the emitting event, and declare that we see the star. Now let's run this backward in time.[282] We find photons moving radially *toward* the star in the future; instead of shining, the star sucks light out of the universe.

So you might think that we could "see" the future star by looking in the opposite direction from where the star actually is, and detecting one of the photons that was headed its way. But that's not right—if we absorb the photon, it never makes it to the star. There is a future boundary condition, which requires that photons be absorbed by the star—not merely that they are headed its way. What we would actually see is our telescopes *emitting* light out into space, in the direction of the future star.[283] If the telescope is pointed in the direction of a future star, it emits light; if it's not, it remains dark. That's the time-reverse of the more conventional idea: "If the telescope is pointed in the direction of a past star, it sees light; if it's not, it doesn't see anything."

All this seems crazy; but that's only because we're not used to thinking about a world with a future boundary condition. "How does the telescope know to emit light when it's pointed in the direction of a star that won't even exist for another trillion years?" That's what future boundary conditions are all about—they pick out the fantastically tiny fraction of microstates within our current macrostate in which such a seemingly unlikely event happens.[284] Deep down, there is nothing stranger about this than there is about the past boundary condition in our actual universe, other than we're used to one but not the other. (By the way, so far nobody has found any experimental evidence for future stars, or any other evidence of a future low-entropy boundary condition. If they had, you probably would have heard about it.)

Meanwhile, the example of the Gold universe serves more as a cautionary tale than as a serious candidate to account for the arrow of time. If you think you have some natural explanation for why the early universe had such a low entropy, but you claim not to invoke any explicit violations of time-reversal symmetry, why shouldn't the late universe look the same way? This thought experiment drives home just how puzzling the low-entropy configuration of the Big Bang really is.

The smart money these days is that the universe won't actually re-collapse. The universe is accelerating; if the dark energy is an absolutely constant vacuum energy (which is the most straightforward possibility), the acceleration will continue forever. We don't know enough to say for sure, but it's most likely that our future is absolutely unlike our past. Which, again, places the unusual circumstances surrounding the Big Bang front and center as a puzzle we would like to solve.

BEFORE THE BIG BANG

We almost seem to have run out of options. If we don't put in time asymmetry by hand (either in the dynamical laws or in a boundary condition), and the Big Bang has a low entropy, but we don't insist on a low-entropy future condition—what is left? We seem to be caught in a viselike grip of logic, with no remaining avenues to reconcile the evolution of entropy in our observable universe with the reversibility of the fundamental laws of physics.

There is a way out: We can accept that the Big Bang had a low entropy, but deny that the Big Bang was the beginning of the universe.

This sounds a bit heretical to anyone who has read about the success of the Big Bang model, or who knows that the existence of an initial singularity is a firm prediction of general relativity. We are often told that there is no such thing as "before the Big Bang"—that time itself (as well as space) doesn't exist prior to the initial singularity. That is, the concept of "prior to the singularity" just doesn't make any sense.

But as I mentioned briefly in Chapter Three, the idea that the Big Bang is truly the beginning of the universe is simply a plausible hypothesis, not a result established beyond reasonable doubt. General relativity doesn't predict that space and time didn't exist before the Big Bang; it predicts that the curvature of spacetime in the very early universe became so large that general relativity itself ceases to be reliable. Quantum gravity, which we can happily ignore when we're talking about the curvature of spacetime in the relatively placid context of the contemporary universe, absolutely must be taken into account. And, sadly, we don't understand quantum gravity well enough to say for sure what actually happens at very early times. It might very well be true that space and time "come into existence" in that era—or not. Perhaps there is a transition from a phase of an irredeemably quantum wave function to the classical spacetime we know and love. But it is equally conceivable that space and time extend beyond the moment that we identify as "the Big Bang." Right now, we simply don't know; researchers are investigating different possibilities, with an open mind about which will eventually turn out to be right.

Some evidence that time doesn't need to have a beginning comes from quantum gravity, and in particular from the holographic principle we talked about in Chapter Twelve.[285] Maldacena showed that a particular theory of gravity in five-dimensional anti–de Sitter space is exactly equivalent to a "dual" four-dimensional theory that doesn't include gravity. There are plenty of questions that are hard to answer in the five-dimensional gravity theory, just like any other model of quantum gravity. But some of these issues become very straightforward from the dual four-dimensional perspective. For example: Does time have a beginning? Answer: no. The four-dimensional theory doesn't involve gravity at all; it's just a field theory that lives in some fixed spacetime, and that spacetime extends infinitely far into the past and the future. That's true even if there are singularities in the five-dimensional gravity theory; somehow, the theory finds a way to continue on beyond them. So we have an explicit example of a complete theory of quantum gravity, where there exists at least one formulation of the theory in which time never begins or ends, but stretches for all eternity. Admittedly, our own universe does not look much like five-dimensional anti–de Sitter space—it has four macroscopic dimensions, and the cosmological constant is positive, not negative. But Maldacena's example demonstrates that it's certainly not necessary that spacetime have a beginning, once quantum gravity is taken into account.

We can also take a less abstract approach to what might have come before the Big Bang. The most obvious strategy is to replace the Bang by some sort of bounce. We imagine that the universe before what we call the Big Bang was actually collapsing and growing denser. But instead of simply continuing to a singular Big Crunch, the universe—somehow—bounced into a phase of expansion, which we experience as the Big Bang.

The question is, what causes this bounce? It wouldn't happen under the usual assumptions made by cosmologists—classical general relativity, plus some reasonable restrictions on the kind of matter and energy in the universe. So we have to somehow change those rules. We could simply wave our hands and say "quantum gravity does it," but that's a little unsatisfying.

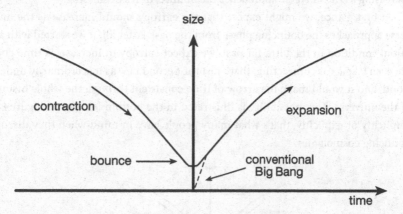

Figure 83: A bouncing-universe cosmology replaces the singularity of the standard Big Bang by a (more or less) smooth crossover between a contracting phase and an expanding phase.

Quite a bit of effort in recent years has gone into developing models that smooth out the Big Bang singularity into a relatively gentle bounce.[286] Each of these proposals offers the possibility of extending the history of the universe beyond the Big Bang, but in every case it's still hard to tell whether the model in question really hangs together. That's life when you're trying to understand the birth of the universe in the absence of a full theory of quantum gravity.

But the crucial point is worth keeping in mind: Even if we don't have one complete and consistent story to tell about how to extend the universe before the Big Bang, cosmologists are hard at work on the problem, and it's very plausible that they will eventually succeed. And the possibility that the Big Bang wasn't really the beginning of the universe has serious consequences for the arrow of time.

AN ARROW FOR ALL TIME

If the Big Bang was the beginning of time, we have a very clear puzzle: why was the entropy so small at that beginning? If the Big Bang was *not* the beginning, we still have a puzzle, but a very different one: why was the entropy small at the bounce, which wasn't even the beginning of the universe? It was just some moment in an eternal history.

For the most part, modern discussions of bouncing cosmologies don't address the question of entropy directly.[287] But it's pretty clear that the addition of a contracting phase before the bounce leaves us with two choices: Either the entropy is increasing as the universe approaches the bounce, or it's decreasing.

At first glance, we might expect that the entropy should increase as the universe approaches the bouncing phase from the past. After all, if we started with an initial condition in the ultra-far past, we expect entropy to increase as time goes on, even if space is contracting; that's just the Second Law as it is ordinarily understood, and it would make the arrow of time consistent through the whole history of the universe. This possibility is illustrated in the bottom left plot of Figure 84. Implicitly or explicitly, that's what many people have in mind when they discuss bouncing cosmologies.

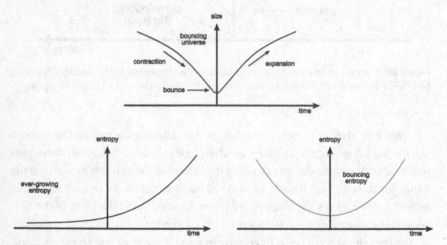

Figure 84: At the top, the size of a bouncing universe through time; at bottom, two possible scenarios for the evolution of entropy. The entropy could simply rise forever, as shown at bottom left, giving rise to a consistent arrow of time through all eternity. Or it could decrease during the contracting phase before beginning to increase in the expanding phase, as shown at bottom right.

But a scenario in which the entropy of our comoving patch increases consistently through a universal bounce faces an incredible problem. In conventional Big Bang cosmology, we have the problem that the entropy is relatively small in the current observable universe, and was substantially smaller in the past. This implies a great deal of hidden fine-tuning in the *present* microstate of the universe, so that entropy would decrease if we used the laws of physics to run it backward in time. But in the bouncing scenario, where we have pushed the "beginning of the universe" infinitely far away, the amount of fine-tuning needed to make this happen becomes infinitely bad. If we believe in reversible laws of physics, we need to imagine a state of the universe today with the property that it could be evolved backward in time forever, with the entropy continually decreasing all the way. That's a lot to ask.[288]

We should also mention a closely related problem. We know that the entropy of our comoving patch immediately after the bounce has to be small—that is, much smaller than it might have been. (From the estimates we made in Chapter Thirteen, it had to be 10^{88} or smaller, while it might have been as large as 10^{120}.) Which implies that the entropy was as small, or smaller, just before the bounce. If the entropy were large, you wouldn't get a bounce; you would get a chaotic mess that would have no hope of coming out the other side as the nice smooth universe from which we emerged. So what we have to imagine is that this comoving patch of space had been contracting for an infinitely long time (from the far past to the moment of the bounce), and in that time the entropy was increasing all along, but managed to increase only a tiny bit. That's not impossible to imagine, but it strikes us as unusual, to say the least.[289]

Even if we do allow ourselves to contemplate the possibility of the extraordinary amount of fine-tuning necessary to let entropy increase consistently for all time, we are left with absolutely no good reason why our universe should actually be that way. We have so far provided no justification for why our universe should be finely tuned at all, and now we are suggesting an infinite amount of fine-tuning. This doesn't really sound like progress.

A MIDDLE HYPOTHESIS

So we are led to consider the alternative, portrayed at bottom right in Figure 84: a bouncing universe where entropy *decreases* during the contracting phase, reaches a minimum value at the bounce, and begins to increase thereafter. Now, perhaps, we are getting somewhere. An explicit model of such a bouncing cosmology was proposed by Anthony Aguirre and Steven Gratton in 2003. They based their construction on inflation and showed that by clever cutting and pasting we could take an inflationary universe that was expanding forward in time and glue it at the beginning to an inflationary universe expanding backward in time, to obtain a smooth bounce.[290]

This alternative comes with a dramatic advantage: The behavior of the universe is symmetric in time. Both the size of the universe, and its entropy, would have a minimum value at the bounce, and increase in either direction. Conceptually, that's a big improvement over any of the other models we've contemplated; the underlying time-reversal symmetry of the laws of physics is reflected in the large-scale behavior of the universe. In particular, we avoid the pitfall of temporal chauvinism—the temptation to treat the "initial" state of the universe differently from the "final" state. It was our wish to sidestep that fallacy that led us to contemplate the Gold universe, which was also symmetric about one moment in time. But now that we allow ourselves to think about a possible universe before the Big Bang, the solution seems more acceptable: The universe is symmetric, not because entropy is low at either end of time, but because it's *high* at either end.

Nevertheless, this is a funny universe. The evolution of entropy is responsible for all the various manifestations of the arrow of time, including our ability to remember the past and our feeling that we move through time. In the bouncing-entropy scenario, the arrow of time *reverses direction* at the bounce. From the perspective of our observable universe, portrayed on the right-hand side of the plots in Figure 84, the past is the low-entropy direction of time, toward the bounce. But observers on the other side of the bounce, which we have (given our own perspective) labeled "contraction" in the plots, would also define the "past" as the direction of time in which entropy was lower—that is, the direction of the bounce. The arrow of time always points in the direction in which entropy is increasing, from the point of view of a local observer. On either side of the bounce, the arrow points toward a "future" in which the universe is expanding and emptying out. To observers on either side, observers on the other side experience time "running backward." But this mismatch of arrows is completely unobservable—people on one side of the bounce can't communicate with people on the other, any more than we can communicate with anyone else in our past. Everyone sees the Second Law of Thermodynamics operating normally in his or her observable part of the universe.

Unfortunately, a bouncing-entropy cosmos is not quite enough to allow us to declare in good conscience that we have solved the problem we set out at the beginning of this chapter. Sure, allowing for a cosmological bounce that is also a minimum point for the entropy of the universe avoids the philosophical pitfall of placing initial conditions and final conditions on a different footing. But it does so at the cost of a new puzzle: Why is the entropy so low in the *middle* of the history of the universe?

In other words, the bouncing-entropy model doesn't, by itself, actually *explain* anything at all about the arrow of time. Rather, it takes the need for a Past Hypothesis and replaces it with the need for a "Middle Hypothesis." There is just as much fine-tuning as ever; we are still stuck trying to explain why the configuration of our comoving patch of space found itself in such a low-entropy state near the cosmological bounce. So it would appear that we still have some work to do.

BABY UNIVERSES

To make an honest attempt at providing a robust dynamical explanation of the low entropy of our early universe, let's take it backward. Put aside for a moment what we know about our actual universe, and return to the question we asked in Chapter Thirteen: What *should* the universe look like? In that discussion, I argued that a natural universe—one that didn't rely on finely tuned low-entropy boundary conditions at any point, past, present, or future—would basically look like empty space. When we have a small positive vacuum energy, empty space takes the form of de Sitter space.

The question that any modern theory of cosmology must therefore answer is: Why don't we live in de Sitter space? It has a high entropy, it lasts forever, and the curvature of spacetime induces a small but nonzero temperature. De Sitter space is empty apart from the thin background of thermal radiation, so for the most part it is completely inhospitable to life; there is no arrow of time, since it's in thermal equilibrium. There will be thermal fluctuations, just as we would expect in a sealed box of gas in a Newtonian spacetime. Such fluctuations can give rise to Boltzmann brains, or entire galaxies, or whatever other macrostate you have in mind, if you wait long enough. But we do not appear to be such a fluctuation—if we were, the world around us would be as high entropy as it could possibly get, which it clearly is not.

There is a way out: De Sitter space might not simply stretch on for all eternity, uninterrupted. Something might happen to it. If that were the case, everything we have said about Boltzmann brains would be out the window. That argument made sense only because we knew exactly what kind of system we were dealing with—a gas at a fixed temperature—and we knew that it would last forever, so that even very improbable events would eventually occur, and we could reliable calculate the relative frequencies of different unreliable events. If we introduce complications into that picture, all bets are off. (Most bets, anyway.)

It's not hard to imagine ways that de Sitter space could fail to last forever. Remember that the "old inflation" model was basically a period of de Sitter space in the early universe, with a very high energy density provided by an inflaton field stuck in a false vacuum state. As long as there is another vacuum state of lower energy, that de Sitter space will eventually decay via the appearance of bubbles of true vacuum. If bubbles appear rapidly, the false vacuum will completely disappear;

if they appear slowly, we'll end up with a fractal mixture of true-vacuum bubbles in a persistent false-vacuum background.

In the case of inflation, a crucial point was that the energy density during the de Sitter phase was very high. Here we are interested in the opposite end of the spectrum—where the vacuum energy is extremely low, as it is in our current universe.

That makes a huge difference. High-energy states naturally like to decay into states of lower energy, but not vice versa. The reason is not because of energy conservation, but because of entropy.[291] The entropy associated with de Sitter space is low when the energy density is high, and high when the energy density is low. The decay of high-energy de Sitter space into a state with lower vacuum energy is just the natural evolution of a low-entropy state into a high-entropy one. But we want to know how we might escape from a situation like the one into which our current universe is evolving: empty de Sitter space with a very small vacuum energy, and a very high entropy. Where do we go from there?

If the correct theory of everything were quantum field theory in a classical de Sitter space background, we'd be pretty much stuck. Space would keep expanding, quantum fields would keep fluctuating, and we'd be more or less in the situation described by Boltzmann and Lucretius. But there is (at least) one possible escape route, courtesy of quantum gravity: the creation of *baby universes*. If de Sitter space gives birth to a continuous stream of baby universes, each of which starts with a low entropy and expands into a high-entropy de Sitter phase of its own, we could have a natural mechanism for creating more and more entropy in the universe.

As we've reiterated at multiple points, there's a lot we don't understand about quantum gravity. But there's a lot that we do understand about classical gravity, and about quantum mechanics; so we have certain reasonable expectations for what should happen in quantum gravity, even if the details remain to be ironed out. In particular, we expect that spacetime itself should be susceptible to quantum fluctuations. Not only should quantum fields in the de Sitter background be fluctuating, but the de Sitter space itself should be fluctuating.

One way in which spacetime might fluctuate was studied in the 1990s by Edward Farhi, Alan Guth, and Jemal Guven.[292] They suggested that spacetime could not only bend and stretch, as in ordinary classical general relativity, but also split into multiple pieces. In particular, a tiny bit of space could branch off from a larger universe and go its own way. The separate bit of space is, naturally, known as a baby universe. (In contrast to the "pocket universes" mentioned in the last chapter, which remained connected to the background spacetime.)

We can be more specific than that. The thermal fluctuations in de Sitter space are really fluctuations of the underlying quantum fields; the particles are just what we see when we observe the fields. Let's imagine that one of those fields has the right properties to be an inflaton—there are places in the potential where the field could sit relatively motionless in a false vacuum valley or a new-inflation plateau. But instead of starting it there, we consider what happens when the field starts at the bottom, where the vacuum energy is very small. Quantum fluctuations will occasionally push the field *up* the potential, from the true vacuum to the false vacuum—not everywhere at once, but in some small region of space.

What happens when a bubble of false vacuum fluctuates into existence in de Sitter space? To be honest once again, we're not sure.[293] One thing seems likely: Most of the time, the field will simply dissipate away back into its thermal surroundings. Inside, where we've fluctuated into the false vacuum, space wants to expand; but the wall separating the inside from the outside of the bubble wants to shrink, and usually it shrinks away quickly before anything dramatic happens.

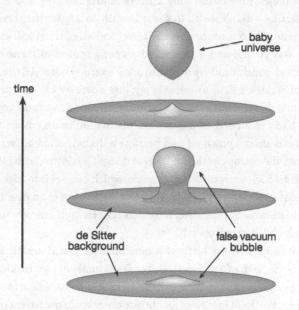

Figure 85: Creation of a baby universe via quantum fluctuation of a false-vacuum bubble.

Every once in a while, however, we could get lucky. The process of getting lucky is portrayed in Figure 85. What we see is a simultaneous fluctuation of the inflaton field, creating a bubble of false vacuum, and of space itself, creating a region that pinches off from the rest of the universe. The tiny throat that connects the two is a wormhole, as we discussed way back in Chapter Six. But this wormhole is unstable and will quickly collapse to nothing, leaving us with two disconnected spacetimes: the original parent universe and the tiny baby.

Now we have a baby universe, dominated by false vacuum energy, all set up to undergo inflation and expand to a huge size. If the properties of the false vacuum are just right, the energy will eventually be converted into ordinary matter and radiation, and we'll have a universe that evolves according to the standard inflation-plus-Big-Bang story. The baby universe can grow to an arbitrarily large size; there is no limitation imposed, for example, by energy conservation. It is a curious feature of general relativity that the total energy of a closed, compact universe is exactly zero, once we account for the energy of the gravitational field as well as everything else. So inflation can take a microscopically tiny ball of space and blow it up to the size of our observable universe, or much larger. As Guth puts it: "Inflation is the ultimate free lunch."

Of course the entropy of the baby universe starts out very small. That might seem like cheating—didn't we go to great lengths to argue that there are many degrees of freedom in our observable universe, and all of them still existed when the universe was young, and if we picked a configuration of them randomly it would be preposterously unlikely to obtain a low-entropy state? All that is true, but the process of making a baby universe is not one where we choose the configuration of our universe randomly. It's chosen in a very specific way: the configuration that is most likely to emerge as a quantum fluctuation in an empty background spacetime that is able to pinch off and become a disconnected universe. Considered as a whole, the entropy of the multiverse doesn't go down during this process; the initial state is high-entropy de Sitter space, which evolves into high-entropy de Sitter space plus a little extra universe. It's not a fluctuation of an equilibrium configuration into a lower-entropy state, but a leakage of a high-entropy state into one with an even higher entropy overall.

You might think that the birth of a new universe is a dramatic and painful event, just like the birth of a new person. But it's actually not so. Inside the bubble, of course, things are pretty dramatic—there's a new universe where there was none before. But from the point of view of an outside observer in the parent universe, the entire process is almost unnoticeable. What it looks like is a fluctuation of thermal particles that come together to form a tiny region of very high density—in fact, a black hole. But it's a microscopic black hole, with a tiny entropy, which then evaporates via Hawking radiation as quickly as it formed. The birth of a baby universe is much less traumatic than the birth of a baby human.

Indeed, if this story is true, a baby universe could be born right in the room where you're reading this book, and you would never notice. It's not very likely; in all the spacetime of the universe we can currently observe, chances are it never happened. If it did, the action would all be on a microscopic scale. The new universe could grow to a tremendous size, but it would be completely disconnected from the original spacetime. Like some other children, there is absolutely no communication between the baby universe and its parent—once they split, they remain separate forever.

A RESTLESS MULTIVERSE

So it's possible that, even when de Sitter space is in a high-entropy true-vacuum state, it's not completely stable. Rather, it can give birth to new baby universes, which grow up into large universes in their own right (and could very well give rise to new babies themselves). The original de Sitter space continues on its way, essentially unperturbed.

The prospect of baby universes makes all the difference in the world to the question of the arrow of time. Remember the basic dilemma: The most natural universe to live in is de Sitter space, empty space with a positive vacuum energy, which acts like an eternal box of gas at a fixed temperature. The gas spends most of its time in thermal equilibrium, with rare fluctuations into states of lower entropy. With that kind of setup, we could fairly reliably quantify what kinds of fluctuations there will be, and how often they will happen. Given any particular thing you would want the fluctuation to contain—a person, or a galaxy, or even a hundred billion galaxies—this scenario strongly predicts that most such fluctuations will look like they are in equilibrium, apart from the presence of the fluctuation itself. Furthermore, most such fluctuations will arise from higher-entropy states, and evolve back into higher-entropy states. So most observers will find themselves alone in the universe, having arisen as random arrangements of molecules out of the surrounding high-entropy gas of particles; likewise most galaxies, and so on. You could potentially fluctuate into something that looks just like the history of our Big Bang cosmology; but the number of observers within such a fluctuation is much smaller than the number of observers who are otherwise alone.

Baby universes change this picture in a crucial way. Now it's no longer true that the only thing that can happen is a thermal fluctuation away from equilibrium and then back again. A baby universe is a kind of fluctuation, but it's one that never comes back—it grows and cools off, but it doesn't rejoin the original spacetime.

What we've done is given the universe a way that it can increase its entropy without limit. In a de Sitter universe, space grows without bound, but the part of space that is visible to any one observer remains finite, and has a finite entropy—the area of the cosmological horizon. Within that space, the fields fluctuate at a fixed temperature that never changes. It's an equilibrium configuration, with every process occurring equally as often as its time-reverse. Once baby universes are added to the game, the system is no longer in equilibrium, for the simple reason that there is no such thing as equilibrium. In the presence of a positive vacuum energy (according to this story), the entropy of the universe never reaches a maximum value and stays there, because there is no maximum value for the entropy of the universe—it can always increase, by creating new universes. That's how the paradox of the Boltzmann-Lucretius scenario can be avoided.

Consider a simple analogy: a ball rolling on a hill. Not a quantum field moving in a potential, an actual down-to-Earth ball. But a very special hill: one that doesn't ever reach a particular bottom, but rolls smoothly away to infinity. And one on which there is absolutely no friction, so the ball can roll forever with the same amount of total energy.

Now let's ask ourselves: What should the ball be doing? That is, if we imagine finding such a ball, which has miraculously been operating as an isolated system for all of eternity, undisturbed by the rest of the universe, what kind of state would we expect the ball to be in?

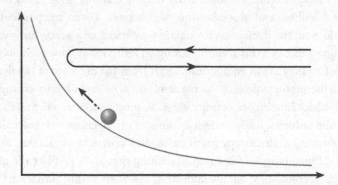

Figure 86: A ball rolling on a hill that doesn't have a bottom. There is only one kind of trajectory such a ball could have: coming in from infinitely far away in the infinite past, rolling up to a turning point, reversing direction, and rolling back out to infinity in the infinite future.

That may or may not be a sensible question, but it's not that hard to answer, because there aren't that many things the ball possibly could be doing. Every allowed trajectory for the ball looks basically the same: It rolls in from infinity, turns around, and rolls back out again. Depending on the total energy that the ball has, the turning point will reach different possible heights up the hill, but the qualitative behavior will be the same. So there will be precisely one moment in the lifetime of the ball when it isn't moving: the point where it turns around. At every other moment, it's either moving to the left, or moving to the right. Therefore, when we observe the ball at some random time, it seems very likely that it will be moving in one direction or the other.

Now imagine further that there is an entire tiny civilization living inside the ball, complete with tiny scientists and philosophers. One of their favorite topics of discussion is what they call the "arrow of motion." These thinkers have noticed that their ball evolves in perfect accord with Newton's laws of motion. Those laws don't distinguish between left and right: They are completely reversible. If a ball were to be placed at the bottom of a valley, it would simply sit there forever, motionless. If it were not quite at the bottom, it would start rolling toward the bottom, and then oscillate back and forth in that vicinity. Yet, their ball seems to be rolling consistently in the same direction for very long periods of time! What can be going on?

In case the terms of this somewhat-off-kilter analogy are not immediately clear, the ball represents our universe, and the position from left to right represents entropy. The reason why it's not surprising to find the ball moving in a consistent direction is that it tends to always be moving in the same direction, with the exception of the one special turnaround point. Despite appearances, the portion of the trajectory where the ball is coming in from right to left is not any different from the portion where it is moving away from left to right; the motion of the ball is time-reversal symmetric around that turning point.

Perhaps the entropy of our universe is like that. The real problem with de Sitter space (without baby universes) is that it's almost always in equilibrium—any particular observer sees a thermal bath that lasts forever, with predictable fluctuations. More generally, if there exists any such thing as "equilibrium" in the context of cosmology, it's hard to understand why we don't find the universe in that state. By suggesting that there is no such thing as equilibrium, we can avoid this dilemma. It becomes natural to observe entropy increasing, simply because entropy can always increase.

This is the scenario suggested by Jennifer Chen and me in 2004.[294] We started by assuming that the universe is eternal—the Big Bang is not the beginning of time—and that de Sitter space was a natural high-entropy state for the universe to be in. That means we can "start" with almost any state you like—pick some favorite distribution of matter and energy throughout space, and let it evolve. We put *start* in quotation marks because we don't want to prejudice initial conditions over

362 FROM ETERNITY TO HERE

conditions at any other time; respecting the reversibility of the laws of physics, we evolve the state both forward and backward in time. As I've argued here, the natural evolution forward in time is for space to expand and empty out, eventually approaching a de Sitter state. But from there, if we wait long enough, we will see the occasional production of baby universes via quantum fluctuations. These baby universes will expand and inflate, and their false vacuum energy will eventually convert into ordinary matter and radiation, which eventually dilutes away until we achieve de Sitter space once again. From there, both the original universe and the new universe can give birth to new babies. This process continues forever. In the parts of spacetime that look like de Sitter, the universe is in equilibrium, and there is no arrow of time. But in baby universes, for the time in between the initial birth and the final cooling off, there is a pronounced arrow of time, as the entropy starts near zero and expands to its equilibrium value.

Most interestingly, the same story can be told backward in time, starting from the initial state, as depicted in Figure 87. If it is not de Sitter already, the universe will empty out backward in time as well as forward. From there it will give birth to baby universes, which expand and cool off. In these baby universes, the arrow of time is oriented in the opposite direction to those in the universes we have put in "the future." The overall direction of the time coordinate is utterly arbitrary, of course. Observers in the universes at the top of the diagram will think of the bottom of the diagram as "the past," while observers in the universes at the bottom of the diagram will think of the top as "the past." Their arrows of time are incompatible, but that doesn't lead to any Benjamin Button unpleasantness; these baby universes are completely separate from one another in time, and their arrows point away from each other, so no communication between them is possible.

In this scenario, the multiverse on ultra-large scales is symmetric about the middle moment; statistically, at least, the far future and the far past are indistinguishable. In that sense this picture resembles the bouncing cosmologies we discussed earlier; entropy increases forever in both directions of time, around a middle point of lowest entropy. There is a crucial difference, however: The moment of "lowest" entropy is not actually a moment of "low" entropy. That middle moment was not finely tuned to some special very-low-entropy initial condition, as in typical bouncing models. It was as high as we could get, for a single connected universe in the presence of a positive vacuum energy. That's the trick: allowing entropy to continue to rise in both directions of time, even though it started out large to begin with. There isn't any state we could possibly have chosen that would have prevented this kind of evolution from happening. An arrow of time is inevitable.[295]

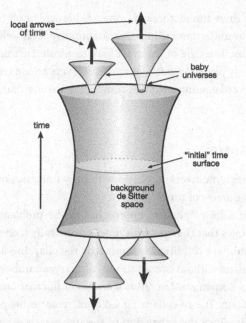

Figure 87: Baby universes created in a background de Sitter space, both to the past and to the future. Each baby universe starts in a dense, low-entropy state, and exhibits a local arrow of time as it expands and cools. The multiverse manifests overall time-reversal symmetry, as baby universes born in the past have an arrow of time pointing in the opposite direction to those in the future.

Having said all that, we may still want to ask why our patch of observable universe has such a low-entropy boundary condition at one end of time—why were our particular degrees of freedom ever found in such an unnatural state? But in this picture, that's not quite the right question to ask. We don't start by knowing which degrees of freedom we are, and then asking why they are (or were) in a certain configuration. Rather, we need to look at the multiverse as a whole, and ask what is most often experienced by observers like us. (If our scenario is going to be useful, the specific definition of "like us" shouldn't matter.)

This version of the multiverse will feature both isolated Boltzmann brains lurking in the empty de Sitter regions, and ordinary observers found in the aftermath of the low-entropy beginnings of the baby universes. Indeed, there should be an infinite number of both types. So which infinity wins? The kinds of fluctuations that create freak observers in an equilibrium background are certainly rare, but the kinds of fluctuations that create baby universes are also very rare. Ultimately, it's not enough to draw fun pictures of universes branching off in both directions of time; we need to understand things at a quantitative level well enough to make reliable predictions. The state of the art, I have to admit, isn't up to that task just yet. But it's certainly plausible that a lot more observers arise as the baby universes grow and cool toward equilibrium than come about through random fluctuations in empty space.

BRINGING IT HOME

Does it work? Does a multiverse scenario with baby universes offer a satisfactory explanation for the arrow of time?

We've covered a lot of possible approaches to the problem of the arrow of time: a space of states that changes with time, intrinsically irreversible dynamical laws, a special boundary condition, a symmetric re-collapsing universe, a bouncing universe with and without overall time-symmetry, an unbounded multiverse, and of course the Boltzmann-Lucretius scenario of fluctuations around an eternal equilibrium state. The re-collapsing Gold universe seems pretty unlikely on empirical grounds, since the expansion of the universe is accelerating; and the Boltzmann-Lucretius universe also seems ruled out by observation, since the Big Bang had a much lower entropy than it had any right to in that picture. But the other possibilities are still basically on the table; we may find them more or less satisfying, but we can't be confident enough to dismiss them out of hand. Not to mention the very real possibility that the right answer is something we simply haven't thought of yet.

It's hard to tell whether baby universes and the multiverse will ultimately play a role in understanding the arrow of time. For one thing, as I've taken pains (perhaps too many) to emphasize, there were many steps along the way where we were fearlessly speculative, to say the least. Our understanding of quantum gravity is not good enough to say for sure whether baby universes really do fluctuate into existence from de Sitter space; there seem to be arguments both for and against. We also don't completely understand the role of the vacuum energy. We've been speaking as if the cosmological constant we observe in our universe today is really the minimum possible vacuum energy, but there is little hard evidence for that assumption. In the context of the string theory landscape, for example, it's easy enough to get states with the right value of the vacuum energy, but it's also easy to get all kinds of states, including ones with negative vacuum energy or precisely zero vacuum energy. A more comprehensive theory of quantum gravity and the multiverse would predict how all of these possible states fit together, including transitions between different numbers of macroscopic dimensions as well as different values of the vacuum energy. Not to mention that we haven't really taken quantum mechanics completely seriously—we've nodded in the direction of quantum fluctuations but have drawn pictures of what are essentially classical space-times. The right answer, whatever it may turn out to be, will more likely be phrased in terms of wave functions, Schrödinger's equation, and Hilbert spaces.

The important point is not the prospects of any particular model, but the crucial clue that the arrow of time provides us as we try to understand the universe on the largest possible scales. If the universe we see is really all there is, with the Big Bang as a low-entropy beginning, we seem to be stuck with an uncomfortable fine-tuning problem. Embedding our observable patch in a wider multiverse alleviates this problem by changing the context: The goal is not to explain why the whole universe has a low-entropy boundary condition at the beginning of time, but why there exist relatively small regions of spacetime, arising within a much larger ensemble, where the entropy dramatically increases. That question, in turn, can be answered if the multiverse doesn't have any state of maximum entropy: The entropy increases because it can always increase, no matter what state we are in. The trick is to set things up so that the mechanism by which entropy increases overall is the production of universes that resemble our own.

The nice thing about a multiverse based on de Sitter space and baby universes is that it avoids all of the standard pitfalls that beset many approaches to the arrow of time: It treats the past and future on an equal footing, doesn't invoke irreversibility at the level of fundamental dynamics, and never assumes an ad hoc low-entropy state for the universe at any moment in time. It serves as a demonstration that such an explanation is at least conceivable, even if we aren't yet able to judge whether this particular one is sensible, much less part of the ultimately correct answer. There's every reason to be optimistic that we will eventually settle on an understanding of how the arrow of time arises naturally and dynamically from the laws of physics themselves.

16

EPILOGUE

Glance into the world just as though time were gone: and everything crooked will become straight to you.

—Friedrich Nietzsche

Unlike many authors, I had no struggle settling on the title for this book.[296] Once I had come up with *From Eternity to Here*, it seemed irresistible. The connotations were perfect: On the one hand, a classic movie (based on a classic novel), with that iconic scene of untamed waves from the Pacific crashing around lovers Deborah Kerr and Burt Lancaster caught in a passionate embrace. On the other hand, the cosmological grandeur implicit in the word *eternity*.

But the title is even more appropriate than those superficial considerations might suggest. This book has not only been about "eternity"; it's also been about "here." The puzzle of the arrow of time doesn't begin with giant telescopes or powerful particle accelerators; it's in our kitchens, every time we break an egg. Or stir milk into coffee, or put an ice cube into warm water, or spill wine onto the carpet, or let aromas drift through a room, or shuffle a new deck of cards, or turn a delicious meal into biological energy, or experience an event that leaves a lasting memory, or give birth to a new generation. All of these commonplace occurrences exhibit the fundamental irreversibility that is the hallmark of the arrow of time.

The chain of reasoning that started with an attempt to understand that arrow led us inexorably to cosmology—to eternity. Boltzmann provided us with an elegant and compelling microscopic understanding of entropy in terms of statistical mechanics. But that understanding does not explain the Second Law of Thermodynamics unless we also invoke a boundary condition—why was the entropy ever low to start with? The entropy of an unbroken egg is much lower than it could be, but such eggs are nevertheless common, because the overall entropy of the universe is much lower than it could be. And that's because it used to be even lower, all the way back to the beginning of what we can observe. What happens here, in our kitchen, is intimately connected with what happens in eternity, at the beginning of the universe.

Figures such as Galileo, Newton, and Einstein are celebrated for proposing laws of physics that hadn't previously been appreciated. But their accomplishments also share a common theme: They illuminate the *universality* of Nature. What happens here happens everywhere—as Richard Feynman put it, "The entire universe is in a glass of wine, if we look at it closely enough."[297] Galileo showed that the heavens were messy and ever changing, just like conditions here on Earth; Newton understood that the same laws of gravity that accounted for falling apples could explain the motions of the planets; and Einstein realized that space and time were different aspects of a single unified spacetime, and that the curvature of spacetime underlies the dynamics of the Solar System and the birth of the universe.

Likewise, the rules governing entropy and time are common to our everyday lives and to the farthest stretches of the multiverse. We don't yet know all the answers, but we're on the threshold of making progress on some big questions.

WHAT'S THE ANSWER?

Over the course of this book, we've lovingly investigated what we know about how time works, both in the smooth deterministic context of relativity and spacetime, and in the messy probabilistic world of statistical mechanics. We finally arrived at cosmology, and explored how our best theories of the universe fall embarrassingly short when confronted with the universe's most obvious feature: the difference in entropy between early times and late times. Then, after fourteen chapters of building up the problems, we devoted a scant single chapter to the possible solutions, and fell short of a full-throated endorsement of any of them.

That may seem frustrating, but the balance was entirely intentional. Understanding a deeply puzzling feature of the natural world is a process that can go through many stages—we may be utterly clueless, we may understand how to state the problem but not have any good ideas about the answer, we may have several reasonable answers at our disposal but not know which (if any) are right, or we may have it all figured out. The arrow of time falls in between the second and third of these options—we can state the problem very clearly but have only a few vague ideas of what the answer might be.

In such a situation, it's appropriate to dwell on understanding the problem, and not become too wedded to any of the prospective solutions. A century from now, most everything we covered in the first three parts of this book should remain standing. Relativity is on firm ground, as is quantum mechanics, and the framework of statistical mechanics. We are even confident in our understanding of the basic evolution of the universe, at least from a minute or so after the Big Bang up to today. But our current ideas about quantum gravity, the multiverse, and what happened at the Big Bang are still very speculative. They may grow into a robust understanding, but many of them may be completely abandoned. At this point it's more important to understand the map of the territory than to squabble over what is the best route to take through it.

Our universe isn't a fluctuation around an equilibrium background, or it would look very different. And it doesn't seem likely that the fundamental laws of physics are irreversible at a microscopic level—or, if they are, it's very hard to see how that could actually account for the evolution of entropy and complexity we observe in our universe. A boundary condition stuck at the beginning of time is impossible to rule out, but also seems to be avoiding the question more than answering it. It may ultimately be the best we can do, but I strongly suspect that the low entropy of our early universe is a clue to something deeper, not just a brute fact we can do no more than accept.

We're left with the possibility that our observable universe is part of a much larger structure, the multiverse. By situating what we see inside a larger ensemble, we open the possibility of explaining our apparently finely tuned beginning without imposing any fine-tuning on the multiverse as a whole. That move isn't sufficient, of course; we need to show why there should be a consistent entropy gradient, and why that gradient should be manifested in a universe that looks like our own, rather than in some other way.

We discussed a specific model of which I am personally fond: a universe that is mostly high-entropy de Sitter space, but which gives birth to disconnected baby universes, allowing the entropy to increase without bound and creating patches of spacetime like the one around us along the way. The details of this model are highly speculative, and rely on assumptions that stretch beyond what the state of the art allows us to reliably compute, to put it mildly. More important, I think, is the general paradigm, according to which entropy is seen to be increasing because entropy can always increase; there is no equilibrium state for the universe. That setup naturally leads to an entropy gradient, and is naturally time-symmetric about some moment of minimal (although not necessarily "small") entropy. It would be interesting to see if there are other ways of possibly carrying out this general program.

There is one other approach lurking in the background, which we occasionally acknowledged but never granted our undivided attention: the idea that "time" itself is simply an approximation that is occasionally useful, including in our local universe, but doesn't have any fundamental meaning. This is a perfectly legitimate possibility. Lessons from the holographic principle, as well as a general feeling that the underlying ingredients of a quantum mechanical theory may appear very different from what shows up in the classical regime, make it quite reasonable to imagine that time might be an emergent phenomenon rather than a necessary part of our ultimate description of the world.

One reason why the time-is-just-an-approximation alternative wasn't emphasized in this book is that there doesn't seem to be too much to say about it, at least within our present state of knowledge. Even by our somewhat forgiving standards, the way in which time might emerge from a more fundamental description is not well understood. But there is a more compelling reason, as well: Even if time is only an approximation, it's an approximation that seems extremely good in the part of the universe we can observe, and that's where the arrow-of-time problem is to be found. Sure, we can imagine that the viability of classical spacetime as a useful concept breaks down completely near the Big Bang. But, all by itself, that doesn't tell us anything at all about why conditions at that end of time (what we call "the past") should be so different from conditions at the other end of time ("the future") within our observable patch. Unless you can say, "Time is only an approximate concept, and therefore entropy should behave as follows in the regime where it's valid to speak about time," this alternative seems more like an evasive maneuver than a viable strategy. But that is largely a statement about our ignorance; it is certainly possible that the ultimate answer might lie in this direction.

THE EMPIRICAL CIRCLE

The pioneers of thermodynamics—Carnot, Clausius, and others—were motivated by practical desires; among other things, they wanted to build better steam engines. We've traveled directly from their insights to grand speculations about universes beyond our own. The crucial question is: How do we get back? Even if our universe does have an arrow of time because it belongs to a multiverse with an unbounded entropy, how would we ever know?

Scientists are fiercely proud of the *empirical* nature of what they do. Scientific theories do not become accepted because they are logical or beautiful, or fulfill some philosophical goal cherished by the scientist. Those might be good reasons why a theory is *proposed*—but being accepted is a much higher standard. Scientific theories must, at the end of the day, fit the data. No matter how intrinsically compelling a theory might be, if it fails to fit the data, it's a curiosity, not an achievement.

But this criterion of "fitting the data" is more slippery than it first appears. For one thing, lots of very different theories might fit the data; for another, a very promising theory might not completely fit the data as it currently stands, even though there is a kernel of truth to it. At a more subtle level, one theory might seem to fit the data perfectly well, but lead to a conceptual dead end, or to an intrinsic inconsistency, while another theory doesn't fit the data well at all, but holds promise for developing into something more acceptable. After all, no matter how much data we collect, we have only ever performed a tiny fraction of all possible experiments. How are we to choose?

The reality of how science is done can't be whittled down to a few simple mottos. The issue of distinguishing "science" from "not science" is sufficiently tricky that it goes by its own name: the *demarcation problem*. Philosophers of science have great fun arguing into the night about the proper way to resolve the demarcation problem.

Despite the fact that the goal of a scientific theory is to fit the data, the worst possible scientific theory would be one that fit *all possible* data. That's because the real goal isn't just to "fit" what we see in the universe; it's to *explain* what we see. And you can explain what we see only if you understand why things are the particular way they are, rather than some other way. In other words, your theory has to say that some things do not ever happen—otherwise you haven't said very much at all.

This idea was put forth most forcefully by Sir Karl Popper, who claimed that the important feature of a scientific theory wasn't whether it was "verifiable," but whether it was "falsifiable."[298] That's not to say that there are data that contradict the theory—only that the theory clearly makes predictions that could, in principle, be contradicted by some experiment we could imagine doing. The theory has to stick its neck out; otherwise, it's not scientific. Popper had in mind Karl Marx's theory of history, and Sigmund Freud's theory of psychoanalysis. These influential intellectual constructs, in his mind, fell far short of the scientific status their proponents liked to claim. Popper felt that you could take anything that happened in the world, or any behavior shown by a human being, and come up with an "explanation" of those data on the basis of Marx or Freud—but you wouldn't ever be able to point to any observed event and say, "Aha, there's no way to make that consistent with these theories." He contrasted these with Einstein's theory of relativity, which sounded equally esoteric and inscrutable to the person on the street, but made very definite predictions that (had the experiments turned out differently) could have falsified the theory.

THE MULTIVERSE IS NOT A THEORY

Where does that leave the multiverse? Here we are, claiming to be engaged in the practice of science, attempting to "explain" the observed arrow of time in our universe by invoking an infinite plethora of unobservable other universes. How is the claim that other universes exist falsifiable? It should come as no surprise that this kind of speculative theorizing about unobservable things leaves a bad taste in the mouths of many scientists. If you can't make a specific prediction that I could imagine doing an experiment to falsify, they say, what you're doing isn't science. It's philosophy at best, and not very good philosophy at that.

But the truth, as is often the case, is a bit more complicated. All this talk of multiverses might very well end up being a dead end. A century from now, our successors might be shaking their heads at all the intellectual effort that was wasted on trying to figure out what came before the Big Bang, as much as we wonder at all that work put into alchemy or the caloric theory of heat. But it won't be because modern cosmologists had abandoned the true path of science; it will (if that's how things turn out) simply be because the theory wasn't correct.

Two points deserve to be emphasized concerning the role of unobservable things in science. First, it's wrong to think of the goal of science as simply to fit the data. The goal of science goes much deeper than that: It's to *understand* the behavior of the natural world.[299] In the early seventeenth century, Johannes Kepler proposed his three laws of planetary motion, which correctly accounted for the voluminous astronomical data that had been collected by his mentor, Tycho Brahe. But we didn't really understand the dynamics of planets within the Solar System until Isaac Newton showed that they could all be explained in terms of a simple inverse-square law for gravity. Similarly, we don't need to look beyond the Big Bang to understand the evolution of our observable universe; all we have to do is specify what conditions were like at early times, and leave it at that. But that's a strategy that denies us any understanding of why things were the way they were.

Similar logic would have argued against the need for the theory of inflation; all inflation did was take things that we already knew were true about the universe (flatness, uniformity, absence of monopoles) and attempt to explain them in terms of simple underlying rules. We didn't need to do that; we could have accepted things as they are. But as a result of our desire to do better, to actually understand the early universe rather than simply accept it, we discovered that inflation provides more than we had even asked for: a theory of the origin and nature of the primordial perturbations that grow into galaxies and large-scale structure. That's the benefit to searching for understanding, rather than being content with fitting the data: True understanding leads you places you didn't know you wanted to go. If we someday understand why the early universe had a low entropy, it is a good bet that the underlying mechanism will teach us more than that single fact.

The second point is even more important, although it sounds somewhat trivial: science is a messy, complicated business. It will never stop being true that the basis of science is empirical knowledge; we are guided by data, not by pure reason. But along the way to being guided by data, we use all sorts of nonempirical clues and preferences in constructing models and comparing them to one another. There's nothing wrong with that. Just because the end product must be judged on the basis of how well it explains the data, doesn't mean that every step along the way must have the benefit of an intimate and detailed contact with experiment.

More specifically: The multiverse is not a "theory." If it were, it would be perfectly fair to criticize it on the basis of our difficulty in coming up with possible experimental tests. The correct way to think about the multiverse is as a *prediction*. The theory—such as it is, in its current underdeveloped state—is the marriage of the principles behind quantum field theory to our basic understanding of how curved spacetime works. Starting from those inputs, we don't simply theorize that the universe could have undergone an early period of superfast acceleration; we *predict* that inflation should occur, if a quantum inflaton field with the right properties finds itself in the right state. Likewise, we don't simply say, "Wouldn't it be cool if there were an infinite number of different universes?" Rather, we predict on the basis of reasonable extrapolations of gravity and quantum field theory that a multiverse really should exist.

The prediction that we live in a multiverse is, as far as we can tell, untestable. (Although, who knows? Scientists have come up with remarkably clever ideas before.) But that misses the point. The multiverse is part of a larger, more comprehensive structure. The question should be not "How can we test whether there is a multiverse?" but "How can we test the theories that predict the multiverse should exist?" Right now we don't know how to use those theories to make a falsifiable prediction. But there's no reason to think that we can't, in principle, do so. It will require a lot more work on the part of theoretical physicists to develop these ideas to the point where we can say what, if any, the testable predictions might be. One might be *impatient* that those predictions aren't laid out before them straightforwardly right from the start—but that's a personal preference, not a principled philosophical stance. Sometimes it takes time for a promising scientific idea to be nurtured and developed to the point where we can judge it fairly.

THE SEARCH FOR MEANING IN A PREPOSTEROUS UNIVERSE

Throughout history, human beings have (quite naturally) tended to consider the universe in human-being-centric terms. That might mean something as literal as putting ourselves at the geographical center of the universe—an assumption that took some effort to completely overcome. Ever since the heliocentric model of the Solar System gained widespread acceptance, scientists have held up the Copernican Principle—"we do not occupy a favored place in the universe"—as a caution against treating ourselves as something special.

But at a deeper level, our anthropocentrism manifests itself as a conviction that human beings somehow *matter* to the universe. This feeling is at the core of much of the resistance in some quarters to accepting Darwin's theory of natural selection as the right explanation for the evolution of life on Earth. The urge to think that we matter can take the form of a straightforward belief that we (or some subset of us) are God's chosen people, or something as vague as an insistence that all this marvelous world around us must be more than just an *accident*.

Different people have different definitions of the word *God*, or different notions of what the nominal purpose of human life might be. God can become such an abstract and transcendental concept that the methods of science have nothing to say about the matter. If God is identified with Nature, or the laws of physics, or our feeling of awe when contemplating the universe, the question of whether or not such a concept provides a useful way of thinking about the world is beyond the scope of empirical inquiry.

There is a very different tradition, however, that seeks evidence for God in the workings of the physical universe. This is the approach of natural theology, which stretches long before Aristotle, through William Paley's watchmaker analogy, up to the present day.[300] It used to be that the best evidence in favor of the argument from design came from living organisms, but Darwin provided an elegant mechanism to explain what had previously seemed inexplicable. In response, some adherents to this philosophy have shifted their focus to a different seemingly inexplicable thing: from the origin of life to the origin of the cosmos.

The Big Bang model, with its singular beginning, seems to offer encouragement to those who would look for the finger of God in the creation of the universe. (Georges Lemaître, the Belgian priest who developed the Big Bang model, refused to enlist it for any theological purposes: "As far as I can see, such a theory remains entirely outside of any metaphysical or religious question."[301]) In Newtonian spacetime, there wasn't even any such thing as the creation of the universe, at least not as an event happening at a particular time; time and space persisted forever. The introduction of a particular beginning to spacetime, especially one that apparently defies easy understanding, creates a temptation to put the responsibility for explaining what went on into the hands of God. Sure, the reasoning goes, you can find dynamical laws that govern the evolution of the universe from moment to moment, but explaining the creation of the universe itself requires an appeal to something outside the universe.

Hopefully, one of the implicit lessons of this book has been that it's not a good idea to bet against the ability of science to explain anything whatsoever about the operation of the natural world, including its beginning. The Big Bang represented a point past which our understanding didn't stretch, back when it was first studied in the 1920s—and it continues to do so today. We don't know exactly what happened 14 billion years ago, but there's no reason whatsoever to doubt that we will eventually figure it out. Scientists are tackling the problem from a variety of angles. The rate at which scientific understanding advances is notoriously hard to predict, but it's not hard to predict that it will be advancing.

Where does that leave us? Giordano Bruno argued for a homogeneous universe with an infinite number of stars and planets. Avicenna and Galileo, with the conservation of momentum, undermined the need for a Prime Mover to explain the persistence of motion. Darwin explained the development of species as an undirected process of descent with random modifications, chosen by natural selection. Modern cosmology speculates that our observable universe could be only one of an infinite number of universes within a grand ensemble multiverse. The more we understand about the world, the smaller and more peripheral to its operation we seem to be.[302]

That's okay. We find ourselves, not as a central player in the life of the cosmos, but as a tiny epiphenomenon, flourishing for a brief moment as we ride a wave of increasing entropy from the Big Bang to the quiet emptiness of the future universe. Purpose and meaning are not to be found in the laws of nature, or in the plans of any external agent who made things that way; it is our job to create them. One of those purposes—among many—stems from our urge to explain the world around us the best we can. If our lives are brief and undirected, at least we can take pride in our mutual courage as we struggle to understand things much greater than ourselves.

NEXT STEPS

It's surprisingly hard to think clearly about time. We're all familiar with it, but the problem might be that we're *too* familiar. We're so used to the arrow of time that it's hard to conceptualize time without the arrow. We are led, unprotesting, to temporal chauvinism, prejudicing explanations of our current state in terms of the past over those in terms of the future. Even highly trained professional cosmologists are not immune.

Despite all the ink that has been spilled and all the noise generated by discussions about the nature of time, I would argue that it's been discussed too little, rather than too much. But people seem to be catching on. The intertwined subjects of time, entropy, information, and complexity bring together an astonishing variety of intellectual disciplines: physics, mathematics, biology, psychology, computer science, the arts. It's about time that we took time seriously, and faced its challenges head-on.

Within physics, that's starting to happen. For much of the twentieth century, the field of cosmology was a bit of a backwater; there were many ideas, and little data to distinguish between them. An era of precision cosmology, driven by large-scale surveys enabled by new technologies, has changed all that; unanticipated wonders have been revealed, from the acceleration of the universe to the snapshot of early times provided by the cosmic microwave background.[303] Now it is the turn for ideas to catch up to the reality. We have interesting suggestions from inflation, from quantum cosmology, and from string theory, to how the universe might have begun and what might have come before. Our task is to develop these promising ideas into honest theories, which can be compared with experiment and reconciled with the rest of physics.

Predicting the future isn't easy. (Curse the absence of a low-entropy future boundary condition!) But the pieces are assembled for science to take dramatic steps toward answering the ancient questions we have about the past and the future. It's time we understood our place within eternity.

APPENDIX: MATH

Lloyd: You mean, not good like one out of a hundred?
Mary: I'd say more like one out of a million.
[pause]
Lloyd: So you're telling me there's a chance.

—Jim Carrey and Lauren Holly, *Dumb and Dumber*

In the main text I bravely included a handful of equations—a couple by Einstein, and a few expressions for entropy in different contexts. An equation is a powerful, talismanic object, conveying a tremendous amount of information in an extraordinarily compact notation. It can be very useful to look at an equation and understand its implications as a rigorous expression of some feature of the natural world.

But, let's face it—equations can be scary. This appendix is a very quick introduction to exponentials and logarithms, the key mathematical ideas used in describing entropy at a quantitative level. Nothing here is truly necessary to comprehending the rest of the book; just bravely keep going whenever the word *logarithm* appears in the main text.

EXPONENTIALS

These two operations—exponentials and logarithms—are exactly as easy or difficult to understand as each other. Indeed, they are opposites; one operation undoes the other one. If we start with a number, take its exponential, and then take the logarithm of the result, we get back the original number we started with. Nevertheless, we tend to come across exponentials more often in our everyday lives, so they seem a bit less intimidating. Let's start there.

Exponentials just take one number, called the *base*, and raise it to the power of another number. By which we simply mean: Multiply the base by itself, a number of times given by the power. The base is written as an ordinary number, and the power is written as a superscript. Some simple examples:

$$2^2 = 2 \bullet 2 = 4,$$

$$2^5 = 2 \bullet 2 \bullet 2 \bullet 2 \bullet 2 = 32,$$

$$4^3 = 4 \bullet 4 \bullet 4 = 64.$$

(We use a dot to stand for multiplication, rather than the × symbol, because that's too easy to confuse with the letter x.) One of the most convenient cases is where we take the base to be 10; in that case, the power simply becomes the number of zeroes to the right of the one.

$$10^1 = 10,$$

$$10^2 = 100,$$

$$10^9 = 1,000,000,000,$$

$$10^{21} = 1,000,000,000,000,000,000,000.$$

That's the idea of exponentiation. When we speak more specifically about the exponential *function*, what we have in mind is fixing a particular base and letting the power to which we raise it be a variable quantity. If we denote the base by a and the power by x, we have

$$a^x = a \bullet a \bullet a \bullet a \bullet a \bullet a \ldots \bullet a, x \text{ times.}$$

This definition, unfortunately, can give you the impression that the exponential function makes sense only when the power x is a positive integer. How can you multiply a number by itself minus-two times, or 3.7 times? Here you will have to have faith that the magic of mathematics allows us to define the exponential for *any* value of x. The result is a smooth function that is very small when x is a negative number, and rises very rapidly when x becomes positive, as shown in Figure 88.

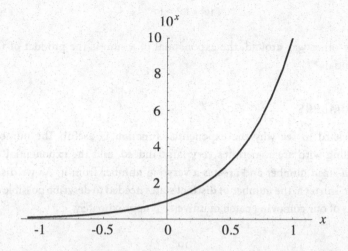

Figure 88: The exponential function 10^x. Note that it goes up very fast, so that it becomes impractical to plot it for large values of x.

There are a couple of things to keep in mind about the exponential function. The exponential of 0 is always equal to 1, for any base, and the exponential of 1 is equal to the base itself. When the base is 10, we have:

$$10^0 = 1,$$

$$10^1 = 10.$$

If we take the exponential of a negative number, it's just the reciprocal of the exponential of the corresponding positive number:

$$10^{-1} = 1/10^1 = 0.1,$$

$$10^{-3} = 1/10^3 = 0.001.$$

These facts are specific examples of a more general set of properties obeyed by the exponential function. One of these properties is of paramount importance: If we *multiply* two numbers that are the same base raised to different powers, that's equal to what we would get by *adding* the two powers and raising the base to that result. That is:

$$10^x \cdot 10^y = 10^{(x+y)}$$

Said the other way around, the exponential of a sum is the product of the two exponentials.[304]

BIG NUMBERS

It's not hard to see why the exponential function is useful: The numbers we are dealing with are sometimes very large indeed, and the exponential takes a medium-sized number and creates a very big number from it. As we discuss in Chapter Thirteen, the number of distinct states needed to describe possible configurations of our comoving patch of universe is approximately

$$10^{10^{120}}$$

That number is just so enormously, unimaginably huge that it would be hard to know how to even begin describing it if we didn't have recourse to exponentiation.

Let's consider some other big numbers to appreciate just how giant this one is. One billion is 10^9, while one trillion is 10^{12}; these have become all too familiar terms in discussions of economics and government spending. The number of particles within our observable universe is about 10^{88}, which was also the entropy at early times. Now that we have black holes, the entropy of the observable universe is something like 10^{101}, whereas it conceivably could have been as high as 10^{120}. (That same 10^{120} is also the ratio of the predicted vacuum energy density to the observed density.)

For comparison's sake, the entropy of a macroscopic object like a cup of coffee is about 10^{25}. That's related to Avogadro's Number, $6.02 \cdot 10^{23}$, which is approximately the number of atoms in a gram of hydrogen. The number of grains of sand in all the Earth's beaches is about 10^{20}. The number of stars in a typical galaxy is about 10^{11}, and the number of galaxies in the observable universe is also about 10^{11}, so the number of stars in the observable universe is about 10^{22}—a bit larger than the number of grains of sand on Earth.

The basic units that physicists use are time, length, and mass, or combinations thereof. The shortest interesting time is the Planck time, about 10^{-43} seconds. Inflation is conjectured to have lasted for about 10^{-30} seconds or less, although that number is extremely uncertain. The universe created helium out of protons and neutrons about 100 seconds after the Big Bang, and it became transparent at the time of recombination, 380,000 years (10^{13} seconds) after that. (One year is about $3 \cdot 10^7$ seconds.) The observable universe now is 14 billion years old, about $4 \cdot 10^{17}$

seconds. In another 10^{100} years or so, all the black holes will have mostly evaporated away, leaving a cold and empty universe.

The shortest length is the Planck length, about 10^{-33} centimeters. The size of a proton is about 10^{-13} centimeters, and the size of a human being is about 10^2 centimeters. (That's a pretty short human being, but we're only being very rough here.) The distance from the Earth to the Sun is about 10^{13} centimeters; the distance to the nearest star is about 10^{18} centimeters, and the size of the observable universe is about 10^{28} centimeters.

The Planck mass is about 10^{-5} grams—that would be extraordinarily heavy for a single particle, but isn't all that much by macroscopic standards. The lightest particles that have more than zero mass are the neutrinos; we don't know for sure what their masses are, but the lightest seem to be about 10^{-36} grams. A proton is about 10^{-24} grams, and a human being is about 10^5 grams. The Sun is about 10^{33} grams, a galaxy is about 10^{45} grams, and the mass within the observable universe is about 10^{56} grams.

LOGARITHMS

The logarithm function is the easiest thing in the world: It undoes the exponential function. That is, if we have some number that can be expressed in the form 10^x—and every positive number can be—then the logarithm of that number is simply

$$\log(10^x) = x.$$

What could be simpler than that? Likewise, the exponential undoes the logarithm:

$$10^{\log(x)} = x.$$

Another way of thinking about it is: If a number is a perfect power of 10 (like 10, 100, 1,000, etc.), the logarithm is simply the number of zeroes to the right of the initial 1:

$$\log(10) = 1,$$

$$\log(100) = 2,$$

$$\log(1,000) = 3.$$

But just as for the exponential, the logarithm is actually a smooth function, as shown in Figure 89. The logarithm of 2.5 is about 0.3979, the logarithm of 25 is about 1.3979, the logarithm of 250 is about 2.3979, and so on. The only restriction is that we can't take the logarithm of a negative number; that makes sense, because the logarithm inverts the exponential function, and we can never *get* a negative number by exponentiating. Roughly speaking, for large numbers the logarithm is simply "the number of digits in the number."

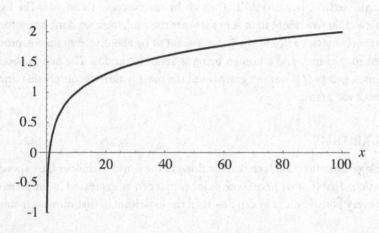

Figure 89: The logarithm function log(x). It is not defined for negative values of x, and as x approaches zero from the right the logarithm goes to minus infinity.

Just like the exponential of a sum is the product of exponentials, the logarithm has a corresponding property: The logarithm of a product is the sum of logarithms. That is:

$$\log(x \bullet y) = \log(x) + \log(y).$$

It's this lovely property that makes logarithms so useful in the study of entropy. As we discuss in Chapter Eight, a physical property of entropy is that the entropy of two systems combined together is equal to the sum of the entropies of the two individual systems. But you get the number of possible states of the combined systems by multiplying the numbers of states of the two individual systems. So Boltzmann concluded that the entropy should be the logarithm of the number of states, not the number of states itself. In Chapter Nine we tell a similar story for information:

Shannon wanted a measure of information for which the total information carried in two independent messages was the sum of the individual informations in each message, so he realized he also had to take the logarithm.

More informally, logarithms have the nice property that they take large numbers and whittle them down to manageable sizes. When we take the logarithm of an unwieldy number like a trillion, we get a nice number like 9. The logarithm is a monotonic function—it always increases as we increase the number we're taking the logarithm of. So the logarithm gives a specific measure of how big a number is, but it collapses huge numbers down to a reasonable size, which is very helpful in fields like cosmology, statistical mechanics, or even economics.

One final crucial detail is that, just like exponentials, logarithms can come in different bases. The "log base b" of a number x is the number to which we would have to raise b in order to get x. That is:

$$\log_2(2^x) = x,$$

$$\log_{12}(12^x) = x,$$

and so on. Whenever we don't write the base explicitly, we take it to be equal to 10, because that's how many fingers most human beings have. But scientists and mathematicians often like to make a seemingly odd choice: they use the *natural logarithm*, often written $\ln(x)$, in which the base is taken to be Euler's constant:

$$\ln(x) = \log_e(x),$$

$$e = 2.7182818284\ldots$$

Euler's constant is an irrational number, like pi or the square root of two, so its explicit form above would go on forever. At first glance that seems like a truly perverse choice to use as a base for one's logarithms. But in fact e has a lot of nice properties, once you get deeper into the math; in calculus, for example, the function e^x is the only one (aside from the trivial function equal to zero everywhere) that is equal to its own derivative, as well as its own integral. In this book all of our logarithms have used base 10, but if you launch yourself into physics and math at a higher level, it will be natural logarithms all the way.

NOTES

PROLOGUE

1 Wikipedia contributors (2009).
2 Let's emphasize the directions here, because they are easily confused: Entropy measures disorder, not order, and it increases with time, not decreases. We informally think "things wind down," but the careful way of saying that is "entropy goes up."

1. THE PAST IS PRESENT MEMORY

3 In an effort not to be too abstract, we will occasionally lapse into a kind of language that assumes the directionality of time—"time passes," we "move into the future," stuff like that. Strictly speaking, part of our job is to explain why that language seems so natural, as opposed to phrasings along the lines of "there is the present, and there is also the future," which seems stilted. But it's less stressful to occasionally give into the "tensed" way of speaking, and question the assumptions behind it more carefully later on.

4 Because the planets orbit in ellipses rather than perfect circles, their velocity around the Sun is not strictly constant, and the actual angle that the Earth describes in its orbit every time Mars completes a single revolution will depend on the time of year. These are details that are easy to take care of when we actually sit down to carefully define units of time.

5 The number of vibrations per second is fixed by the size and shape of the crystal. In a watch, the crystal is tuned to vibrate 32,768 times per second, which happens to be equal to 2 to the 15th power. That number is chosen so that it's easy for the watch's inner workings to divide successively by 2 to obtain a frequency of exactly once per second, appropriate for driving the second hand of a watch.

6 Alan Lightman's imaginative novel *Einstein's Dreams* presents a series of vignettes that explore what the world would be like if time worked very differently than it does in the real world.

7 See for example Barbour (1999) or Rovelli (2008).

8 There is a famous joke, attributed to Einstein: "When a man sits with a pretty girl for an hour, it seems like a minute. But let him sit on a hot stove for a minute and it's longer than any hour. That's relativity." I don't know whether Einstein actually ever said those words. But I do know that's not relativity.

9 Here is a possible escape clause, if we were really committed to restoring the scientific integrity of Baker's fantasy: Perhaps time in the rest of the world didn't completely stop, but just slowed down by a tremendous factor, and still ticked along at a sufficient rate that light could travel from the objects Arno was looking at to his eyes. Close, but no cigar. Even if that happened, the fact that the light was slowed down would lead to an enormous redshift—what looked like visible light in the ordinary world would appear to Arno as radio waves, which his poor eyes wouldn't be able to see. Perhaps X-rays would be redshifted down to visible wavelengths, but X-ray flashlights are hard to come by. (It does, admittedly, provoke one into thinking how interesting a more realistic version of this scenario might be.)

10 *Temporal*: of or pertaining to time. It's a great word that we'll be using frequently. Sadly, an alternative meaning is "pertaining to the present life or this world"—and we'll be roaming very far away from that meaning.

11 As a matter of historical accuracy, while Einstein played a central role in the formulation of special relativity, it was legitimately a collaborative effort involving the work of a number of physicists and mathematicians, including George FitzGerald, Hendrik Lorentz, and Henri Poincaré. It was eventually Hermann Minkowski who took Einstein's final theory and showed that it could be understood in terms of a four-dimensional spacetime, which is often now called "Minkowski space." His famous 1909 quote was "The views of space and time which I wish to lay before you have sprung from the soil of

experimental physics, and therein lies their strength. They are radical. Henceforth space by itself, and time by itself, are doomed to fade away into mere shadows, and only a kind of union of the two will preserve an independent reality" (Minkowski, 1909).

12 Pirsig (1974), 375.
13 Price (1996), 3.
14 Vonnegut (1969), 34. Quoted in Lebowitz (2008).
15 Augustine (1998), 235.
16 Good discussions of these issues can be found in Callender (2005), Lockwood (2005), and Davies (1995).
17 Philosophers often discuss different conceptions of time in terms laid out by J. M. E. McTaggart in his famous paper "The Unreality of Time" (1908). There, McTaggart distinguished between three different notions of time, which he labeled as different "series" (see also Lockwood, 2005). The A-series is a series of events measured relative to now, that move through time—"one year ago" doesn't denote a fixed moment, but one that changes as time passes. The B-series is the sequence of events with permanent temporal labels, such as "October 12, 2009." And the C-series is simply an ordered list of events—"*x* happens before *y* but after *z*"—without any time stamps at all. McTaggart argued—very roughly—that the B-series and C-series are fixed arrays, lacking the crucial element of change, and therefore insufficient to describe time. But the A-series itself is incoherent, as any specific event will be classified simultaneously as "past," "present," and "future," from the point of view of different moments in time. (The moment of your birth is in the past to you now but was in the future to your parents when they first met.) Therefore, he concludes, time doesn't exist.
    If you get the feeling that this purported contradiction seems more like a problem with language than one with the nature of time, you are on the right track. To a physicist, there seems to be no contradiction between stepping outside the universe and thinking of all of spacetime at once, and admitting that from the point of view of any individual inside the universe time seems to flow.

2. THE HEAVY HAND OF ENTROPY

18 Amis (1991), 11.
19 Fitzgerald (1922).
20 Carroll, L. (2000), 175.
21 Obviously.
22 Diedrick (1995) lists a number of stories that feature time reversals in one form or another, in addition to the ones mentioned here: Lewis Carroll's *Sylvie and Bruno*, Jean Cocteau's *Le Testament d'Orphee*, Brian Aldiss's *An Age*, and Philip K. Dick's *Counter-Clock World*. In T. H. White's *The Once and Future King*, the character of Merlyn experiences time backward, although White doesn't try very hard to consistently maintain the conceit. More recently, the technique has been used by Dan Simmons in *Hyperion*, and serves as a major theme in Andrew Sean Greer's *The Confessions of Max Tivoli* and in Greg Egan's short story "The Hundred-Year Diary." Vonnegut's *Slaughterhouse-Five* includes a brief description of the firebombing of Dresden in reversed order, which Amis credits in the Afterword to *Time's Arrow*.
23 Stoppard (1999), 12.
24 In addition to the First Law of Thermodynamics ("the total energy remains constant in any physical process") and the Second Law ("the entropy of a closed system never decreases"), there is also a Third Law: As the temperature of a system is lowered, there is a minimum value (absolute zero) for which the entropy is also a minimum. These three laws have been colorfully translated as: "You can't win; you can't break even; and you can't even get out of the game." There is also a Zeroth Law: If two systems are both in thermal equilibrium with a third system, they are in thermal equilibrium with each other. Feel free to invent your own whimsical sporting analogies.
25 Eddington (1927), 74.
26 Snow (1998), 15.
27 In fact, it would be fair to credit Sadi Carnot's father, French mathematician and military officer Lazare Carnot, with the first glimmerings of this concept of entropy and the Second Law. In 1784, Lazare Carnot wrote a treatise on mechanics in which he argued that perpetual motion was impossible, because any realistic machine would dissipate useful energy through the rattling and shaking of its component parts. He later became a successful leader of the French Revolutionary Army.
28 Not strictly true, actually. Einstein's general theory of relativity, which explains gravitation in terms of the curvature of spacetime, implies that what we ordinarily call "energy" is not really conserved, for

example, in an expanding universe. We'll talk about that in Chapter Five. But for the purposes of most combustion engines, the expansion of the universe can be neglected, and energy really is conserved.

29   Specifically, by "measures the number of ways we can rearrange the individual parts," we mean "is proportional to the logarithm of the number of ways we can rearrange the individual parts." See the Appendix for a discussion of logarithms, and Chapter Nine for a detailed discussion of the statistical definition of entropy.

30   The temperature of the surface of the Sun is approximately 5,800 Kelvin. (One Kelvin is the same as one degree Celsius, except that zero Kelvin corresponds to -273 degrees C—absolute zero, the lowest possible temperature.) Room temperature is approximately 300 Kelvin. Space—or, more properly, the cosmic background radiation that suffuses space—is at about 3 Kelvin. There is a nice discussion of the role of the Sun as a hot spot in a cold sky in Penrose (1989).

31   You will sometimes hear claims by creationists to the effect that evolution according to Darwinian natural selection is incompatible with the growth of entropy, since the history of life on Earth has involved increasingly complex organisms purportedly descending from less complex forms. This is crazy on several levels. The most basic level is simply: The Second Law refers to closed systems, and an organism (or a species, or the biosphere) is not a closed system. We'll discuss this a bit more in Chapter Nine, but that's basically all there is to it.

32   Thompson (1862).

33   Pynchon (1984), 88.

3. THE BEGINNING AND END OF TIME

34   In fact there was a literal debate—the "Great Debate" between astronomers Harlow Shapley and Heber Curtis was held in 1920 at the Smithsonian in Washington, D.C. Shapley defended the position that the Milky Way was the entirety of the universe, while Curtis argued that the nebulae (or at least some of them, and in particular the Andromeda nebula M31) were galaxies like our own. Although Shapley ended up on the losing side of the big question, he did correctly understand that the Sun was not at the center of the Milky Way.

35   That's a bit of poetic license. As we will explain later, the cosmological redshift is conceptually distinct from the Doppler effect, despite their close similarity. The former arises from the expansion of space through which the light is traveling, while the latter arises from the motion of the sources through space.

36   After decades of heroic effort, modern astronomers have finally been able to pin down the actual value of this all-important cosmological parameter: 72 kilometers per second per Megaparsec (Freedman et al., 2001). That is, for every million parsecs of distance between us and some galaxy, we will observe an apparent recession velocity of 72 km/sec. For comparison, the current size of the observable universe is about 28 billion parsecs across. A parsec is about 3.26 light years, or 30 trillion kilometers.

37   Strictly speaking, we should say "every sufficiently distant galaxy . . ." Nearby galaxies could be bound into pairs or groups or clusters under the influence of their mutual gravitational attraction. Such groups, like any bound systems, do not expand along with the universe; we say that they have "broken away from the Hubble flow."

38   Admittedly, it's a bit subtle. Just two footnotes prior, we said the observable universe was "28 billion parsecs" across. It's been 14 billion years since the Big Bang, so you might think there are 14 billion light-years from here to the edge of the observable universe, which we can multiply by two to get the total diameter—28 billion light years, or about 9 billion parsecs, right? Was there a typo, or how can these be reconciled? The point is that distances are complicated by the fact that the universe is expanding, and in particular because it is being accelerated by dark energy. The physical distance today to the most distant galaxies within our observable universe is actually larger than 14 billion light-years. If you go through the math, the farthest point that was ever within our observable patch of universe is now 46 billion light-years, or 14 billion parsecs, distant.

39   The idea that particles aren't created out of empty space should be clearly labeled as an assumption, although it seems to be a pretty good one—at least, within the current universe. (Later we'll see that particles can very rarely appear from the vacuum in an accelerating universe, in a process analogous to Hawking radiation around black holes.) The old Steady State theory explicitly assumed the opposite, but had to invoke new kinds of physical processes to make it work (and it never really did).

40   To be careful about it, the phrase *Big Bang* is used in two different ways. One way is as we've just defined it—the hypothetical moment of infinite density at the beginning of the universe, or at least conditions in the universe very, very close to that moment in time. But we also speak of the "Big Bang model,"

which is simply the general framework of a universe that expands from a hot, dense state according to the rules of general relativity; and sometimes we drop the *model*. So you might read newspaper stories about cosmologists "testing the predictions of the Big Bang." You can't test the predictions of some moment in time; you can only test predictions of a model. Indeed, the two concepts are fairly independent—we will be arguing later in the book that a complete theory of the universe will have to replace the conventional Big Bang singularity by something better, but the Big Bang model of the evolution of the universe over the last 14 billion years is well established and not going anywhere.

41   The microwave background has a messy history. George Gamow, Ralph Alpher, and Robert Herman wrote a series of papers in the late 1940s and early 1950s that clearly predicted the existence of relic microwave radiation from the Big Bang, but their work was subsequently largely forgotten. In the 1960s, Robert Dicke at Princeton and A. G. Doroshkevich and Igor Novikov in the Soviet Union independently recognized the existence and detectability of the radiation. Dicke went so far as to assemble a talented group of young cosmologists (including David Wilkinson and P. J. E. Peebles, who would go on to become leaders in the field) to build an antenna and search for the microwave background themselves. They were scooped by Penzias and Wilson, just a few miles away, who were completely unaware of their work. Gamow passed away in 1968, but it remains mysterious why Alpher and Herman never won the Nobel Prize for their predictions. They told their side of the story in a book, *Genesis of the Big Bang* (Alpher and Herman, 2001). In 2006, John Mather and George Smoot were awarded the Prize for their measurements of the blackbody spectrum and temperature anisotropies in the microwave background, using NASA's Cosmic Background Explorer (COBE) satellite.

42   The full story is told by Farrell (2006).

43   Bondi and Gold (1948); Hoyle (1948).

44   See for example Wright (2008).

45   Needless to say, that's making a long story very short. Type Ia supernovae are believed to be the result of the catastrophic gravitational collapse of white dwarf stars. A white dwarf is a star that has used up all of its nuclear fuel and just sits there quietly, supported by the basic fact that electrons take up space. But some white dwarfs have companion stars, from which matter can slowly dribble onto the dwarf. Eventually the white dwarf hits a point—the Chandrasekhar Limit, named after Subrahmanyan Chandrasekhar—where the outward pressure due to electrons cannot compete with the gravitational pull, and the star collapses into a neutron star, ejecting its outer layers as a supernova. Because the Chandrasekhar Limit is approximately the same for every white dwarf in the universe, the brightness of the resulting explosions is approximately the same for every Type Ia supernova. (There are other types of supernovae, which don't involve white dwarfs at all.) But astronomers have learned how to correct for the differences in brightness by using the empirical fact that brighter supernovae take longer to decline in brightness after the peak luminosity. The story of how astronomers search for such supernovae, and how they eventually discovered the acceleration of the universe, is told in Goldsmith (2000), Kirshner (2004), and Gates (2009); the original papers are Riess et al. (1998) and Perlmutter et al. (1999).

46   Another subtle point needs to be explained. The expansion rate of the universe is measured by the Hubble constant, which relates distance to redshift. It's not really a "constant"; in the early universe the expansion was much faster, and what we might call the Hubble "parameter" was a lot larger than our current Hubble constant. We might expect that the phrase *the universe is accelerating* means "the Hubble parameter is increasing," but that's not true—it just means "it's not decreasing very fast." The "acceleration" refers to an increase in the apparent velocity of any particular galaxy over time. But that velocity is equal to the Hubble parameter times the distance, and the distance is increasing as the universe expands. So an accelerating universe is not necessarily one in which the Hubble parameter is increasing, just one in which the product of the Hubble parameter with the distance to any particular galaxy is increasing. It turns out that, even with a cosmological constant, the Hubble parameter never actually increases; it decreases more slowly as the universe expands and dilutes, until it approaches a fixed constant value after all the matter has gone away and there's nothing left but cosmological constant.

47   We're being careful to distinguish between two forms of energy that are important for the evolution of the contemporary universe: "matter," made of slowly moving particles that dilute away as the universe expands, and "dark energy," some mysterious stuff that doesn't dilute away at all, but maintains a constant energy density. But matter itself comes in different forms: "ordinary matter," including all of the kinds of particles we have ever discovered in experiments here on Earth, and "dark matter," some other kind of particle that can't be anything we've yet directly seen. The mass (and therefore energy) in ordinary matter is mostly in the form of atomic nuclei—protons and neutrons—but electrons also

contribute. So ordinary matter includes you, me, the Earth, the Sun, stars, and all the gas and dust and rocks in space. We know how much of that stuff there is, and it's not nearly enough to account for the gravitational fields observed in galaxies and clusters. So there must be dark matter, and we've ruled out all known particles as candidates; theorists have invented an impressive menu of possibilities, including "axions" and "neutralinos" and "Kaluza-Klein particles." All told, ordinary matter makes up about 4 percent of the energy in the universe, dark matter makes up about 22 percent, and dark energy makes up about 74 percent. Trying to create or detect dark matter directly is a major goal of modern experimental physics. See Hooper (2007), Carroll (2007), or Gates (2009) for more details.

48   So how much energy is there in the dark energy, anyway? It's about one billion-billionth of a calorie per cubic centimeter. (That's using the calories used to measure the energy content of food, or 1,000 standard calories.) So if we took all of the cubic centimeters composing the volume of Lake Michigan, their total dark energy content is roughly equal to one food calorie. Seen another way, if we converted all of the dark energy in all the cubic centimeters within the volume of the Earth into electricity, it would be roughly equal to the electricity usage of an average American over one year. The point is, there's not all that much dark energy per cubic centimeter—it's spread very thinly throughout the universe. Of course, we *cannot* convert dark energy into useful energy of this form—dark energy is completely useless. (Why? Because it's in a high-entropy state.)

49   Planck wasn't really doing quantum gravity. In 1899, in attempting to understand some mysteries of blackbody radiation, he had hit upon the need for a new fundamental constant of nature, now known as "Planck's constant," h. Taking that new quantity and multiplying and dividing in appropriate ways by the speed of light c and Newton's constant of gravitation G, Planck invented a system of fundamental units that we now think of as characteristic of quantum gravity: the Planck length $L_p = 1.6 \times 10^{-35}$ meters, the Planck time $t_p = 5.4 \times 10^{-44}$ seconds, and the Planck mass $M_p = 2.2 \times 10^{-8}$ kilograms, along with the Planck energy. Interestingly, Planck's first thought was that the universal nature of these quantities—based in physics, rather than determined by human convention—could someday help us communicate with extraterrestrial civilizations.

50   Fred Adams and Greg Laughlin devoted an entire book to the subject, well worth reading (Adams and Laughlin, 1999).

51   Huw Price has diagnosed this tendency very convincingly (Price, 1996). He accuses cosmologists of an implicit double standard, applying criteria of naturalness to the early universe that they would never apply to the late universe, and vice versa. Price suggests that a consistent cosmology governed by time-symmetric laws should have time-symmetric evolution. Given that the Big Bang has a low entropy, this implies that the future should feature eventual re-collapse to a Big Crunch that also has low entropy—the Gold universe, first contemplated by Thomas Gold (of Steady State fame). In such a universe, the arrow of time would reverse when the universe reached its maximum size, and entropy would begin to decrease toward the Crunch. This kind of scenario seems less likely now that we have discovered dark energy. (The way we will meet Price's challenge in this book is to imagine that the universe is indeed time-symmetric on large scales, with *high* entropy toward both the far past and the far future, which can obviously be achieved only if the Big Bang is not really the beginning.)

52   The universe is not actually going to collapse into one big black hole. As discussed, it's going to empty out. Remarkably, however, in the presence of dark energy even empty space has entropy, and we obtain the same number (10^{120}) for the maximum entropy of the observable universe. Note that 10^{120} was also the discrepancy between the theoretical estimate of the vacuum energy and its observed value. This apparent coincidence of two different numbers is actually the same coincidence as that between the current density of matter (which is related to the maximum entropy) and the energy density in a vacuum. In both cases, the numbers work out to be given by taking the size of the observable universe—roughly 10 billion light years—dividing by the Planck length, and squaring the result.

4. TIME IS PERSONAL

53   On the other hand, the achievements for which Paris Hilton is famous are also pretty mysterious.

54   Einstein's "miraculous year" was 1905, when he published a handful of papers that individually would have capped the career of almost any other scientist: the definitive formulation of special relativity, the explanation of the photoelectric effect (implying the existence of photons and laying the groundwork for quantum mechanics), proposing a theory of Brownian motion in terms of random collisions at the atomic level, and uncovering the equivalence between mass and energy. For most of the next decade he concentrated on the theory of gravity; his ultimate answer, the general theory of relativity, was completed in 1915, when Einstein was thirty-six years old. He died in 1955 at the age of seventy-six.

55 We should also mention Dutch physicist Hendrik Antoon Lorentz, who beginning in 1892 developed the idea that times and distances were affected when objects moved near the speed of light, and derived the "Lorentz transformations," relating measurements obtained by observers moving with respect to each other. To Lorentz, velocities were measured with respect to a background of aether; Einstein was the one who first realized that the aether was an unnecessary fiction.

56 Galison (2003). One gets the impression from Galison's book that he finds the case of Poincaré to actually be more interesting than that of Einstein. However, when an author has a chance to put Einstein in a book title, his name will generally go first. Einstein is box office.

57 George Johnson (2008), in reviewing Leonard Susskind's book *The Black Hole Wars* (2008), laments the fate of the modern reader of popular physics books.

> I was eager to learn how, in the end, Susskind and company showed that Hawking was probably wrong—that information is indeed conserved. But first I had to get through a sixty-six-page crash course on relativity and quantum mechanics. Every book about contemporary physics seems to begin this way, which can be frustrating to anyone who reads more than one. (Imagine if every account of the 2008 presidential campaign had to begin with the roots of Athenian democracy and the heritage of the French Enlightenment.)

The solution is obvious: The basics of relativity and quantum mechanics should be a regular part of secondary education, just like the roots of Athenian democracy and the heritage of the French Enlightenment. In the meantime, this chapter will be part of the inevitable crash course, but by concentrating in particular on the role of "time" we'll hopefully be able to avoid the most shopworn ways of explaining things.

58 Science fiction movies and television shows tend to flagrantly disregard this feature of reality, mostly for the practical reason that it's very hard to fake weightlessness. (*Star Trek: Enterprise* did feature one amusing scene in which the ship "lost its gravity" while Captain Archer was taking a shower.) The artificial gravity you need to make the captain and crew stride purposefully about the ship's bridge doesn't seem compatible with the laws of physics as we know them. If you're not accelerating, the only way to make that much gravity is to carry around a small planet's worth of mass, which doesn't seem practical.

59 Velocity is just the rate of change of position, and acceleration is the rate of change of velocity. In terms of calculus, velocity is the first derivative of the position, and acceleration is the second derivative. It is a deep feature of classical mechanics that the information one can specify about the state of a particle is its position and velocity; the acceleration is then determined by the local conditions and the appropriate laws of physics.

60 Left as exercises for the reader: Can we imagine a world in which absolute orientation in space were observable? What about a world in which position, velocity, and acceleration were all unobservable, but the rate of change of acceleration were observable?

61 Don't get lost in the hypotheticals here. Today we strongly believe that there is *not* any medium pervading space, with respect to which we could measure our velocity. But they did believe that in the late nineteenth century; that's the aether we'll be talking about. On the other hand, we do believe that there are *fields* defined at every point in space, and some of those fields (such as the hypothetical Higgs field) might even have nonzero values in empty space. We now believe that waves, electromagnetic and otherwise, are propagating oscillations in these fields. But a field doesn't really count as a "medium," both because it can have a zero value, and because we can't measure our velocity with respect to it.

On the third hand, it's possible that we don't know everything, and some imaginative theoretical physicists have been wondering whether there actually might be fields that do define a rest frame, and with respect to which we could imagine measuring our velocities (see, for example, Mattingly, 2005). Such fields have been whimsically dubbed "aether," but they are not really the kind of aether that was being proposed in the nineteenth century. In particular, they have nothing to do with the propagation of electromagnetic waves, and are perfectly consistent with the underlying principles of relativity.

62 For some of the historical background, see Miller (1981). Many of the original papers concerning relativity are reprinted in Einstein (1923).

63 To actually experience length contraction or time dilation, we need either to have incredibly exquisite measuring devices, or to be moving at velocities close to the speed of light. Neither such devices nor such velocities are part of our everyday lives, which is why special relativity seems so counterintuitive to us. Of course, the fact that most objects around us have relative velocities that are small compared to the speed of light is an interesting fact about the world, which a complete theory of the universe should try to explain.

64 You might be suspicious that this argument doesn't really demonstrate the impossibility of moving faster than light, only the impossibility of taking something moving slower than light and *accelerating* it to move faster than light. We might imagine that there exist objects that are *always* moving faster

than light, so they don't have to be accelerated. And that certainly is a logical possibility; such hypothetical particles are known as "tachyons." But as far as we know, tachyons do not exist in the real world, and it's a good thing, too; the ability to send signals faster than light would entail the ability to send signals backward in time, and that would wreak havoc with our notions of causality.

65   You will sometimes hear that special relativity is unable to deal with accelerating bodies, and you need general relativity to take acceleration into account. That is complete rubbish. General relativity is required when (and only when) gravity becomes important and spacetime is curved. Far away from any gravitational fields, where spacetime is flat, special relativity applies, no matter what is going on—including accelerating bodies. It's true that freely falling (unaccelerated) trajectories have a special status in special relativity, as they are all created equal. But it is entirely incorrect to leap from there to the idea that accelerated trajectories cannot even be described within the language of special relativity.

66   Apologies for the sloppy lapse into temporal chauvinism (by presuming that one moves forward in time), not to mention giving in to the metaphor of "moving" through time. Rather than saying "Every object moves through spacetime," it would be less prejudicial to say "The history of every object describes a world line that extends through spacetime." But sometimes it's just too tedious to be so pedantically precise all the time.

67   One way of relating relativity to Newtonian spacetime is to imagine "letting the speed of light get infinitely large." Then the light cones we draw would become wider and wider, and the spacelike region would be squeezed down to a single surface, just as in the Newtonian setup. This is a suggestive picture but not terribly respectable. For one thing, we can always choose units in which the speed of light is unity; just measure time in years, and distance in light-years. So what we would actually try to do is change all of the constants of nature so that other velocities diminished with respect to the speed of light. Even if we did that, the process is highly non-unique; we have made an arbitrary choice about how to take the limit so that the light cones converge to some particular surfaces of constant time.

68   That is, at least three dimensions of space. It is quite possible, and taken for granted in certain corners of the theoretical-physics community, that there exist additional dimensions of space that for some reason are invisible to us, at least at the low energies to which we have ready access. There are a number of ways in which extra spatial dimensions could be hidden; see Greene (2000), or Randall (2005). Extra hidden timelike dimensions are considered much less likely, but you never know.

69   Both are reprinted in Einstein (1923).

5. TIME IS FLEXIBLE

70   Special relativity grew out of the incompatibility of Newtonian mechanics and Maxwellian electrodynamics, while general relativity grew out of the incompatibility of special relativity and Newtonian gravity. Right now, physics faces another troublesome incompatibility: general relativity and quantum mechanics. We are all hopeful that someday they will be united into a theory of quantum gravity. String theory is the leading candidate at present, but matters are not yet settled.

71   It might seem crazy that tension, which pulls things together, is responsible for the acceleration of the universe, which pushes things apart. The point is that the tension from dark energy is equal at every point throughout space, and precisely cancels, so there is no direct pulling. Instead, we are left with the indirect effect of the dark energy on the curvature of spacetime. That effect is to impart a perpetual push to the universe, because the dark energy density does not dilute away.

72   Here is another way of thinking about it. The fact that energy is conserved in Newtonian mechanics is a reflection of an underlying symmetry of the theory: time-translation invariance. The background spacetime in which particles move is fixed once and for all. But in general relativity that's no longer true; the background is dynamical, pushing things around, changing their energies.

73   See Michell (1784); Laplace's essay is reprinted as an appendix in Hawking and Ellis (1974). It is occasionally pointed out, with great raising of eyebrows and meaningful murmurs, that the radius of a "black star" as calculated according to Newtonian gravity is precisely the same size as the predicted Schwarzschild radius of a black hole in general relativity ($2GM/c^2$, where G is Newton's constant of gravitation, M is the mass of the object, and c is the speed of light). This coincidence is completely accidental, due primarily to the fact that there aren't many ways you can create a quantity with units of length out of G, M, and c.

74   For purposes of this chapter, we are assuming the validity of classical general relativity, even though we know that it must be replaced by a better theory when it comes to singularities. For more on these issues, see Hawking (1988) or Thorne (1994).

75   Feel free to construct your own moral lessons.

6. LOOPING THROUGH TIME

76 Referring, of course, to the time machines in George Pal's 1960 movie version of H. G. Wells's *The Time Machine*; Robert Zemeckis's 1985 film *Back to the Future*; and the long-running BBC serial *Doctor Who*, respectively.

77 In the interest in getting on with our story, we're not being completely fair to the subject of tachyons. Allowing objects that travel faster than light opens the door to paradoxes, but that doesn't necessarily force us to walk through the door. We might imagine models that allowed for tachyons, but only in self-consistent ways. For some discussion see Feinberg (1967) or Nahin (1999). To make things more confusing, in quantum field theory the word "tachyon" often simply refers to a momentarily unstable configuration of a field, where nothing is actually traveling faster than light.

78 Gödel (1949). In doing research for their massive textbook *Gravitation* (1973), Charles Misner, Kip Thorne, and John Wheeler visited Gödel to talk about general relativity. What Gödel wanted to ask them, however, was whether contemporary astronomical observations had provided any hints for an overall rotation in the universe. He remained interested in the possible relevance of his solution to the real world.

79 Kerr (1963). The Kerr solution is discussed at a technical level in any modern textbook on general relativity, and at a popular level in Thorne (1994). Thorne relates the story of how Kerr presented his solution at the first Texas Symposium on Relativistic Astrophysics, only to be completely (and somewhat rudely) ignored by the assembled astrophysicists, who were busily arguing about quasars. To be fair, at the time Kerr found his solution he didn't appreciate that it represented a black hole, although he knew it was a spinning solution to Einstein's equation. Later on, astrophysicists would come to understand that quasars are powered by spinning black holes, described by Kerr's spacetime.

80 Tipler (1974). The solution for the curvature of spacetime around an infinite cylinder was actually found by Willem Jacob van Stockum, Dutch physicist (and bomber pilot), in 1937, but Van Stockum didn't notice that his solution contained closed timelike curves. An excellent overview of both research into time machines in general relativity, and the appearance of time travel in fiction, can be found in Nahin (1999).

81 Erwin Schrödinger, one of the pioneers of quantum mechanics, proposed a famous thought experiment to illustrate the bizarre nature of quantum superposition. He imagined placing the cat in a sealed box containing a radioactive source that, in some fixed time interval, had a 50 percent chance of decaying and activating a source that would release poison gas into the box. According to the conventional view of quantum mechanics, the resulting system is in an equal superposition of "alive cat" and "dead cat," at least until someone observes the cat; see Chapter Eleven for discussion.

82 Kip Thorne has pointed out the "grandfather paradox" seems a bit squeamish, with the introduction of the extra generation and all, not to mention that it's somewhat patriarchal. He suggests we should be contemplating the "matricide" paradox.

83 This rule is sometimes raised to the status of a principle; see discussions in Novikov (1983) or Horwich (1987). Philosophers such as Hans Reichenbach (1958) and Hilary Putnam (1962) have also emphasized that closed timelike curves do not necessitate the introduction of paradoxes, so long as the events in spacetime are internally consistent. Really, it's just common sense. It's perfectly obvious that there are no paradoxes in the real world; the interesting question is how Nature manages to avoid them.

84 In Chapter Eleven we'll backtrack from this statement just a bit, when we discuss quantum mechanics. In quantum mechanics, the real world may include more than one classical history. David Deutsch (1997) has suggested that we might take advantage of multiple histories to include one in which you were in the Ice Age, and one in which you were not. (And an infinite number of others.)

85 *Back to the Future* was perhaps the least plausible time-travel movie ever. Marty McFly travels from the 1980s back to the 1950s, and commences to change the past right and left. What is worse, whenever he interferes with events that supposedly already happened, ramifications of those changes propagate "instantaneously" into the future, and even into a family photograph that Marty has carried with him. It is hard to imagine how that notion of "instantaneous" could be sensibly defined. Although perhaps not impossible—one would have to posit the existence of an additional dimension with many of the properties of ordinary time, through which Marty's individual consciousness was transported by the effects of his actions. There is probably a good Ph.D. thesis in there somewhere: "Toward a Consistent Ontology of Time and Memory in *Back to the Future, et seq.*" I'm not sure what department it would belong to, however.

86 More or less the final word in consistent histories in the presence of closed timelike curves was explored in Robert A. Heinlein's story "All You Zombies—" (1959). Through a series of time jumps and one sex-change operation, the protagonist manages to be his/her own father, mother, and recruiter into

the Temporal Corps. Note that the life story is not, however, a self-contained closed loop; the character ages into the future.

87 For a discussion of this point see Friedman et al. (1990).

88 Actually, we are committed determinists. Human beings are made of particles and fields that rigidly obey the laws of physics, and in principle (although certainly not in practice) we could forget that we are human and treat ourselves as complicated collections of elementary particles. But that doesn't mean we should shrink from facing up to how bizarre the problem of free will in the presence of closed timelike curves really is.

89 This is a bit more definitive-sounding than what physicists are able to actually prove. Indeed, in some extremely simplified cases we can show that the future can be predicted from the past, even in the presence of closed timelike curves; see Friedman and Higuchi (2006). It seems very likely (to me, anyway), that in more realistically complicated models this will no longer be the case; but a definitive set of answers has not yet been obtained.

90 We might be able to slice spacetime into moments of constant time, even in the presence of closed timelike curves—for example, we can do that in the simple circular-time universe. But that's a very special case, and in a more typical spacetime with closed timelike curves it will be impossible to find any slicing that consistently covers the entire universe.

91 The exception, obviously, is the rotating black hole. We can certainly imagine creating such a hole by the collapse of a rotating star, but there is a different problem: The closed timelike curves are hidden behind an event horizon, so we can't actually get there without leaving the external world behind once and for all. We'll discuss later in the chapter whether that should count as an escape hatch. Perhaps more important, the solution found by Kerr that describes a rotating black hole is valid only in the ideal case where there is absolutely no matter in spacetime; it is a black hole all by itself, not one that is created by the collapse of a star. Most experts in general relativity believe that a real-world collapsing star would never give rise to closed timelike curves, even behind an event horizon.

92 Abbott (1899); see also Randall (2005).

93 The original paper was Gott (1991); he also wrote a popular-level book on the subject (2001). Almost every account you will read of this work will not talk about "massive bodies moving in Flatland," but rather "perfectly straight, parallel cosmic strings moving in four-dimensional spacetime." That's because the two situations are precisely equivalent. A cosmic string is a hypothetical relic from the early universe that can be microscopically thin but stretch for cosmological distances; an idealized version would be perfectly straight and stretch forever, but in the real world cosmic strings would wiggle and curve in complicated ways. But if such a string were perfectly straight, nothing at all would depend on the direction of spacetime along that string; in technical terms, the entire spacetime would be invariant with respect to both translations and boosts along the string. Which means, in effect, that the direction along the string is completely irrelevant, and we are free to ignore it. If we simply forget that dimension, an infinitely long string in three-dimensional space becomes equivalent to a point particle in two-dimensional space. The same goes for a collection of several strings, as long as they are all perfectly straight and remain absolutely parallel to one another. Of course, the idea of pushing around infinitely long and perfectly straight strings is almost as bizarre as imagining that we live in a three-dimensional spacetime. That's okay; we're just making unrealistic assumptions because we want to push our theories to the edge of what is conceivable, to distinguish what is impossible in principle from what is merely a daunting technical challenge.

94 Soon after Gott's paper appeared, Curt Cutler (1992) showed that the closed timelike curves extended to infinity, another signal that this solution didn't really count as building a time machine (as we think of "building" as something that can be accomplished in a local region). Deser, Jackiw, and 't Hooft (1992) examined Gott's solution and found that the total momentum corresponded to that of a tachyon. I worked with Farhi, Guth, and Olum (1992, 1994) to show that an open Flatland universe could never contain enough energy to create a Gott time machine starting from scratch. 't Hooft (1992) showed that a closed Flatland universe would collapse to a singularity before a closed timelike curve would have a chance to form.

95 Farhi, Guth, and Guven (1990).

96 Think of a plane, seen from the perspective of some particular point, as stretching around for 360 degrees of angle. What happens in Flatland is that every bit of energy decreases the total angle around you; we say that every mass is associated with a "deficit angle," which is removed from the space by its presence. The more mass, the more angle is removed. The resulting geometry looks like a cone at large distances, rather than like a flat piece of paper. But there are only 360 degrees available to be removed, so there is an upper limit on the total amount of energy we can have in an open universe.

97 "Something like" because we are speaking of the topology of space, not its geometry. That is, we're not saying that the curvature of space is everywhere perfectly spherical, just that you could smoothly deform it into a sphere. A spherical topology accommodates a deficit angle of exactly 720 degrees, twice the upper limit available in an open universe. Think of a cube (which is topologically equivalent to a sphere). It has eight vertices, each of which corresponds to a deficit angle of 90 degrees, for a total of 720.

98 Sagan (1985). The story of how Sagan's questions inspired Kip Thorne's work on wormholes and time travel is related in Thorne (1994).

99 As should be obvious from the dates, the work on wormhole time machines actually predates the Flatland explorations. But it involves a bit more exotic physics than Gott's idea, so it's logical to discuss the proposals in this order. The original wormhole-as-time-machine paper was Morris, Thorne, and Yurtsever (1988). A detailed investigation into the possible consistency of time travel in wormhole spacetimes was Friedman et al. (1990), and the story is related at a popular level in Thorne (1994).

100 I once introduced Bob Geroch for a talk he was giving. It's useful in these situations to find an interesting anecdote to relate about the speaker, so I Googled around and stumbled on something perfect: a Star Trek site featuring a map of our galaxy, prominently displaying something called the "Geroch Wormhole." (Apparently it connects the Beta Quadrant to the Delta Quadrant, and was the source of a nasty spat with the Romulans.) So I printed a copy of the map on a transparency and showed it during my introduction, to great amusement all around. Later Bob told me he assumed I had made it up myself, and was pleased to hear that his work on wormholes had produced a beneficial practical effect on the outside world. The paper that showed you would have to make a closed timelike curve in order to build a wormhole is Geroch (1967).

101 Hawking (1991). In his conclusion, Hawking also claimed that there was observational evidence that travel backward in time was impossible, based on the fact that we had not been invaded by historians from the future. He was joking (I'm pretty sure). Even if it were possible to construct closed timelike curves from scratch, they could never be used to travel backward to a time before the closed curves had been constructed. So there is no observational evidence against the possibility of building a time machine, just evidence that no one has built one yet.

7. RUNNING TIME BACKWARD

102 See O'Connor and Robertson (1999), Rouse Ball (1908). You'll remember Laplace as one of the people who were speculating about black holes long before general relativity came along.

103 Apparently Napoleon found this quite amusing. He related Laplace's quip to Joseph Lagrange, another distinguished physicist and mathematician of the time. Lagrange responded with, "Ah, but it is a fine hypothesis; it explains so many things." Rouse Ball (1908), 427.

104 Laplace (2007).

105 There is no worry that Laplace's Demon exists out there in the universe, smugly predicting our every move. For one thing, it would have to be as big as the universe, and have a computational power equal to that of the universe itself.

106 Stoppard (1999), 103–4. Valentine, one presumes, is referring to the idea that the phenomenon of chaos undermines the idea of determinism. Chaotic dynamics, which is very real, happens when small changes in initial conditions lead to large differences in later evolution. As a practical matter, this makes the future extremely difficult to predict for systems that are chaotic (not everything is)—there will always be some tiny error in our understanding of the present state of a system. I'm not sure that this argument carries much force with respect to Laplace's Demon. As a practical matter, there was no danger that we were ever going to know the entire state of the universe, much less use it to predict the future; this conception was always a matter of principle. And the prospect of chaos doesn't change that at all.

107 Granted, physicists couldn't actually live on any of our checkerboards, for essentially anthropic reasons: The setups are too simplistic to allow for the formation and evolution of complex structures that we might identify with intelligent observers. This stifling simplicity can be traced to an absence of interesting "interactions" between the different elements. In the checkerboard worlds we will look at, the entire description consists of just a single kind of thing (such as a vertical or diagonal line) stretching on without alteration. An interesting world is one in which things can persist more or less for an extended period of time, but gradually change via the influence of interactions with other things in the world.

108 This "one moment at a time" business isn't perfectly precise, as the real world is not (as far as we know) divided up into discrete steps of time. Time is continuous, flowing smoothly from one time to another

while going through every possible moment in between. But that's okay; calculus provides exactly the right set of mathematical tools to make sense of "chugging forward one moment at a time" when time itself is continuous.

109   Note that translations in space and spatial inversions (reflections between left and right) are also perfectly good symmetries. That doesn't seem as obvious, just from looking at the picture, but that's only because the states themselves (the patterns of o's and 1's) are not invariant under spatial shifts or reflections.

Lest you think these statements are completely vacuous, there are some symmetries that might have existed, but don't. We cannot, for example, exchange the roles of time and space. As a general rule, the more symmetries you have, the simpler things become.

110   This whole checkerboard-worlds idea sometimes goes by the name of *cellular automata*. A cellular automaton is just some discrete grid that follows a rule for determining the next row from the state of the previous row. They were first investigated in the 1940s, by John von Neumann, who is also the guy who figured out how entropy works in quantum mechanics. Cellular automata are fascinating for many reasons having little to do with the arrow of time; they can exhibit great complexity and can function as universal computers. See Poundstone (1984) or Shalizi (2009).

Not only are we disrespecting cellular automata by pulling them out only to illustrate a few simple features of time reversal and information conservation, but we are also not speaking the usual language of cellular-automaton cognoscenti. For one thing, computer scientists typically imagine that time runs from top to bottom. That's crazy; everyone knows that time runs from bottom to top on a diagram. More notably, even though we are speaking as if each square is either in the state "white" or the state "gray," we just admitted that you have to keep track of more information than that to reliably evolve into the future in what we are calling example B. That's no problem; it just means that we're dealing with an automaton where the "cells" can take on more than two different states. One could imagine going beyond white and gray to allow squares to have any of four different colors. But for our current purposes that's a level of complexity we needn't explicitly introduce.

111   If the laws of physics are not completely deterministic—if they involve some random, stochastic element—then the "specification" of the future evolution will involve probabilities, rather than certainties. The point is that the state includes all of the information that is required to do as well as we can possibly do, given the laws of physics that we are working with.

112   Sometimes people count relativity as a distinct theory, distinguishing between "classical mechanics" and "relativistic mechanics." But more often they don't. It makes sense, for most purposes, to think of relativity as introducing a particular *kind* of classical mechanics, rather than a completely new way of thinking. The way we specify the state of a system, for example, is pretty much the same in relativity as it would be in Newtonian mechanics. Quantum mechanics, on the other hand, really is quite different. So when we deploy the adjective *classical*, it will usually denote a contrast with *quantum*, unless otherwise specified.

113   It is not known, at least to me, whether Newton himself actually played billiards, although the game certainly existed in Britain at the time. Immanuel Kant, on the other hand, is known to have made pocket money as a student playing billiards (as well as cards).

114   So the momentum is not just a number; it's a vector, typically denoted by a little arrow. A vector can be defined as a magnitude (length) and a direction, or as a combination of sub-vectors (components) pointing along each direction of space. You will hear people speak, for example, of "the momentum along the x-direction."

115   This is a really good question, one that bugged me for years. At various points when one studies classical mechanics, there are times when one hears one's teachers talk blithely about momenta that are completely inconsistent with the actual trajectory of the system. What is going on?

The problem is that, when we are first introduced to the concept of "momentum," it is typically *defined* as the mass times the velocity. But somewhere along the line, as you move into more esoteric realms of classical mechanics, that idea ceases to be a definition and becomes something that you can *derive* from the underlying theory. In other words, we start conceiving of the essence of momentum as "some vector (magnitude and direction) defined at each point along the path of the particle," and then derive equations of motion that insist the momentum will be equal to the mass times the velocity. (This is known as the Hamiltonian approach to dynamics.) That's the way we are thinking in our discussion of time reversal. The momentum is an independent quantity, part of the state of the system; it is equal to the mass times the velocity only when the laws of physics are being obeyed.

116   David Albert (2000) has put forward a radically different take on all this. He suggests that we should define a "state" to be just the positions of particles, *not* the positions and momenta (which he would

call the "dynamical condition"). He justifies this by arguing that states should be logically independent at each moment of time—the states in the future should not depend on the present state, which
clearly they do in the way we defined them, as that was the entire point. But by redefining things in
this way, Albert is able to live with the most straightforward definition of time-reversal invariance: "A
sequence of states played backward in time still obeys the same laws of physics," without resorting to
any arbitrary-sounding transformations along the way. The price he pays is that, although Newtonian
mechanics is time-reversal invariant under this definition, almost no other theory is, including classical electromagnetism. Which Albert admits; he claims that the conventional understanding that electromagnetism is invariant under time reversal, handed down from Maxwell to modern textbooks, is
simply wrong. As one might expect, this stance invited a fusillade of denunciations; see, for example,
Earman (2002), Arntzenius (2004), or Malament (2004).
      Most physicists would say that it just doesn't matter. There's no such thing as the one true meaning of
time-reversal invariance, which is out there in the world waiting for us to capture its essence. There are
only various concepts, which we may or may not find useful in thinking about how the world works.
Nobody disagrees on how electrons move in the presence of a magnetic field; they just disagree on the
words to use when describing that situation. Physicists tend to express bafflement that philosophers
care so much about the words. Philosophers, for their part, tend to express exasperation that physicists
can use words all the time without knowing what they actually mean.

117   Elementary particles come in the form of "matter particles," called "fermions," and "force particles,"
called "bosons." The known bosons include the photon carrying electromagnetism, the gluons carrying the strong nuclear force, and the W and Z bosons carrying the weak nuclear force. The known
fermions fall neatly into two types: six different kinds of "quarks," which feel the strong force and get
bound into composite particles like protons and neutrons, and six different kinds of "leptons," which
do not feel the strong force and fly around freely. These two groups of six are further divided into collections of three particles each; there are three quarks with electric charge $+2/3$ (the up, charm, and
top quarks), three quarks with electric charge $-1/3$ (the down, strange, and bottom quarks), three leptons with electric charge -1 (the electron, the muon, and the tau), and three leptons with zero charge
(the electron neutrino, the muon neutrino, and the tau neutrino). To add to the confusion, every type
of quark and lepton has a corresponding antiparticle with the opposite electric charge; there is an
anti-up-quark with charge $-2/3$, and so on.
      All of which allows us to be a little more specific about the decay of the neutron (two down quarks
and one up): it actually creates a proton (two up quarks and one down), an electron, and an electron *antineutrino*. It's important that it's an antineutrino, because that way the net number of leptons
doesn't change; the electron counts as one lepton, but the antineutrino counts as minus one lepton, so
they cancel each other out. Physicists have never observed a process in which the net number of leptons or the net number of quarks changes, although they suspect that such processes must exist. After
all, there seem to be a lot more quarks than antiquarks in the real world. (We don't know the net number of leptons very well, since it's very hard to detect most neutrinos in the universe, and there could be
a lot of antineutrinos out there.)

118   "Easiest" means "lowest in mass," because it takes more energy to make higher-mass particles, and
when you do make them they tend to decay more quickly. The lightest two kinds of quarks are the up
(charge $+2/3$) and the down (charge $-1/3$), but combining an up with an anti-down does not give a neutral particle, so we have to look at higher-mass quarks. The next heaviest is the strange quark, with
charge $-1/3$, so it can be combined with a down to make a kaon.

119   Angelopoulos et al. (1998). A related experiment, measuring time-reversal violation by neutral kaons
in a slightly different way, was carried out by the KTeV collaboration at Fermilab, outside Chicago
(Alavi-Harati et al. 2000).

120   Quoted in Maglich (1973). The original papers were Lee and Yang (1956) and Wu et al. (1957). As
Wu had suspected, other physicists were able to reproduce the result very rapidly; in fact, another
group at Columbia performed a quick confirmation experiment, the results of which were published
back-to-back with the Wu et al. paper (Garwin, Lederman, and Weinrich, 1957).

121   Christenson et al. (1964). Within the Standard Model of particle physics, there is an established method
to account for CP violation, developed by Makoto Kobayashi and Toshihide Maskawa (1973), who generalized an idea due to Nicola Cabbibo. Kobayashi and Maskawa were awarded the Nobel Prize in 2008.

122   We're making a couple of assumptions here: namely, that the laws are time-translation invariant (not
changing from moment to moment), and that they are deterministic (the future can be predicted with
absolute confidence, rather than simply with some probability). If either of these fails to be true, the
definition of whether a particular set of laws is time-reversal invariant becomes a bit more subtle.

8. ENTROPY AND DISORDER

123   Almost the same example is discussed by Wheeler (1994), who attributes it to Paul Ehrenfest. In what Wheeler calls "Ehrenfest's Urn," exactly one particle switches side at every step, rather than every particle having a small chance of switching sides.

124   When we have 2 molecules on the right, the first one could be any of the 2,000, and the second could be any of the remaining 1,999. So you might guess there are 1,999 × 2,000 = 3,998,000 different ways this could happen. But that's overcounting a bit, because the two molecules on the right certainly don't come in any particular *order*. (Saying "molecules 723 and 1,198 are on the right" is exactly the same statement as "molecules 1,198 and 723 are on the right.") So we divide by two to get the right answer: There are 1,999,000 different ways we can have 2 molecules on the right and 1,998 on the left. When we have 3 molecules on the right, we take 1,998 × 1,999 × 2,000 and divide by 3 × 2 different orderings. You can see the pattern; for 4 particles, we would divide 1,997 × 1,998 × 1,999 × 2,000 by 4 × 3 × 2, and so on. These numbers have a name—"binomial coefficients"—and they represent the number of ways we can choose a certain set of objects out of a larger set.

125   We are assuming the logarithm is "base 10," although any other base can be used. The "logarithm base 2" of 8 is 2^3 is 3; the logarithm base 2 of 2,048 = 2^{11} is 11. See Appendix for fascinating details.

126   The numerical value of k is about 3.2 × 10^{-16} ergs per Kelvin; an erg is a measure of energy, while Kelvin of course measures temperature. (That's not the value you will find in most references; this is because we are using base-10 logarithms, while the formula is more often written using natural logarithms.) When we say "temperature measures the average energy of moving molecules in a substance," what we mean is "the average energy per degree of freedom is one-half times the temperature times Boltzmann's constant."

127   The actual history of physics is so much messier than the beauty of the underlying concepts. Boltzmann came up with the idea of "$S = k \log W$," but those are not the symbols he would have used. His equation was put into that form by Max Planck, who suggested that it be engraved on Boltzmann's tomb; it was Planck who first introduced what we now call "Boltzmann's constant." To make things worse, the equation on the tomb is *not* what is usually called "Boltzmann's equation"—that's a different equation discovered by Boltzmann, governing the evolution of a distribution of a large number of particles through the space of states.

128   One requirement of making sense of this definition is that we actually know how to *count* the different kinds of microstates, so we can quantify how many of them belong to various macrostates. That sounds easy enough when the microstates form a discrete set (like distributions of particles in one half of a box or the other half) but becomes trickier when the space of states is continuous (like real molecules with specific positions and momenta, or almost any other realistic situation). Fortunately, within the two major frameworks for dynamics—classical mechanics and quantum mechanics—there is a perfectly well-defined "measure" on the space of states, which allows us to calculate the quantity W, at least in principle. In some particular examples, our understanding of the space of states might get a little murky, in which case we need to be careful.

129   Feynman (1964), 119–20.

130   I know what you're thinking. "I don't know about you, but when I dry myself off, most of the water goes onto the towel; it's not fifty-fifty." That's true, but the reason why is because the fiber structure of a nice fluffy towel provides many more places for the water to be than your smooth skin does. That's also why your hair doesn't dry as efficiently, and why you can't dry yourself very well with pieces of paper.

131   At least in certain circumstances, but not always. Imagine we had a box of gas, where every molecule on the left side was "yellow" and every molecule on the right was "green," although they were otherwise identical. The entropy of that arrangement would be pretty low and would tend to go up dramatically if we allowed the two colors to mix. But we couldn't get any useful work out of it.

132   The ubiquity of friction and noise in the real world is, of course, due to the Second Law. When two billiard balls smack into each other, there are only a very small number of ways that all the molecules in each ball could respond precisely so as bounce off each other without disturbing the outside world in any way; there are a much larger number of ways that those molecules can interact gently with the air around them to create the noise of the two balls colliding. All of the guises of dissipation in our everyday lives—friction, air resistance, noise, and so on—are manifestations of the tendency of entropy to increase.

133   Thought of yet another way: The next time you are tempted to play the Powerball lottery, where you pick five numbers between 1 and 59 and hope that they come up in a random drawing, pick the numbers "1, 2, 3, 4, 5." That sequence is precisely as likely as any other "random-looking" sequence. (Of course,

a nationwide outcry would ensue if you won, as people would suspect that someone had rigged the drawing. So you'd probably never collect, even if you got lucky.)

134   Strictly speaking, since there are an infinite number of possible positions and an infinite number of possible momenta for each particle, the number of microstates per macrostate is also infinite. But the possible positions and momenta for a particle on the left side of the box can be put into one-to-one correspondence with the possible positions and momenta on the right side; even though both are infinite, they're "the same infinity." So it's perfectly legitimate to say that there are an equal number of possible states per particle on each side of the box. What we're really doing is counting "the volume of the space of states" corresponding to a particular macrostate.

135   To expand on that a little bit, at the risk of getting hopelessly abstract: As an alternative to averaging within a small region of space, we could imagine averaging over a small region in *momentum* space. That is, we could talk about the average position of particles with a certain value of momentum, rather than vice versa. But that's kind of crazy; that information simply isn't accessible via macroscopic observation. That's because, in the real world, particles tend to interact (bump into one another) when they are *nearby in space*, but nothing special happens when two distant particles have the same momentum. Two particles that are close to each other in position can interact, no matter what their relative velocities are, but the converse is not true. (Two particles that are separated by a few light years aren't going to interact noticeably, no matter what their momentum is.) So the laws of physics pick out "measuring average properties within a small region of space" as a sensible thing to do.

136   A related argument has been given by mathematician Norbert Wiener in *Cybernetics* (1961), 34.

137   There is a loophole. Instead of starting with a system that had delicately tuned initial conditions for which the entropy would decrease, and then letting it interact with the outside world, we could just ask the following question: "Given that this system will go about interacting with the outside world, what state do I need to put it in right now so that its entropy will decrease in the future?" That kind of future boundary condition is not inconceivable, but it's a little different than what we have in mind here. In that case, what we have is not some autonomous system with a naturally reversed arrow of time, but a conspiracy among every particle in the universe to permit some subsystem to decrease in entropy. That subsystem would not look like the time-reverse of an ordinary object in the universe; it would look like the rest of the world was conspiring to nudge it into a low-entropy state.

138   Note the caveat "at room temperature." At a sufficiently high temperature, the velocity of the individual molecules is so high that the water doesn't stick to the oil, and once again a fully mixed configuration has the highest entropy. (At that temperature the mixture will be vapor.) In the messy real world, statistical mechanics is complicated and should be left to professionals.

139   Here is the formula: For each possible microstate x, let p_x be the probability that the system is in that microstate. The entropy is then the sum over all possible microstates x of the quantity $-kp_x \log p_x$, where k is Boltzmann's constant.

140   Boltzmann actually calculated a quantity H, which is essentially the difference between the maximum entropy and the actual entropy, thus the name of the theorem. But that name was attached to the theorem only later on, and in fact Boltzmann himself didn't even use the letter H; he called it E, which is even more confusing. Boltzmann's original paper on the H-Theorem was 1872; an updated version, taking into account some of the criticisms by Loschmidt and others, was 1877. We aren't coming close to doing justice to the fascinating historical development of these ideas; for various different points of view, see von Baeyer (1998), Lindley (2001), and Cercignani (1998); at a more technical level, see Uffink (2004) and Brush (2003). Any Yale graduates, in particular, will lament the short shrift given to the contributions of Gibbs; see Rukeyser (1942) to redress the balance.

141   Note that Loschmidt is *not* saying that there are equal numbers of increasing-entropy and decreasing-entropy evolutions that start with the same initial conditions. When we consider time reversal, we switch the initial conditions with the final conditions; all Loschmidt is pointing out is that there are equal numbers of increasing-entropy and decreasing-entropy evolutions overall, when we consider every possible initial condition. If we confine our attention to the set of low-entropy initial conditions, we can successfully argue that entropy will usually increase; but note that we have sneaked in time asymmetry by starting with low-entropy *initial* conditions rather than final ones.

142   Albert (2000); see also (among many examples) Price (2004). Although I have presented the need for a Past Hypothesis as (hopefully) perfectly obvious, its status is not uncontroversial. For a dash of skepticism, see Callender (2004) or Earman (2006).

143   Readers who have studied some statistical mechanics may wonder why they don't recall actually doing this. The answer is simply that it doesn't matter, as long as we are trying to make predictions about the future. If we use statistical mechanics to predict the future behavior of a system, the predictions we

get based on the Principle of Indifference plus the Past Hypothesis are indistinguishable from those we would get from the Principle of Indifference alone. As long as there is no assumption of any special *future* boundary condition, all is well.

9. INFORMATION AND LIFE

144 Quoted in Tribus and McIrvine (1971).

145 Proust (2004), 47.

146 We are, however, learning more and more all the time. See Schacter, Addis, and Buckner (2007) for a recent review of advances in neuroscience that have revealed how the way actual brains reconstruct memories is surprisingly similar to the way they go about imagining the future.

147 Albert (2000).

148 Rowling (2005).

149 Callender (2004). In Callender's version, it's not that you die; it's that the universe ends, but I didn't want to get confused with Big Crunch scenarios. But really, it would be nice to see more thought experiments in which the future boundary condition was "you fall in love" or "you win the lottery."

150 Davis (1985, 11) writes: "I will lay out four rules, but each is really only a special application of the great principle of causal order: *after cannot cause before* . . . there is no way to change the past . . . one-way arrows flow with time."

151 There are a number of references that go into the story of Maxwell's Demon in greater detail than we will here. Leff and Rex (2003) collect a number of the original papers. Von Baeyer (1998) uses the Demon as a theme to trace the history of thermodynamics; Seife (2006) gives an excellent introduction to information theory and its role in unraveling this puzzle. Bennett and Landauer themselves wrote about their work in *Scientific American* (Bennett and Landauer, 1985; Bennett, 1987).

152 This scenario can be elaborated on further. Imagine that the box was embedded in a bath of thermal gas at some temperature T, and that the walls of the box conducted heat, so that the molecule inside was kept in thermal equilibrium with the gas outside. If we could continually renew our information about which side of the box the molecule was on, we could keep extracting energy from it, by cleverly inserting the piston on the appropriate side; after the molecule lost energy to the piston, it would gain the energy back from the thermal bath. What we've done is to construct a perpetual motion machine, powered only by our hypothetical limitless supply of information. (Which drives home the fact that information never just comes for free.) Szilárd could even quantify precisely how much energy could be extracted from a single bit of information: $kT \log 2$, where k is Boltzmann's constant.

153 It's interesting how, just as much of the pioneering work on thermodynamics in the early nineteenth century was carried out by practical-minded folks who were interested in building better steam engines, much of the pioneering work on information theory in the twentieth century has been carried out by practical-minded folks who were interested in building better communications systems and computers.

154 We can go further than this. Just as Gibbs came up with a definition of entropy that referred to the probability that a system was in various different states, we can define the "information entropy" of a space of possible messages in terms of the probability that the message takes various forms. The formulas for the Gibbs entropy and the information entropy turn out to be identical, although the symbols in them have slightly different meanings.

155 For recent overviews, see Morange (2008) or Regis (2009).

156 The argument that follows comes from Bunn (2009), which was inspired by Styer (2008). See also Lineweaver and Egan (2008) for details and additional arguments.

157 Crick (1990).

158 Schrödinger (1944), 69.

159 *From Being to Becoming* is the title of a popular book (1980) by Belgian Nobel Laureate Ilya Prigogine, who helped pioneer the study of "dissipative structures" and self-organizing systems in statistical mechanics. See also Prigogine (1955), Kauffman (1993), and Avery (2003).

160 A good recent book is Nelson (2007).

161 He would have been even more wary in modern times; a Google search on "free energy" returns a lot of links to perpetual-motion schemes, along with some resources on clean energy.

162 Informally speaking, the concepts of "useful" and "useless" energy certainly predate Gibbs; his contribution was to attach specific formulas to the ideas, which were later elaborated on by German physicist Hermann von Helmholtz. In particular, what we are calling the "useless" energy is (in Helmholtz's formulation) simply the temperature of the body times its entropy. The free energy is then the total internal energy of the body minus that quantity.

163 In the 1950s, Claude Shannon built "The Ultimate Machine," based on an idea by Marvin Minsky. In its resting state, the machine looked like a box with a single switch on one face. If you were to flip the switch, the box would buzz loudly. Then the lid would open and a hand would reach out, flipping the switch back to its original position, and retreat back into the box, which became quiet once more. One possible moral of which is: Persistence can be a good in its own right.

164 Specifically, more massive organisms—which typically have more moving parts and are correspondingly more complex—consume free energy at a higher rate per unit mass than less massive organisms. See, for example, Chaisson (2001).

165 This and other quantitative measures of complexity are associated with the work of Andrey Kolmogorov, Ray Solomonoff, and Gregory Chaitin. For a discussion, see, for example, Gell-Mann (1994).

166 For some thoughts on this particular question, see Dyson (1979) or Adams and Laughlin (1999).

10. RECURRENT NIGHTMARES

167 Nietzsche (2001), 194. What is it with all the demons, anyway? Between Pascal's Demon, Maxwell's Demon, and Nietzsche's Demon, it's beginning to look more like Dante's *Inferno* than a science book around here. Earlier in *The Gay Science* (189), Nietzsche touches on physics explicitly, although in a somewhat different context: "We, however, want to *become who we are*—human beings who are new, unique, incomparable, who give themselves laws, who create themselves! To that end we must become the best students and discoverers of everything lawful and necessary in the world: we must become *physicists* in order to be creators in this sense—while hitherto all valuations and ideals have been built on *ignorance* of physics or in *contradiction* to it. So, long live physics! And even more, long live what compels us to it—our honesty!"

168 Note that, if each cycle were truly a perfect copy of the previous cycles, you would have no memory of having experienced any of the earlier versions (since you didn't have such a memory before, and it's a perfect copy). It's not clear how different such a scenario would be than if the cycle occurred only once.

169 For more of the story, see Galison (2003). Poincaré's paper is (1890).

170 Another subtlety is that, while the system is guaranteed to return to its starting configuration, it is not guaranteed to attain *every* possible configuration. The idea that a sufficiently complicated system does visit every possible state is equivalent to the idea that the system is ergodic, which we discussed in Chapter Eight in the context of justifying Boltzmann's approach to statistical mechanics. It's true for some systems, but not for all systems, and not even for all interesting ones.

171 It's my book, so Pluto still counts.

172 Roughly speaking, the recurrence time is given by the exponential of the maximum entropy of the system, in units of the typical time it takes for the system to evolve from one state to the next. (We are assuming some fixed definition of when two states are sufficiently different as to count as distinguishable.) Remember that the entropy is the logarithm of the number of states, and an exponential undoes a logarithm; in other words, the recurrence time is simply proportional to the total number of possible states the system can be in, which makes perfect sense if the system spends roughly equal amounts of time in each allowed state.

173 Poincaré (1893).

174 Zermelo (1896a).

175 Boltzmann (1896).

176 Zermelo (1896b); Boltzmann (1897).

177 Boltzmann (1897).

178 "At least" three ways, because the human imagination is pretty clever. But there aren't that many choices. Another one would be that the underlying laws of physics are intrinsically irreversible.

179 Boltzmann (1896).

180 We're imagining that the spirit of the recurrence theorem is valid, not the letter of it. The proof of the recurrence theorem requires that the motions of particles be bounded—perhaps because they are planets moving in closed orbits around the Sun, or because they are molecules confined to a box of gas. Neither case really applies to the universe, nor is anyone suggesting that it might. If the universe consisted of a finite number of particles moving in an infinite space, we would expect some of them to simply move away forever, and recurrences would not happen. However, if there are an *infinite* number of particles in an infinite space, we can have a fixed finite average *density*—the number of particles per (for example) cubic light-year. In that case, fluctuations of the form illustrated here are sure to occur, which look for all the world like Poincaré's recurrences.

181   Boltzmann (1897). He made a very similar suggestion in a slightly earlier paper (1895), where he attributed it to his "old assistant, Dr. Schuetz." It is unclear whether this attribution should be interpreted as a generous sharing of credit, or a precautionary laying of blame.

182   Note that Boltzmann's reasoning actually goes past the straightforward implications of the recurrence theorem. The crucial point now is not that any particular low-entropy starting state will be repeated infinitely often in the future—although that's true—but that anomalously low-entropy states of all sorts will eventually appear as random fluctuations.

183   Epicurus is associated with Epicureanism, a philosophical precursor to utilitarianism. In the popular imagination, "epicurean" conjures up visions of hedonism and sensual pleasure, especially where food and drink are concerned; while Epicurus himself took pleasure as the ultimate good, his notion of "pleasure" was closer to "curling up with a good book" than "partying late into the night" or "gorging yourself to excess."

      Much of the original writing by the Atomists has been lost; Epicurus, in particular, wrote a thirty-seven-volume treatise on nature, but his only surviving writings are three letters reproduced in Diogenes Laertius's *Lives of the Philosophers*. The atheistic implications of their materialist approach to philosophy were not always popular with later generations.

184   Lucretius (1995), 53.

185   A careful quantitative understanding of the likelihood of different kinds of fluctuations was achieved only relatively recently, in the form of something called the "fluctuation theorem" (Evans and Searles, 2002). But the basic idea has been understood for a long time. The probability that the entropy of a system will take a random jump downward is proportional to the exponential of minus the change in entropy. That's a fancy way of saying that small fluctuations are common, and large fluctuations are extremely rare.

186   It's tempting to think, *But it's incredibly unlikely for a featureless collection of gas molecules in equilibrium to fluctuate into a pumpkin pie, while it's not that hard to imagine a pie being created in a world with a baker and so forth.* True enough. But as hard as it is to fluctuate a pie all by itself, it's much more difficult to fluctuate a baker and a pumpkin patch. Most pies that come to being *under these assumptions*—an eternal universe, fluctuating around equilibrium—will be all by themselves in the universe. The fact that the world with which we are familiar doesn't seem to work that way is evidence that something about these assumptions is not right.

187   Eddington (1931). Note that what really matters here is not so much the likelihood of significant dips in the entropy of the entire universe, but the conditional question: "Given that one subset of the universe has experienced a dip in entropy, what should we expect of the rest of the universe?" As long as the subset in question is coupled weakly to everything else, the answer is what you would expect, and what Eddington indicated: The entropy of the rest of the universe is likely to be as high as ever. For discussions (at a highly mathematical level) in the context of classical statistical mechanics, see the books by Dembo and Zeitouni (1998) or Ellis (2005). For related issues in the context of quantum mechanics, see Linden et al. (2008).

188   Albrecht and Sorbo (2004).

189   Feynman, Leighton, and Sands (1970).

190   This discussion draws from Hartle and Srednicki (2007). See also Olum (2002), Neal (2006), Page (2008), Garriga and Vilenkin (2008), and Bousso, Freivogel, and Yang (2008).

191   There are a couple of closely related questions that arise when we start comparing different kinds of observers in a very large universe. One is the "simulation argument" (Bostrom 2003), which says that it should be very easy for an advanced civilization to make a powerful computer that simulates a huge number of intelligent beings, and therefore we are most likely to be living in a computer simulation. Another is the "doomsday argument" (Leslie, 1990; Gott, 1993), which says that the human race is unlikely to last for a very long time, because if it did, those of us (now) who live in the early days of human civilization would be very atypical observers. These are very provocative arguments; their persuasive value is left up to the judgment of the reader.

192   See Neal (2006), who calls this approach "Full Non-indexical Conditioning." "Conditioning" means that we make predictions by asking what the rest of the universe looks like when certain conditions hold (e.g., that we are an observer with certain properties); "full" means that we condition over every single piece of data we have, not only coarse features like "we are an observer"; and "non-indexical" means that we consider absolutely every instance in which the conditions are met, not just one particular instance that we label as "us."

193   Boltzmann's travelogue is reprinted in Cercignani (1998), 231. For more details of his life and death, see that book as well as Lindley (2001).

11. QUANTUM TIME

194 Quoted in von Baeyer (2003), 12–13.

195 This is not to say that the ancient Buddhists weren't wise, but their wisdom was not based on the failure of classical determinism at atomic scales, nor did they anticipate modern physics in any meaningful way, other than the inevitable random similarities of word choice when talking about grand cosmic
concepts. (I once heard a lecture claming that the basic ideas of primordial nucleosynthesis were pre-
figured in the Torah; if you stretch your definitions enough, eerie similarities are everywhere.) It is dis-
respectful to both ancient philosophers and modern physicists to ignore the real differences in their
goals and methods in an attempt to create tangible connections out of superficial resemblances.

196 More recently, dogs have also been recruited for the cause. See Orzel (2009).

197 We're still glossing over one technicality—the truth is actually one step more complex (as it were) than
this description would have you believe, but it's not a complication that is necessary for our present
purposes. Quantum amplitudes are really *complex numbers*, which means they are combinations of
two numbers: a real number, plus an imaginary number. (Imaginary numbers are what you get when
you take the square root of a negative real number; so "imaginary two" is the square root of minus four,
and so on.) A complex number looks like $a + bi$, where a and b are real numbers and "i" is the square
root of minus one. If the amplitude associated with a certain option is $a + bi$, the probability it corresponds to is simply $a^2 + b^2$, which is guaranteed to be greater than or equal to zero. You will have to
trust me that this extra apparatus is extremely important to the workings of quantum mechanics—
either that, or start learning some of the mathematical details of the theory. (I can think of less reward-
ing ways of spending your time, actually.)

198 The fact that any particular sequence of events assigns positive or negative amplitudes to the two final
possibilities is an assumption we are making for the purposes of our thought experiment, not a deep
feature of the rules of quantum mechanics. In any real-world problem, details of the system being considered will determine what precisely the amplitudes are, but we're not getting our hands quite that
dirty at the moment. Note also that the particular amplitudes in these examples take on the numerical
values of plus or minus 0.7071—that's the number which, when squared, gives you 0.5.

199 At a workshop attended by expert researchers in quantum mechanics in 1997, Max Tegmark took an
admittedly highly unscientific poll of the participants' favored interpretation of quantum mechanics
(Tegmark, 1998). The Copenhagen interpretation came in first with thirteen votes, while the many-worlds
interpretation came in second with eight. Another nine votes were scattered among other alternatives.
Most interesting, eighteen votes were cast for "none of the above/undecided." And these are the experts.

200 So what does happen if we hook up a surveillance camera but then don't examine the tapes? It doesn't
matter whether we look at the tapes or not; the camera still counts as an observation, so there will
be a chance to observe Ms. Kitty under the table. In the Copenhagen interpretation, we would say,
"The camera is a classical measuring device whose influence collapses the wave function." In the
many-worlds interpretation, as we'll see, the explanation is "the wave function of the camera becomes
entangled with the wave function of the cat, so the alternative histories decohere."

201 Many people have thought about changing the rules of quantum mechanics so that this is no longer
the case; they have proposed what are called "hidden variable theories" that go beyond the standard
quantum mechanical framework. In 1964, theoretical physicist John Bell proved a remarkable theorem: No local theory of hidden variables can possibly reproduce the predictions of quantum mechanics. This hasn't stopped people from investigating nonlocal theories—ones where distant events can
affect each other instantaneously. But they haven't really caught on; the vast majority of modern physicists believe that quantum mechanics is simply correct, even if we don't yet know how to interpret it.

202 There is a slightly more powerful statement we can actually make. In classical mechanics, the state is
specified by both position and velocity, so you might guess that the quantum wave function assigns
probabilities to every possible combination of position and velocity. But that's not how it works. If
you specify the amplitude for every possible position, you are done—you've completely determined
the entire quantum state. So what happened to the velocity? It turns out that you can write the same
wave function in terms of an amplitude for every possible velocity, completely leaving position out of
the description. These are not two different states; they are just two different ways of writing exactly the
same state. Indeed, there is a cookbook recipe for translating between the two choices, known in the
trade as a "Fourier transform." Given the amplitude for every possible position, you can do a Fourier
transform to determine the amplitude for any possible velocity, and vice versa. In particular, if the
wave function is an eigenstate, concentrated on one precise value of position (or velocity), its Fourier
transform will be completely spread out over all possible velocities (or positions).

203 Einstein, Podolsky, and Rosen (1935).

204 Everett (1957). For discussion from various viewpoints, see Deutsch (1997), Albert (1992), or Ouellette (2007).
205 Note how crucial entanglement is to this story. If there were no entanglement, the outside world would still exist, but the alternatives available to Miss Kitty would be completely independent of what was going on out there. In that case, it would be perfectly okay to attribute a wave function to Miss Kitty all by herself. And thank goodness; that's the only reason we are able to apply the formalism of quantum mechanics to individual atoms and other simple isolated systems. Not everything is entangled with everything else, or it would be impossible to say much about any particular subsystem of the world.

12. BLACK HOLES: THE ENDS OF TIME

206 Bekenstein (1973).
207 Hawking (1988), 104. Or, as Dennis Overbye (1991, 107) puts it: "In Cambridge Bekenstein's break-through was greeted with derision. Hawking was outraged. He knew this was nonsense."
208 For discussion of observations of stellar-mass black holes, see Casares (2007); for supermassive black holes in other galaxies, see Kormendy and Richstone (1995). The black hole at the center of our galaxy is associated with a radio source known as "Sagittarius A*"; see Reid (2008).
209 Okay, for some people the looking is even more fun.
210 Way more than that, actually. As of January 2009, Hawking's original paper (1975) had been cited by more than 3,000 other scientific papers.
211 As of this moment, we have never detected gravitational waves directly, although indirect evidence for their existence (as inferred from the energy lost by a system of two neutron stars known as the "binary pulsar") was enough to win the Nobel Prize for Joseph Taylor and Russell Hulse in 1993. Right now, several gravitational-wave observatories are working to discover such waves directly, perhaps from the coalescence of two black holes.
212 The area of the event horizon is proportional to the square of the mass of the black hole; in fact, if the area is A and the mass is M, we have $A = 16\pi G^2 M^2 / c^4$, where G is Newton's constant of gravitation and c is the speed of light.
213 The analogy between black hole mechanics and thermodynamics was spelled out in Bardeen, Carter, and Hawking (1973).
214 One way to think about why the surface gravity is not infinite is to take seriously the caveat "as measured by an observer very far away." The force right near the black hole is large, but when you measure it from infinity it undergoes a gravitational redshift, just as an escaping photon would. The force is infinitely strong, but there is an infinite redshift from the point of view of a distant observer, and the effects combine to give a finite answer for the surface gravity.
215 More carefully, Bekenstein suggested that the entropy was proportional to the area of the event horizon. Hawking eventually worked out the constant of proportionality.
216 Hawking (1988), 104–5.
217 You may wonder why it seems natural to think of the electromagnetic and gravitational fields, but not the electron field or the quark field. That's because of the difference between fermions and bosons. Fermions, like electrons and quarks, are matter particles, distinguished by the fact that they can't pile on top of one another; bosons, like photons and gravitons, are force particles that pile on with abandon. When we observe a macroscopic, classical-looking field, that's a combination of a huge number of boson particles. Fermions like electrons and quarks simply can't pile up that way, so their field vibrations only ever show up as individual particles.
218 Overbye (1991), 109.
219 For reference purposes, the Planck length is equal to $(G\hbar/c^3)^{1/2}$, where G is Newton's constant of gravitation, $\hbar$ is Planck's constant from quantum mechanics, and c is the speed of light. (We've set Boltzmann's constant equal to 1.) So the entropy can be expressed as $S = (c^3/4\hbar G)A$. The area of the event horizon is related to the mass M of the black hole by $A = 8\pi G^2 M^2$. Putting it all together, the entropy is related to the mass by as $S = (4\pi G c^3/\hbar) M^2$.
220 Particles and antiparticles are all "particles," if that makes sense. Sometimes the word *particle* is used specifically to contrast with *antiparticle*, but more often it just refers to any pointlike elementary object. Nobody would object to the sentence "the positron is a particle, and the electron is its antiparticle."
221 "Known" is an important caveat. Cosmologists have contemplated the possibility that some unknown process, perhaps in the very early universe, might have created copious amounts of very small black holes, perhaps even related to the dark matter. If these black holes were small enough, they wouldn't be all that dark; they'd be emitting increasing amounts of Hawking radiation, and the final explosions might even be detectable.

222 One speculative but intriguing idea is that we could *make* a black hole in a particle accelerator, and then observe it decaying through Hawking radiation. Under ordinary circumstances, that's hopelessly unrealistic; gravity is such an incredibly weak force that we'll never be able to build a particle accelerator powerful enough to make even a microscopic black hole. But some modern scenarios, featuring hidden dimensions of spacetime, suggest that gravity becomes much stronger than usual at short distances (see Randall, 2005). In that case, the prospect of making and observing small black holes gets upgraded from "crazy" to "speculative, but not completely crazy." I'm sure Hawking is rooting for it to happen.

　Unfortunately, the prospect of microscopic black holes has been seized on by a group of fearmongers to spin scenarios under which the Large Hadron Collider, a new particle accelerator at the CERN laboratory in Geneva, is going to destroy the world. Even if the chances are small, destroying the world is pretty bad, so we should be careful, right? But careful reviews of the possibilities (Ellis et al., 2008) have concluded that there's nothing the LHC will do that hasn't occurred many times already elsewhere in the universe; if something disastrous were going to happen, we should have seen signs of it in other astrophysical objects. Of course, it's always possible that everyone involved in these reviews is making some sort of unfortunate math mistake. But lots of things are possible. The next time you open a jar of tomato sauce, it's possible that you will unleash a mutated pathogen that will wipe out all life on Earth. It's possible that we are being watched and judged by a race of super-intelligent aliens, who will think badly of us and destroy the Earth if we allow ourselves to be cowed by frivolous lawsuits and *don't* turn on the LHC. When possibilities become as remote as what we're speaking about here, it's time to take the risks and get on with our lives.

223 You might be tempted to pursue ideas along exactly those lines—perhaps the information is copied, and is contained simultaneously in the book falling into the singularity and in the radiation leaving the black hole. A result in quantum mechanics—the "No-Cloning Theorem"—says that can't happen. Not only can information not be destroyed, but it can't be duplicated.

224 Preskill's take on the black hole bets can be found at his Web page: http://www.theory.caltech.edu/people/preskill/bets.html. For an in-depth explanation of the black hole information loss paradox, see Susskind (2008).

225 You might think we could sidestep this conclusion by appealing to photons once again, because photons are particles that have zero mass. But they do have energy; the energy of a photon is larger when its wavelength is smaller. Because we're dealing with a box of a certain fixed size, each photon inside will have a minimum allowed energy; otherwise, it simply wouldn't fit. And the energy of all those photons, through the miracle of $E = mc^2$, contributes to the mass of the box. (Each photon is massless, but a box of photons has a mass, given by the sum of the photon energies divided by the speed of light squared.)

226 The area of a sphere is equal to 4π times its radius squared. The area of a black hole event horizon, logically enough, is 4π times the Schwarzschild radius squared. This is actually the *definition* of the Schwarzschild radius, since the highly curved spacetime inside the hole makes it difficult to sensibly define the distance from the singularity to the horizon. (Remember—that distance is timelike!) So the area of the event horizon is proportional to the square of the mass of the black hole. This is all for black holes with zero rotation and no net electric charge; if the hole is spinning or charged, the formulas are slightly more complicated.

227 The holographic principle is discussed in Susskind (2008); for technical details, see Bousso (2002).

228 Maldacena (1998). The title of Maldacena's paper, "The Large N Limit of Superconformal Field Theories and Supergravity," doesn't immediately convey the excitement of his result. When Juan came to Santa Barbara in 1997 to give a seminar, I stayed in my office to work, having not been especially intrigued by his title. Had the talk been advertised as "An Equivalence Between a Five-Dimensional Theory with Gravity and a Four-Dimensional Theory Without Gravity," I probably would have attended the seminar. Afterward, it was easy to tell from the conversations going on in the hallway—excited, almost frantic, scribbling on blackboards to work out implications of these new ideas—that I had missed something big.

229 The good thing about string theory is that it seems to be a unique theory; the bad thing is that this theory seems to have many different phases, which look more or less like completely different theories. Just like water can take the form of ice, liquid, or water vapor, depending on the circumstances, in string theory spacetime itself can come in many different phases, with different kinds of particles and even different numbers of observable dimensions of spacetime. And when we say "many," we're not kidding—people throw around numbers like 10^{500} different phases, and it could very well be an infinite number. So the theoretical uniqueness of string theory seems to be of little practical help in understanding the particles and interactions of our particular world. See Greene (2000) or Musser (2008) for

overviews of string theory, and Susskind (2006) for a discussion (an optimistic one) of the problem of many different phases.
230 Strominger and Vafa (1996). For a popular-level account, see Susskind (2008).
231 While the Strominger-Vafa work implies that the space of states for a black hole in string theory has the right size to account for the entropy, it doesn't quite tell us what those states look like when gravity is turned on. Samir Mathur and collaborators have suggested that they are "fuzzballs"—configurations of oscillating strings that fill up the volume of the black hole inside the event horizon (Mathur, 2005).

13. THE LIFE OF THE UNIVERSE

232 In the eighteenth century, Gottfried Wilhelm Leibniz posed the Primordial Existential Question: "Why is there something rather than nothing?" (One might answer, "Why not?") Subsequently, some philosophers have tried to argue that the very existence of the universe should be surprising to us, on the grounds that "nothing" is simpler than "something" (e.g., Swinburne, 2004). But that presupposes a somewhat dubious definition of "simplicity," as well as the idea that this particular brand of simplicity is something a universe ought to have—neither of which is warranted by either experience or logic. See Grünbaum (2004) for a discussion.
233 Some would argue that God plays the role of the Universal Chicken, creating the universe in a certain state that accounts for the low-entropy beginning. This doesn't seem like a very parsimonious explanatory framework, as it's unclear why the entropy would be quite so low, and why (for one thing among many) there should be a hundred billion galaxies in the universe. More important, as scientists we want to explain the most with the least, so if we can come up with naturalistic theories that account for the low entropy of our observed universe without recourse to anything other than the laws of physics, that would be a triumph. Historically, this has been a very successful strategy; pointing at "gaps" in naturalistic explanations of the world and insisting that only God can fill them has, by contrast, had a dismal track record.
234 This isn't exactly true, although it's a pretty good approximation. If a certain kind of particle couples very weakly to the rest of the matter and radiation in the universe, it can essentially stop interacting, and drop out of contact with the surrounding equilibrium configuration. This is a process known as "freeze-out," and it is crucially important to cosmologists—for example, when they would like to calculate the abundance of dark matter particles, which plausibly froze out at a very early time. In fact, the matter and radiation in the late universe (today) has frozen out long ago, and we are no longer in equilibrium even when you ignore gravity. (The temperature of the cosmic microwave background is about 3 Kelvin, so if we were in equilibrium, everything around you would be at a temperature of 3 Kelvin.)
235 The speed of light divided by the Hubble constant defines the "Hubble length," which works out to about 14 billion light-years in the current universe. For not-too-crazy cosmologies, this quantity is almost the same as the age of the universe times the speed of light, so they can be used interchangeably. Because the universe expands at different rates at different times, the current size of our comoving patch can actually be somewhat larger than the Hubble length.
236 See, for example, Kofman, Linde, and Mukhanov (2002). That paper was written in response to a paper by Hollands and Wald (2002) that raised some similar issues to those we're exploring in this chapter, in the specific context of inflationary cosmology. For a popular-level discussion that takes a similar view, see Chaisson (2001).
237 Indeed, Eric Schneider and Dorion Sagan (2005) have argued that the "purpose of life" is to accelerate the rate of entropy production by smoothing out gradients in the universe. It's hard to make a proposal like that rigorous, for various reasons; one is that, while the Second Law says that entropy tends to increase, there's no law of nature that says entropy tends to increase as fast as it can.
238 Also in contrast with the gravitational effects of sources of energy density other than "particles." This loophole is relevant to the real world because of dark energy. The dark energy isn't a collection of particles; it's a smooth field that pervades the universe, and its gravitational impact is to push things apart. Nobody ever said things would be simple.
239 Other details are also important. In the early universe, ordinary matter is ionized—electrons are moving freely, rather than being attached to atomic nuclei. The pressure in an ionized plasma is generally larger than in a collection of atoms.
240 Penrose (2005), 706. An earlier version of this argument can be found in Penrose (1979).
241 Most of the matter in the universe—between 80 percent and 90 percent by mass—is in the form of dark matter, not the ordinary matter of atoms and molecules. We don't know what the dark matter is, and

it's conceivable that it takes the form of small black holes. But there are problems with that idea, including the difficulty of making so many black holes in the first place. So most cosmologists tend to believe that the dark matter is very likely to be some sort of new elementary particle (or particles) that hasn't yet been discovered.

242   Black-hole entropy increases rapidly as the black hole gains mass—it's proportional to the mass squared. (Entropy goes like area, which goes like radius squared, and the Schwarzschild radius is proportional to the mass.) So a black hole of 10 million solar masses would have 100 times the entropy of one coming in at 1 million solar masses.

243   Penrose (2005), 707.

244   The argument here closely follows a paper I wrote in collaboration with Jennifer Chen (Carroll and Chen, 2004).

245   See, for example, Zurek (1982).

246   It's also very far from being accepted wisdom among physicists. Not that there is any accepted answer to the question "What do the highest-entropy states look like when gravity is taken into account?" other than "We don't know." But hopefully you'll become convinced that "empty space" is the best answer we have at the moment.

247   This is peeking ahead a bit, but note that we could also play this game backward in time. That is: start from some configuration of matter in the universe, a slice of spacetime at one moment in time. In some places we'll see expansion and dilution, in others contraction and collapse and ultimately evaporation. But we can also ask what would happen if we evolved that "initial" state backward in time, using the same reversible laws of physics. The answer, of course, is that we would find the same kind of behavior. The regions that are expanding toward the future are contracting toward the past, and vice versa. But ultimately space would empty out as the "expanding" regions took over. The very far past looks just like the very far future: empty space.

248   Here in our own neighborhood, NASA frequently uses a similar effect—the "gravitational slingshot"—to help accelerate probes to the far reaches of the Solar System. If a spacecraft passes by a massive planet in just the right way, it can pick up some of the planet's energy of motion. The planet is so heavy that it hardly notices, but the spacecraft gets flung away at a much higher velocity.

249   Wald (1983).

250   In particular, we can define a "horizon" around every observable patch of de Sitter space, just as we can with black holes. Then the entropy formula for that patch is precisely the same formula as the entropy of a black hole—it's the area of that horizon, measured in Planck units, divided by four.

251   If H is the Hubble parameter in de Sitter space, the temperature is $T = (\hbar/2\pi k)H$, where $\hbar$ is Planck's constant and k is Boltzmann's constant. This was first worked out by Gary Gibbons and Stephen Hawking (1977).

252   You might think this prediction is a bit too bold, relying on uncertain extrapolations into regimes of physics that we don't really understand. It's undeniably true that we don't have direct experimental access to an eternal de Sitter universe, but the scenario we have sketched out relies only on a few fairly robust principles: the existence of thermal radiation in de Sitter space, and the relative frequency of different kinds of random fluctuations. In particular, it's tempting to wonder whether there is some special kind of fluctuation that makes a Big Bang, and that kind of fluctuation is more likely than a fluctuation that makes a Boltzmann brain. That might be what actually happens, according to the ultimately correct laws of physics—indeed, we'll propose something much like that later in the book— but it's absolutely not what happens under the assumptions we are making here. The nice thing about thermal fluctuations in eternal de Sitter space is that we understand thermal fluctuations very well, and we can calculate with confidence how frequently different fluctuations occur. Specifically, fluctuations involving large changes in entropy are enormously less likely than fluctuations involving small changes in entropy. It will always be easier to fluctuate into a brain than into a universe, unless we depart from this scenario in some profound way.

253   Dyson, Kleban, and Susskind (2002); Albrecht and Sorbo (2004).

14. INFLATION AND THE MULTIVERSE

254   Toulmin (1988), 393.

255   See Guth (1997), also Overbye (1991).

256   Space can be curved even if spacetime is flat. A space with negative curvature, expanding with a size proportional to time, corresponds to a spacetime that is completely flat. Likewise, space can be flat even if spacetime is curved; if a spatially flat universe is expanding (or contracting) in time, the spacetime will certainly be curved. (The point is that the expansion contributes to the total curvature of spacetime,

and the curvature of space also contributes. That's why an expanding negatively curved space can correspond to a spacetime with zero curvature; the contribution from spatial curvature is negative and can precisely cancel the positive contribution from the expansion.) When cosmologists refer to "a flat universe" they mean a *spatially* flat universe, and likewise for positive or negative curvature.

257   They add up to less than 180 degrees.

258   One way of measuring the curvature of the universe is indirectly, using Einstein's equation. General relativity implies a relationship between the curvature, the expansion rate, and the amount of energy in the universe. For a long time, astronomers measured the expansion rate and the amount of matter in the universe (which they assumed was the most important part of the energy), and kept finding that the universe was pretty close to flat, but it should have a tiny amount of negative curvature. The discovery of dark energy changed all that; it provided exactly the right amount of energy to make the universe flat. Subsequently, astronomers have been able to measure the curvature directly, by using the pattern of temperature fluctuations in the cosmic microwave background as a kind of giant triangle (Miller et al., 1999; de Bernardis et al., 2000; Spergel et al., 2003). This method indicates strongly that the universe really is spatially flat, which is a nice consistency check with the indirect reasoning.

259   Nobody else calls it that. Because this form of dark energy serves the purpose of driving inflation, it is usually postulated to arise from a hypothetical field dubbed the "inflaton." It would be nice if the inflaton field served some other purpose, or fit snugly into some more complete theory of particle physics, but as yet we don't know enough to say.

260   You might think that, because the Big Bang itself is a point, the past light cones of any event in the universe must necessarily meet at the Big Bang. But that's misleading. For one thing, the Big Bang is not a point in space—it's a moment in time. More important, the Big Bang in classical general relativity is a singularity and shouldn't even be included in the spacetime; we should talk only about what happens after the Big Bang. And even if we included moments immediately after the Big Bang, the past light cones would not overlap.

261   The original papers are by Andrei Linde (1981) and Andreas Albrecht and Paul Steinhardt (1982). See Guth (1997) for an accessible discussion.

262   See, for example, Spergel et al. (2003).

263   See Vilenkin (1983), Linde (1986), Guth (2007).

264   This scenario was invented under the slightly misleading name of "open inflation" (Bucher, Goldhaber, and Turok, 1995). At the time, before the discovery of dark energy, cosmologists had begun to get a bit nervous—inflation seemed to robustly predict that the universe should be spatially flat, but observations of the density of matter kept implying that there wasn't enough energy to make it work out. Some people panicked, and tried to invent models of inflation that didn't necessarily predict a flat universe. That turned out not to be necessary—the dark energy has exactly the right amount of energy density to make the universe flat, and observations of the cosmic microwave background strongly indicate that it really is flat (Spergel et al., 2003). But that's okay, because out of the panic came a clever idea—how to make a realistic universe inside a bubble embedded in a false-vacuum background.

265   In fact, the early papers on eternal inflation were set in the context of new inflation, not old-inflation-with-new-inflation-inside-the-bubbles. In new inflation it is actually more surprising that inflation is eternal, as you would think the field would just roll down the hill defined by its potential energy. But we should remember that the rolling field has quantum fluctuations; if conditions are right, those fluctuations can be quite large. In fact, they can be large enough that in some regions of space the field actually moves *up* the hill, even though on average it is rolling down. Regions where it rolls up are rare, but they expand faster because the energy density is larger. We end up with a picture similar to the old-inflation story; lots of the universe sees an inflaton roll down and convert to matter and radiation, but an increasing volume stays stuck in the inflating stage, and inflation never ends.

266   See Susskind (2006), or Vilenkin (2006). An earlier, related version of a landscape of different vacuum states was explored by Smolin (1993).

267   In the original papers about inflation, it was implicitly assumed that the particles in the early universe were close to thermal equilibrium. The scenario described here, which seems a bit more robust, goes under the name of "chaotic inflation," and was originally elucidated by Andrei Linde (1983, 1986).

268   See for example Penrose (2005), Hollands and Wald (2002).

269   This is not to imply that choosing a configuration of the universe randomly from among all possible allowed states is something we are ordered to do, or that there is some reason to believe that it's actually what happens. Rather, that if the state of the universe is clearly *not* chosen randomly, then there must be something that determines how it is chosen; that's a clue we would like to use to help understand how the universe works.

270 You may object that there is another candidate for a "high-entropy state"—the chaotic mess into which our universe would evolve if we let it collapse. (Or equivalently, if we started with a typical microstate consistent with the current macrostate of the universe, and ran the clock backward.) It's true that such a state is much lumpier than the current universe, as singularities and black holes would form in the process of collapse. But that's exactly the point; even among states that pack the entire current universe into a very small region, an incredibly small fraction take the form of a smooth patch dominated by dark super-energy, as required by inflation. Most such states, on the contrary, are in a regime where quantum field theory doesn't apply, because quantum gravity is absolutely necessary to describe them. But "we don't know how to describe such states" is a very different statement than "such states don't exist" or even "we can ignore such states when we enumerate the possible initial states of the universe." If the dynamics are reversible, we have no choice but to take those states very seriously.

271 For example, Guth (1997).

15. THE PAST THROUGH TOMORROW

272 Pascal (1995), 66.

273 What would be even better is if some young person read this book, became convinced that this was a serious problem worthy of our attention, and went on to solve it. Or an older person, it doesn't really matter. In either case, if you end up finding an explanation for the arrow of time that becomes widely accepted within the physics community, please let me know if this book had anything to do with it.

274 Perhaps the closest to something along these lines is the "Holographic Cosmology" scenario advocated by Tom Banks and Willy Fischler (2005; also Banks, 2007). They suggest that the effective dynamical laws of quantum gravity could be very different in different spacetimes. In other words, the laws of physics themselves could be time-dependent. This is a speculative scenario, but worth paying attention to.

275 A related strategy is to posit a particular form for the wave function of the universe, as advocated by James Hartle and Stephen Hawking (1983). They rely on a technique known as "Euclidean quantum gravity"; attempting to do justice to the pros and cons of this approach would take us too far afield from our present concerns. It has been suggested that the Hartle-Hawking wave function implies that the universe must be smooth near the Big Bang, which would help explain the arrow of time (Halliwell and Hawking, 1985), but the domain of validity of the approximations used to derive this result is a bit unclear. My own suspicion is that the Hartle-Hawking wave function predicts that we should live in empty de Sitter space, just as a straightforward contemplation of entropy would lead us to expect.

276 Penrose (1979). When you dig deeply into the mathematics of spacetime curvature, you find that it comes in two different forms: "Ricci curvature," named after Italian mathematician Gregorio Ricci-Curbastro, and "Weyl curvature," named after German mathematician Hermann Weyl. Ricci curvature is tied directly to the matter and energy in spacetime—where there's stuff, the Ricci curvature is nonzero, and where there's not, the Ricci curvature vanishes. Weyl curvature, on the other hand, can exist all by itself; a gravitational wave, for example, propagates freely through space, and leads to Weyl curvature but no Ricci curvature. The Weyl curvature hypothesis states that singularities in one direction of time always have vanishing Weyl curvature, while those at the other are unconstrained. We would assign the descriptive adjectives *initial* and *final* after the fact, since the low-Weyl-curvature direction would have a lower entropy.

277 Another problem is the apparent danger of Boltzmann brains if the universe enters an eternal de Sitter phase in the future. Also, the concept of a "singularity" from classical general relativity is unlikely to survive intact in a theory of quantum gravity. A more realistic version of the Weyl curvature hypothesis would have to be phrased in quantum-gravity language.

278 Gold (1962).

279 For a brief while, Stephen Hawking believed that his approach to quantum cosmology predicted that the arrow of time would actually reverse if the universe re-collapsed (Hawking, 1985). Don Page convinced him that this was not the case—the right interpretation was that the wave function had two branches, oriented oppositely in time (Page, 1985). Hawking later called this his "greatest blunder," in a reference to Einstein's great blunder of suggesting the cosmological constant rather than predicting the expansion of the universe (Hawking, 1988).

280 Price (1996).

281 See, for example, Davies and Twamley (1993), Gell-Mann and Hartle (1996). A different form of future boundary condition, which does not lead to a reversal of the arrow of time, has been investigated in particle physics; see Lee and Wick (1970), Grinstein, O'Connell, and Wise (2009).

282 Once again, the English language lacks the vocabulary for nonstandard arrows of time. We will choose the convention that the "direction of time" is defined by us, here in the "ordinary" post-Big-Bang phase of the universe; with respect to this choice, entropy decreases "toward the future" in the collapsing phase. Of course, organisms that actually live in that phase will naturally define things in the opposite sense; but it's our book, and the choice is simply a matter of convention, so we can make the rules.

283 Greg Egan worked through the dramatic possibilities of this scenario, in his short story "The Hundred Light-Year Diary" (reprinted in Egan, 1997).

284 Cf. Callender's Fabergé eggs, discussed in Chapter Nine.

285 See also Carroll (2008).

286 One of the first bouncing scenarios was simply called the "Pre-Big-Bang scenario." It makes use of a new field called the "dilaton" from string theory, which affects the strength of gravity as it changes (Gasperini and Veneziano, 1993). A related example is the "ekpyrotic universe" scenario, which was later adapted into the "cyclic universe." In this picture, the energy that powers what we see as the "Bang" comes when a hidden, compact dimension squeezes down to zero size. The cyclic universe idea is discussed in depth in a popular book by Paul Steinhardt and Neil Turok (2007); its predecessor, the ekpyrotic universe, was proposed by Khoury et al. (2001). There are also bouncing cosmologies that don't rely on strings or extra dimensions, but on the quantum properties of spacetime itself, under the rubric of "loop quantum cosmology" (Bojowald, 2006).

287 Hopefully, after the appearance of this book, that will all change.

288 The same argument holds for Steinhardt and Turok's cyclic universe. Despite the label, their model is not recurrent in the way that the Boltzmann-Lucretius model would be. In an eternal universe with a finite-sized state space, allowed sequences of events happen both forward and backward in time, equally often. But in the Steinhardt-Turok model, the arrow of time always points in the same direction; entropy grows forever, requiring an infinite amount of fine-tuning at any one moment. Interestingly, Richard Tolman (1931) long ago discussed problems of entropy in a cyclic universe, although he talked about only the entropy of matter, not including gravity. See also Bojowald and Tavakol (2008).

289 This discussion assumes that the assumptions we previously made in discussing the entropy of our comoving patch remain valid—in particular, that it makes sense to think of the patch as an autonomous system. That is certainly not necessarily correct, but it is usually implicitly assumed by people who study these scenarios.

290 Aguirre and Gratton (2003). Hartle, Hawking, and Hertog (2008) also investigated universes with high entropy in the past and future and low entropy in the middle, in the context of Euclidean quantum gravity.

291 This is true even in ordinary nongravitational situations, where the total energy is strictly conserved. When a high-energy state decays into a lower-energy one, like a ball rolling down a hill, energy isn't created or destroyed; it's just transformed from a useful low-entropy form into a useless high-entropy form.

292 Farhi, Guth, and Guven (1990). See also Farhi and Guth (1987), and Fischler, Morgan, and Polchinski (1990a, 1990b). Guth writes about this work in his popular-level book (1997).

293 The most comprehensive recent work on this question was carried out by Anthony Aguirre and Matthew Johnson (2006). They catalogued all the different ways that baby universes might be created by quantum tunneling, but in the end were unable to make a definitive statement about what actually happens. ("The unfortunate bottom line, then, is that while the relation between the various nucleation processes is much clearer, the question of which ones actually occur remains open.") From a completely different perspective, Freivogel et al. (2006) considered inflation in an anti–de Sitter background, using Maldacena's correspondence. They concluded that baby universes were not created. But our interest is in de Sitter backgrounds, not anti–de Sitter backgrounds; it's unclear whether the results can be extended from one context to the other. For one more take on the evolution of de Sitter space, see Bousso (1998).

294 Carroll and Chen (2004).

295 One assumption here is that the de Sitter space is in a true vacuum state; in particular, that there is no other state of the theory where the vacuum energy vanishes, and spacetime could look like Minkowski space. To be honest, that is not necessarily a realistic assumption. In string theory, for example, we are pretty sure that 10-dimensional Minkowski space is a good solution of the theory. Unlike de Sitter, Minkowski space has zero temperature, so can plausibly avoid the creation of baby universes. To make the scenario described here work, we have to imagine either that there are no states with zero vacuum energy, or that the amount of spacetime that is actually in such a state is sufficiently small compared to the de Sitter regions.

16. EPILOGUE

296   And that's despite the fact that, just as the manuscript was being completed, another book with exactly
      the same title appeared on the market! (Viola, 2009). His subtitle is quite different, however: "Redis-
      covering the Ageless Purpose of God." I do hope nobody orders the wrong book by accident.

297   Feynman, Leighton, and Sands (1970), 46–8.

298   Popper (1959). Note that Popper went a bit further than the demarcation problem; he wanted to under-
      stand all of scientific progress as a series of falsified conjectures. Compared to how science is actually
      done, this is a fairly impoverished way of understanding the process; ruling out conjectures is impor-
      tant, but there's a lot more that goes into the real workings of science.

299   See Deutsch (1997) for more on this point.

300   For one example among many, see Swinburne (2004).

301   Lemaître (1958).

302   Steven Weinberg put it more directly: "The more the universe seems comprehensible, the more it also
      seems pointless" (Weinberg 1977, 154).

303   I regret that this book has paid scant attention to current and upcoming new experiments in funda-
      mental physics. The problem is that, as fascinating and important as those experiments are, it's very
      hard to tell ahead of time what we are going to learn from them, especially about a subject as deep and
      all-encompassing as the arrow of time. We're not going to build a telescope that will use tachyons to
      peer into other universes, unfortunately. What we might do is build particle accelerators that reveal
      something about supersymmetry, which in turn teaches us something about string theory, which we
      can use to understand more about quantum gravity. Or we might gather data from giant telescopes—
      collecting not only photons of light, but also cosmic rays, neutrinos, gravitational waves, or even
      particles of dark matter—that reveal something surprising about the evolution of the universe. The
      real world surprises us all the time: dark matter and dark energy are obvious examples. As a theoreti-
      cal physicist, I've written this book from a rather theoretical perspective, but as a matter of history it's
      often new experiments that end up awakening us from our dogmatic slumbers.

APPENDIX: MATH

304   These properties are behind the "magic of mathematics" appealed to above. For example, suppose we
      wanted to figure out what was meant by 10 to the power 0.5. I know that, whatever that number is, it has
      to have the property that $10^{0.5} \cdot 10^{0.5} = 10^{(0.5 + 0.5)} = 10^1 = 10$.

      In other words, the number $10^{0.5}$ times itself gives us 10; that means that $10^{0.5}$ must simply be the
      square root of 10. (And likewise for any other base raised to the power 0.5.) By similar tricks, we can
      figure out the exponential of any number we like.

BIBLIOGRAPHY

Note: Many of these bibliography entries are referenced explicitly in the text, but many are not. Some were influential in shaping the point of view presented here; others were foils to be argued against. Some are research articles that fill in technical details on the topics of this book, while others provide additional background reading at an accessible level. All of them are interesting.

My favorite modern books about the arrow of time include David Albert's *Time and Chance*, Huw Price's *Time's Arrow and Archimedes' Point*, Brian Greene's *The Fabric of the Cosmos*, and Michael Lockwood's *The Labyrinth of Time*. Any of those will give you a complementary viewpoint to that presented here. Etienne Klein's *Chronos*, Craig Callender's *Introducing Time*, and Paul Davies's *About Time* all discuss the subject of time more broadly. For background in general relativity I suggest Kip Thorne's *Black Holes and Time Warps*, and for black holes and information loss in particular Leonard Susskind's *The Black Hole War*. For cosmology I recommend Dennis Overbye's *Lonely Hearts of the Cosmos* or Alan Guth's *The Inflationary Universe*. David Lindley's *Boltzmann's Atom* and Hans Christian von Baeyer's *Warmth Disperses and Time Passes* provide fascinating historical background, as does Stephen Brush's anthology of original papers on kinetic theory. Zeh's *The Physical Basis of the Direction of Time* tackles the subject at a technical level.

Many of the modern (post-1992) research articles can be downloaded for free from the arXiv physics preprint server http://arxiv.org/.

More information and links to further resources can be found at the book's Web site, http://eternitytohere.com.

Abbot, E. A. *Flatland: A Romance of Many Dimensions*. Cambridge: Perseus, 1899.
Adams, F., and Laughlin, G. *The Five Ages of the Universe: Inside the Physics of Eternity*. New York: Free Press, 1999.
Aguirre, A., and Gratton, S. "Inflation Without a Beginning: A Null Boundary Proposal." *Physical Review D* 67 (2003): 083515.
Aguirre, A., and Johnson, M. C. "Two Tunnels to Inflation." *Physical Review D* 73 (2006): 123529.
Alavi-Harati, A., et al. (KTeV Collaboration). "Observation of CP Violation in $KL \to \pi^+\pi^- e^+e^-$ Decays." *Physical Review Letters* 84 (2000): 408–11.
Albert, D. Z. *Quantum Mechanics and Experience*. Cambridge, MA: Harvard University Press, 1992.
Albert, D. Z. *Time and Chance*. Cambridge, MA: Harvard University Press, 2000.
Albert, D. Z., and Loewer, B. "Interpreting the Many Worlds Interpretation." *Synthese* 77 (1988): 195–213.
Albrecht, A. "Cosmic Inflation and the Arrow of Time." In *Science and Ultimate Reality: From Quantum to Cosmos*, edited by Barrow, J. D., Davies, P. C. W., and Harper, C. L. Cambridge: Cambridge University Press, 2004.
Albrecht, A., and Sorbo, L. "Can the Universe Afford Inflation?" *Physical Review D* 70 (2004): 63528.
Albrecht, A., and Steinhardt, P. J. "Cosmology for Grand Unified Theories with Radiatively Induced Symmetry Breaking." *Physical Review Letters* 48 (1982): 1220–23.
Ali, A., Ellis, J., and Randjbar-Daemi, S., eds. *Salamfestschrift: A Collection of Talks*. Singapore: World Scientific,. 1993.
Alpher, R. A., and Herman, R. *Genesis of the Big Bang*. Oxford: Oxford University Press, 2001.
Amis, M. *Time's Arrow*. New York: Vintage, 1991.
Angelopoulos, A., et al. (CPLEAR Collaboration). "First Direct Observation of Time Reversal Noninvariance in the Neutral Kaon System." *Physics Letters B* 444 (1998): 43–51.
Arntzenius, F. "Time Reversal Operations, Representations of the Lorentz Group, and the Direction of Time." *Studies in History and Philosophy of Science Part B* 35 (2004): 31–43.
Augustine, Saint. *Confessions*. Translated by H. Chadwick. Oxford: Oxford University Press, 1998.
Avery, J. *Information Theory and Evolution*. Singapore: World Scientific, 2003.

Baker, N. *The Fermata*. New York: Random House, 2004.

Banks, T. "Entropy and Initial Conditions in Cosmology" (2007). http://arxiv.org/abs/ hep-th/0701146.

Banks, T., and Fischler, W. "Holographic Cosmology 3.0." *Physica Scripta* T117 (2005): 56–63.

Barbour, J. *The End of Time: The Next Revolution in Physics*. Oxford University Press, 1999.

Bardeen, J. M., Carter, B., and Hawking, S. W. "The Four Laws of Black Hole Mechanics." *Communications in Mathematical Physics* 31 (1973): 161–70.

Barrow, J. D., Davies, P. C. W., and Harper, C. L. *Science and Ultimate Reality: From Quantum to Cosmos*, honoring John Wheeler's 90th birthday. Cambridge: Cambridge University Press, 2004.

Barrow, J. D., and Tipler, F. J. *The Anthropic Cosmological Principle*. Oxford: Oxford University Press, 1988.

Baum, E. B. *What Is Thought?* Cambridge, MA: MIT Press, 2004.

Bekenstein, J. D. "Black Holes and Entropy." *Physical Review D* 7 (1973): 2333–46.

Bekenstein, J. D. "Statistical Black Hole Thermodynamics." *Physical Review D* 12 (1975): 3077–85.

Bennett, C. H. "Demons, Engines, and the Second Law." *Scientific American* 257, no. 5 (1987): 108–16.

Bennett, C. H., and Landauer, R. "Fundamental Limits of Computation." *Scientific American* 253, no. 1 (1985): 48–56.

Bojowald, M. "Loop Quantum Cosmology." *Living Reviews in Relativity* 8 (2006): 11.

Bojowald, M., and Tavakol, R. "Recollapsing Quantum Cosmologies and the Question of Entropy." *Physical Review D* 78 (2008): 23515.

Boltzmann, L. "Weitere Studien über das Wärmegleichgewicht unter Gasmoleculen" [Further studies on the thermal equilibrium of gas molecules]. *Sitzungsberichte Akad. Wiss.* 66 (1872): 275–370.

Boltzmann, L. "Über die Beziehung eines allgemeine mechanischen Satzes zum zweiten Hauptsatze der Warmetheorie" [On the relation of a general mechanical theorem to the Second Law of Thermodynamics]. *Sitzungsberichte Akad. Wiss.* 75 (1877): 67–73.

Boltzmann, L. "On Certain Questions of the Theory of Gases." *Nature* 51 (1895): 413–15.

Boltzmann, L. "Entgegnung auf die wärmetheoretischen Betrachtungen des Hern. E. Zermelo" [Reply to Zermelo's remarks on the Theory of Heat]. *Annalen der Physik* 57 (1896): 773.

Boltzmann, L. "Zu Hrn. Zermelo's Abhandlung 'Über die mechanische Erklärung irreversibler Vorgänge'" [On Zermelo's paper "On the Mechanical Explanation of Irreversible Processes"]. *Annalen der Physik* 60 (1897): 392.

Bondi, H., and Gold, T. "The Steady-State Theory of the Expanding Universe." *Monthly Notices of the Royal Astronomical Society* 108 (1948): 252–70.

Bostrom, N. "Are You Living in a Computer Simulation?" *Philosophical Quarterly* 53 (2003): 243–55.

Bousso, R. "Proliferation of de Sitter Space." *Physical Review D* 58 (1998): 083511.

Bousso, R. "A Covariant Entropy Conjecture." *Journal of High Energy Physics* 9907 (1999): 4.

Bousso, R. "The Holographic Principle." *Reviews of Modern Physics* 74 (2002): 825–74.

Bousso, R., Freivogel, B., and Yang, I.-S. "Boltzmann Babies in the Proper Time Measure." *Physical Review D* 77 (2008): 103514.

Brush, S. G., ed. *The Kinetic Theory of Gases: An Anthology of Classic Papers with Historical Commentary*. London: Imperial College Press, 2003.

Bucher, M., Goldhaber, A. S., and Turok, N. "An Open Universe from Inflation," *Physical Review D* 52 (1995): 3314–37.

Bunn, E. F. "Evolution and the Second Law of Thermodynamics" (2009). http://arxiv.org/abs/0903.4603.

Bunn, E. F., and Hogg, D. W. "The Kinematic Origin of the Cosmological Redshift" (2008). http://arxiv.org/abs/0808.1081.

Callender, C. "There Is No Puzzle About the Low Entropy Past." In *Contemporary Debates in Philosophy of Science*, edited by C. Hitchcock, 240–55. Malden: Wiley-Blackwell, 2004.

Callender, C. *Introducing Time*. Illustrated by Ralph Edney. Cambridge: Totem Books, 2005.

Carroll, L. *Alice's Adventures in Wonderland and Through the Looking Glass*. New York: Signet Classics, 2000.

Carroll, S. M. *Spacetime and Geometry: An Introduction to General Relativity*. New York: Addison-Wesley, 2003.

Carroll, S. M. *Dark Matter and Dark Energy: The Dark Side of the Universe*. DVD Lectures. Chantilly, VA: Teaching Company, 2007.

Carroll, S. M. "What If Time Really Exists?" (2008). http://arxiv.org/abs/0811.3772.

Carroll, S. M., and Chen, J. "Spontaneous Inflation and the Origin of the Arrow of Time" (2004). http://arxiv.org/abs/hep-th/0410270.

Carroll, S. M., Farhi, E., and Guth, A. H. "An Obstacle to Building a Time Machine." *Physical Review Letters* 68 (1992): 263–66; Erratum: Ibid., 68 (1992): 3368.

Carroll, S. M., Farhi, E., Guth, A. H., and Olum, K. D. "Energy Momentum Restrictions on the Creation of Gott Time Machines." *Physical Review D* 50 (1994): 6190–6206.

Carter, B. "The Anthropic Principle and Its Implications for Biological Evolution." *Philosophical Transactions of the Royal Society of London* A310 (1983): 347–63.

Casares, J. "Observational Evidence for Stellar-Mass Black Holes." In *Black Holes from Stars to Galaxies—Across the Range of Masses*, edited by V. Karas and G. Matt. Proceedings of IAU Symposium #238, 3–12. Cambridge: Cambridge University Press, 2007.

Cercignani, C. *Ludwig Boltzmann: The Man Who Trusted Atoms*. Oxford: Oxford University Press, 1998.

Chaisson, E. J. *Cosmic Evolution: The Rise of Complexity in Nature*. Cambridge, MA: Harvard University Press, 2001.

Christenson, J. H., Cronin, J. W., Fitch, V. L., and Turlay, R. "Evidence for the 2π Decay of the $K_2^{\,0}$ Meson." *Physical Review Letters* 13 (1964): 138–40.

Coveney, P., and Highfield, R. *The Arrow of Time: A Voyage Through Science to Solve Time's Greatest Mystery*. New York: Fawcett Columbine, 1990.

Crick, F. *What Mad Pursuit: A Personal View of Scientific Discovery*. New York: Basic Books, 1990.

Cutler, C. "Global Structure of Gott's Two-String Spacetime." *Physical Review D* 45 (1992): 487–94.

Danielson, D. R., ed. *The Book of the Cosmos: Imagining the Universe from Heraclitus to Hawking*. Cambridge: Perseus Books, 2000.

Darwin, C. *On the Origin of Species*. London: John Murray, 1859.

Davies, P. C. W. *The Physics of Time Asymmetry*. London: Surrey University Press, 1974.

Davies, P. C. W. "Inflation and Time Asymmetry in the Universe." *Nature* 301 (1983): 398–400.

Davies, P. C. W. *About Time: Einstein's Unfinished Revolution*. New York: Simon & Schuster, 1995.

Davies, P. C. W., and Twamley, J. "Time Symmetric Cosmology and the Opacity of the Future Light Cone." *Classical and Quantum Gravit* 10 (1993): 931–45.

Davis, J. A. *The Logic of Causal Order*. Thousand Oaks, CA: Sage Publications, 1985.

Dawkins, R. *The Blind Watchmaker*. New York: W. W. Norton, 1987.

de Bernardis, P. et al., BOOMERanG Collaboration. "A Flat Universe from High-Resolution Maps of the Cosmic Microwave Background Radiation." *Nature* 404 (2000): 955–59.

Dembo, A., and Zeitouni, O. *Large Deviations Techniques and Applications*. New York: Springer-Verlag, 1998.

Deser, S., Jackiw, R., and 't Hooft, G. "Physical Cosmic Strings Do Not Generate Closed Timelike Curves." *Physical Review Letters* 68 (1992): 267–69.

Deutsch, D. *The Fabric of Reality: The Science of Parallel Universes—And Its Implications*. New York: Allen Lane, 1997.

Dicke, R. H., and Peebles, P. J. E. "The Big Bang Cosmology—Enigmas and Nostrums." In *General Relativity: An Einstein Centenary Survey*, edited by S. W. Hawking and W. Israel, 504–17. Cambridge: Cambridge University Press, 1979.

Diedrick, J. *Understanding Martin Amis*. Charleston: University of South Carolina Press, 1995.

Dieks, D. "Doomsday—or: The Dangers of Statistics." *Philosophical Quarterly* 42 (1992): 78–84.

Dodelson, S. *Modern Cosmology*. San Diego, CA: Academic Press, 2003.

Dugdale, J. S. *Entropy and Its Physical Meaning*. London: Taylor and Francis, 1996.

Dyson, F. J. "Time Without End: Physics and Biology in an Open Universe." *Reviews of Modern Physics* 51 (1979): 447–60.

Dyson, L., Kleban, M., and Susskind, L. "Disturbing Implications of a Cosmological Constant." *Journal of High Energy Physics* 210 (2002): 11.

Earman, J. "What Time Reversal Is and Why It Matters." *International Studies in the Philosophy of Science* 16 (2002): 245–64.

Earman, J. "The 'Past Hypothesis': Not Even False." *Studies in History and Philosophy of Modern Physics* 37 (2006): 399–430.

Eddington, A. S. *The Nature of the Physical World* (Gifford Lectures). Brooklyn: AMS Press, 1927.

Eddington, A. S. *Nature* 127 (1931): 3203. Reprinted in Danielson (2000): 406.

Egan, G. *Axiomatic*. New York: Harper Prism, 1997.

Einstein, A., ed. *The Principle of Relativity*. Translated by W. Perrett and G. B. Jeffrey. Mineola. Dover, 1923.

Einstein, A., Podolsky, B., and Rosen, N. "Can Quantum-Mechanical Description of Physical Reality Be Considered Complete?" *Physical Review* 47 (1935): 777–80.

Ellis, J., Giudice, G., Mangano, M. L., Tkachev, I., and Wiedemann, U. "Review of the Safety of LHC Collisions." *Journal of Physics G* 35 (2008): 115004.

Ellis, R. S. *Entropy, Large Deviations, and Statistical Mechanics*. New York: Springer-Verlag, 2005.

Evans, D. J., and Searles, D. J. "The Fluctuation Theorem." *Advances in Physics* 51 (2002): 1529–89.
Everett, H. "Relative State Formulation of Quantum Mechanics." *Reviews of Modern Physics* 29 (1957): 454–62.
Falk, D. *In Search of Time: The Science of a Curious Dimension*. New York: Thomas Dunne Books, 2008.
Farhi, E., and Guth, A. H. "An Obstacle to Creating a Universe in the Laboratory." *Physics Letters B* 183 (1987): 149.
Farhi, E., Guth, A. H., and Guven, J. "Is It Possible to Create a Universe in the Laboratory by Quantum Tunneling?" *Nuclear Physics B* 339 (1990): 417–90.
Farrell, J. *The Day Without Yesterday: Lemaître, Einstein, and the Birth of Modern Cosmology*. New York: Basic Books, 2006.
Feinberg, G. "Possibility of Faster-Than-Light Particles." *Physical Review* 159 (1967): 1089–1105.
Feynman, R. P. *The Character of Physical Law*. Cambridge, MA: MIT Press, 1964.
Feynman, R. P., Leighton, R., and Sands, M. *The Feynman Lectures on Physics*. New York: Addison Wesley Longman, 1970.
Fischler, W., Morgan, D., and Polchinski, J. "Quantum Nucleation of False Vacuum Bubbles." *Physical Review D* 41 (1990a): 2638.
Fischler, W., Morgan, D., and Polchinski, J. "Quantization of False Vacuum Bubbles: A Hamiltonian Treatment of Gravitational Tunneling." *Physical Review D* 42 (1990b): 4042–55.
Fitzgerald, F. S. "The Curious Case of Benjamin Button." *Collier's Weekly* (May 1922): 27.
Freedman, W. L., et al. "Final Results from the Hubble Space Telescope Key Project to Measure the Hubble Constant." *Astrophysical Journal* 553, no. 1 (2001): 47–72.
Freivogel, B., Hubeny, V. E., Maloney, A., Myers, R. C., Rangamani, M., and Shenker, S. "Inflation in AdS/CFT." *Journal of High Energy Physics* 603 (2006): 7.
Friedman, J., et al. "Cauchy Problem in Space-times with Closed Timelike Curves." *Physical Review D* 42 (1990): 1915–30.
Friedman, J., and Higuchi, A. "Topological Censorship and Chronology Protection." *Annalen der Physik* 15 (2006): 109–28.
Galison, P. *Einstein's Clocks, Poincaré's Maps: Empires of Time*. New York: W.W. Norton, 2003.
Gamow, G. *One Two Three . . . Infinity: Facts and Speculations of Science*. New York: Viking Press, 1947.
Garriga, J., and Vilenkin, A. "Recycling Universe." *Physical Review D* 57 (1998): 2230.
Garriga, J., and Vilenkin, A. "Prediction and Explanation in the Multiverse." *Physical Review D* 77 (2008): 043526.
Garwin, R. L., Lederman, L. L., and Weinrich, M. "Observation of the Failure of Conservation of Parity and Charge Conjugation in Meson Decays: The Magnetic Moment of the Free Muon." *Physical Review* 105 (1957): 1415–17.
Gasperini, M., and Veneziano, G. "Pre-Big-Bang in String Cosmology." *Astroparticle Physics* 1 (1993): 317–39.
Gates, E. I. *Einstein's Telescope*. New York: W.W. Norton, 2009.
Gell-Mann, M. *The Quark and the Jaguar: Adventures in the Simple and Complex*. New York: W. H. Freeman, 1994.
Gell-Mann, M., and Hartle, J. B. "Time Symmetry and Asymmetry in Quantum Mechanics and Quantum Cosmology." In *Physical Origins of Time Asymmetry*, edited by Halliwell, J. J., Pérez-Mercader, J., and Zurek, W. H. 311–45. Cambridge: Cambridge University Press, 1996.
Geroch, R. P. "Topology Change in General Relativity." *Journal of Mathematical Physics* 8 (1967): 782.
Gibbons, G. W., and Hawking, S. W. "Cosmological Event Horizons, Thermodynamics, and Particle Creation." *Physical Review D* 15 (1977): 2738–51.
Gödel, K. "An Example of a New Type of Cosmological Solution of Einstein's Field Equations of Gravitation." *Reviews of Modern Physics* 21 (1949): 447–50.
Gold, T. "The Arrow of Time." *American Journal of Physics* 30 (1962): 403–10.
Goldsmith, D. *The Runaway Universe: The Race to Find the Future of the Cosmos*. New York: Basic Books, 2000.
Goncharov, A. S., Linde, A. D., and Mukhanov, V. F. "The Global Structure of the Inflationary Universe." *International Journal of Modern Physics A* 2 (1987): 561–91.
Gott, J. R. "Closed Timelike Curves Produced by Pairs of Moving Cosmic Strings: Exact Solutions." *Physical Review Letters* 66 (1991): 1126–29.
Gott, J. R. "Implications of the Copernican Principle for Our Future Prospects." *Nature* 363 (1993): 315–19.
Gott, J. R. *Time Travel in Einstein's Universe: The Physical Possibilities of Travel Through Time*. Boston: Houghton Mifflin, 2001.

Gould, S. J. *Time's Arrow, Time's Cycle: Myth and Metaphor in the Discovery of Geological Time.* Cambridge, MA: Harvard University Press, 1987.

Greene, B. *The Elegant Universe: Superstrings, Hidden Dimensions, and the Quest for the Ultimate Theory.* New York: Vintage, 2000.

Greene, B. *The Fabric of the Cosmos: Space, Time, and the Texture of Reality.* New York: Knopf, 2004.

Grinstein, B., O'Connell, D., and Wise, M. B. "Causality as an Emergent Macroscopic Phenomenon: The Lee-Wick O(N) Model." *Physical Review D* 79 (2009): 105019.

Grünbaum, A. *Philosophical Problems of Space and Time.* Dortrecht: Reidel, 1973.

Grünbaum, A. "The Poverty of Theistic Cosmology." *British Journal for the Philosophy of Science* 55 (2004): 561–614.

Guth, A. H. "The Inflationary Universe: A Possible Solution to the Horizon and Flatness Problems." *Physical Review D* 23 (1981): 347–56.

Guth, A. H. *The Inflationary Universe: The Quest for a New Theory of Cosmic Origins.* Reading: Addison-Wesley, 1997.

Guth, A. H. "Eternal Inflation and Its Implications." *Journal of Physics A* 40 (2007): 6811–26.

Halliwell, J. J., and Hawking, S. W. "Origin of Structure in the Universe." *Physical Review D* 31 (1985): 1777.

Halliwell, J. J., Pérez-Mercader, J., and Zurek, W. H. *Physical Origins of Time Asymmetry.* Cambridge: Cambridge University Press, 1996.

Hartle, J. B., and Hawking, S. W. "Wave Function of the Universe." *Physical Review D* 28 (1983): 2960–75.

Hartle, J. B., Hawking, S. W., and Hertog, T. "The Classical Universes of the No-Boundary Quantum State." *Physical Review D* 77 (2008): 123537.

Hartle, J. B., and Srednicki, M. "Are We Typical?" *Physical Review D* 75 (2007): 123523.

Hawking, S. W. "Particle Creation by Black Holes." *Communications in Mathematical Physics* 43 (1975): 199–220; Erratum: Ibid., 46 (1976): 206.

Hawking, S. W. "The Arrow of Time in Cosmology." *Physical Review D* 32 (1985): 2489.

Hawking, S. W. *A Brief History of Time: From the Big Bang to Black Holes.* New York: Bantam, 1988.

Hawking, S. W. "The Chronology Protection Conjecture." *Physical Review D* 46 (1991): 603.

Hawking, S. W. "The No Boundary Condition and the Arrow of Time." In Halliwell et al. (1996): 346–57.

Hawking, S. W., and Ellis, G. F. R. *The Large-Scale Structure of Spacetime.* Cambridge: Cambridge University Press, 1974.

Hedman, M. *The Age of Everything: How Science Explores the Past.* Chicago: University of Chicago Press, 2007.

Heinlein, R. A. "All You Zombies—." *Magazine of Fantasy and Science Fiction* (March 1959).

Hollands, S., and Wald, R. M. "An Alternative to Inflation." *General Relativity and Gravitation* 34 (2002): 2043–55.

Holman, R., and Mersini-Houghton, L. "Why the Universe Started from a Low Entropy State." *Physical Review D* 74 (2006): 123510.

Hooper, D. *Dark Cosmos: In Search of Our Universe's Missing Mass and Energy.* New York: HarperCollins, 2007.

Horwich, P. *Asymmetries in Time: Problems in the Philosophy of Science.* Cambridge, MA: MIT Press, 1987.

Hoyle, F. "A New Model for the Expanding Universe." *Monthly Notices of the Royal Astronomical Society* 108 (1948): 372–82.

Jaynes, E. T. "Gibbs vs. Boltzmann Entropies." *American Journal of Physics* 33 (1965): 391–98.

Jaynes, E. T. *Probability Theory: The Logic of Science.* Cambridge: Cambridge University Press, 2003.

Johnson, G. "The Theory That Ate the World." *New York Times,* August 22, 2008, BR16.

Kauffman, S. A. *The Origins of Order: Self-Organization and Selection in Evolution.* Oxford: Oxford University Press, 1993.

Kauffman, S. A. *At Home in the Universe: The Search for the Laws of Self-Organization and Complexity.* Oxford: Oxford University Press, 1996.

Kauffman, S. A. *Reinventing the Sacred: A New View of Science, Reason, and Religion.* New York: Basic Books, 2008.

Kerr, R. P. "Gravitational Field of a Spinning Mass as an Example of Algebraically Special Metrics." *Physical Review Letters* 11 (1963): 237–38.

Khoury, J., Ovrut, B. A., Steinhardt, P. J., and Turok, N. "The Ekpyrotic Universe: Colliding Branes and the Origin of the Hot Big Bang." *Physical Review D* 64 (2001): 123522.

Kirshner, R. P. *The Extravagant Universe: Exploding Stars, Dark Energy, and the Accelerating Cosmos.* Princeton, NJ: Princeton University Press, 2004.

Klein, E. *Chronos: How Time Shapes Our Universe.* New York: Thunders Mouth Press, 2005.

Kobayashi, M., and Maskawa, T. "CP-Violation in the Renormalizable Theory of Weak Interaction." *Progress of Theoretical Physics* 49 (1973): 652-57.

Kofman, L., Linde, A., and Mukhanov, V. "Inflationary Theory and Alternative Cosmology." *Journal of High Energy Physics* 210 (2002): 57.

Kolb, R. *Blind Watchers of the Sky: The People and Ideas That Shaped Our View of the Universe.* New York: Addison Wesley, 1996.

Kormendy, J., and Richstone, D. "Inward Bound—The Search for Supermassive Black Holes in Galactic Nuclei." *Annual Review of Astronomy and Astrophysics* 33 (1995): 581.

Laplace, P.-S. *A Philosophical Essay on Probabilities.* Translated by F. W. Tuscott and F. L. Emory, reprinted. New York: Cosimo Classics, 2007.

Lebowitz, J. L. "Statistical Mechanics: A Selective Review of Two Central Issues." *Reviews of Modern Physics* 71 (1999): S346-57.

Lebowitz, J. L. "Time's Arrow and Boltzmann's Entropy." *Scholarpedia* 3, no. 4 (2008): 3448.

Lee, T. D., and Wick, G. C. "Finite Theory of Quantum Electrodynamics." *Physical Review D* 2 (1970): 1033-48.

Lee, T. D., and Yang, C. N. "Question of Parity Conservation in Weak Interactions." *Physical Review* 104 (1956): 254-58.

Leff, H. S., and Rex, A. F., eds. *Maxwell's Demon 2: Entropy, Classical and Quantum Information, Computing.* Bristol: Institute of Physics, 2003.

Lemaître, G. "The Primeval Atom Hypothesis and the Problem of the Clusters of Galaxies." In *La Structure et l'Evolution de l'Univers*, edited by R. Stoops, 1-32. Brussels: Coudenberg, 1958.

Leslie, J. "Is the End of the World Nigh?" *Philosophical Quarterly* 40 (1990): 65-72.

Linde, A. D. "A New Inflationary Universe Scenario: A Possible Solution of the Horizon, Flatness, Homogeneity, Isotropy and Primordial Monopole Problems." *Physics Letters B* 108 (1981): 389-93.

Linde, A. D. "Chaotic Inflation." *Physics Letters B* 129 (1983): 177-81.

Linde, A. D. "Eternally Existing Selfreproducing Chaotic Inflationary Universe." *Physics Letters B* 175 (1986): 395-400.

Linden, N., Popescu, S., Short, A. J., and Winter, A. "Quantum Mechanical Evolution Towards Thermal Equilibrium" (2008). http://arxiv.org/abs/0812.2385.

Lindley, D. *Boltzmann's Atom: The Great Debate That Launched a Revolution in Physics.* New York: Free Press, 2001.

Lineweaver, C. H., and Egan, C. A. "Life, Gravity, and the Second Law of Thermodynamics." *Physics of Life Reviews* 5 (2008): 225-42.

Lippincott, K. *The Story of Time.* With U. Eco, E. H. Gombrich, and others. London: Merrell Holberton, 1999.

Lloyd, S. *Programming the Universe: A Quantum Computer Scientist Takes On the Cosmos.* New York: Knopf, 2006.

Lockwood, M. *The Labyrinth of Time: Introducing the Universe.* Oxford: Oxford University Press, 2005.

Lucretius. *De Rerum Natura (On the Nature of Things).* Edited and translated by A. M. Esolen. Baltimore: Johns Hopkins University Press, 1995.

Maglich, B. *Adventures in Experimental Physics, Gamma Volume.* Princeton, NJ: World Science Communications, 1973.

Malament, D. B. "On the Time Reversal Invariance of Classical Electromagnetic Theory." *Studies in History and Philosophy of Science Part B* 35 (2004): 295-315.

Maldacena, J. M. "The Large N Limit of Superconformal Field Theories and Supergravity." *Advances in Theoretical and Mathematical Physics* 2 (1998): 231-52.

Mathur, S. D. "The Fuzzball Proposal for Black Holes: An Elementary Review." *Fortschritte der Physik* 53 (2005): 793-827.

Mattingly, D. "Modern Tests of Lorentz Invariance." *Living Reviews in Relativity* 8 (2005): 5.

McTaggart, J. M. E. "The Unreality of Time." *Mind: A Quarterly Review of Psychology and Philosophy* 17 (1908): 456.

Michell, J. *Philosophical Transactions of the Royal Society* (London) 74 (1784): 35-57.

Miller, A. D., et al., TOCO Collaboration. "A Measurement of the Angular Power Spectrum of the CMB from l = 100 to 400." *Astrophysical Journal Letters* 524 (1999): L1-L4.

Miller, A. I. *Albert Einstein's Special Theory of Relativity. Emergence (1905) and Early Interpretation (1905-1911).* Reading: Addison-Wesley, 1981.

Minkowski, H. "Raum und Zeit" [Space and Time]. *Phys. Zeitschrift* 10 (1909): 104.

Misner, C. W., Thorne, K. S., and Wheeler, J. A. *Gravitation*. San Francisco: W. H. Freeman, 1973.

Morange, M. *Life Explained*. Translated by M. Cobb and M. DeBevoise. New Haven, CT: Yale University Press, 2008.

Morris, M. S., Thorne, K. S., and Yurtsever, U. "Wormholes, Time Machines, and the Weak Energy Condition." *Physical Review Letters* 61 (1988): 1446–49.

Musser, G. *The Complete Idiot's Guide to String Theory*. New York: Alpha Books, 2008.

Mustonen, V., and Lässig, M. "From Fitness Landscapes to Seascapes: Non-Equilibrium Dynamics of Selection and Adaptation." *Trends in Genetics* 25 (2009): 111–19.

Nahin, P. J. *Time Machines: Time Travel in Physics, Metaphysics, and Science Fiction*. New York: Springer-Verlag, 1999.

Neal, R. M. "Puzzles of Anthropic Reasoning Resolved Using Full Non-indexical Conditioning" (2006). http://arxiv.org/abs/math/0608592.

Nelson, P. *Biological Physics: Energy, Information, Life*. Updated edition. New York: W. H. Freeman, 2007.

Nielsen, H. B. "Random Dynamics and Relations Between the Number of Fermion Generations and the Fine Structure Constants." *Acta Physica Polonica* B20 (1989): 427–68.

Nielsen, M. A., and Chuang, I. L. *Quantum Computation and Quantum Information*. Cambridge: Cambridge University Press, 2000.

Nietzsche, F. W. *Die Fröhliche Wissenshaft*. Translated (2001) as *The Gay Science: With a Prelude in German Rhymes and an Appendix of Songs*. Edited by B. A. O. Williams, translated by J. Nauckhoff, poems translated by A. Del Caro. Cambridge: Cambridge University Press, 1882.

Novikov, I. D. *Evolution of the Universe*. Cambridge: Cambridge University Press, 1983.

Novikov, I. D. *The River of Time*. Cambridge: Cambridge University Press, 1998.

O'Connor, J. J., and Robertson, E. F. "Pierre-Simon Laplace." *MacTutor History of Mathematics Archive* (1999). http://www-groups.dcs.st-and.ac.uk/~history/Biographies/ Laplace.html.

Olum, K. D. "The Doomsday Argument and the Number of Possible Observers." *Philosophical Quarterly* 52 (2002): 164–84.

Orzel, C. *How to Teach Physics to Your Dog*. New York: Scribner, 2009.

Ouellette, J. *The Physics of the Buffyverse*. New York: Penguin, 2007.

Overbye, D. *Lonely Hearts of the Cosmos*. New York: HarperCollins, 1991.

Page, D. N. "Inflation Does Not Explain Time Asymmetry." *Nature* 304 (1983): 39–41.

Page, D. N. "Will Entropy Decrease If the Universe Recollapses?" *Physical Review D* 32 (1985): 2496.

Page, D. N. "Typicality Derived." *Physical Review D* 78 (2008): 023514.

Pascal, B. *Pensées*. Translated by A.J. Krailsheimer. New York: Penguin Classics, 1995.

Penrose, R. "Singularities and Time-Asymmetry." In *General Relativity, and Einstein Centenary Survey*, edited by S. W. Hawking and W. Israel, 581–638. Cambridge: Cambridge University Press, 1979.

Penrose, R. *The Emperor's New Mind: Concerning Computers, Minds, and the Laws of Physics*. Oxford: Oxford University Press, 1989.

Penrose, R. *The Road to Reality: A Complete Guide to the Laws of the Universe*. New York: Knopf, 2005.

Perlmutter, S., et al., Supernova Cosmology Project. "Measurements of Omega and Lambda from 42 High Redshift Supernovae." *Astrophysical Journal* 517 (1999): 565–86.

Pirsig, R. M. *Zen and the Art of Motorcycle Maintenance*. New York: Bantam, 1974.

Poincaré, H. "Sur les problème des trois corps et les équations de la dynamique." *Acta Mathematica* 13 (1890): 1–270. Excerpts translated in Brush (2003, vol. 2) as "On the Three-Body Problem and the Equations of Dynamics," 194–202.

Poincaré, H. "Le mécanisme et l'expérience." *Revue de Metaphysique et de Morale* 4 (1893): 534. Translated in Brush (2003, vol. 2) as "Mechanics and Experience," 203–7.

Popper, Karl R. *The Logic of Scientific Discovery*. London: Routledge, 1959.

Poundstone, W. *The Recursive Universe: Cosmic Complexity and the Limits of Scientific Knowledge*. New York: W. W. Norton, 1984.

Price, H. *Time's Arrow and Archimedes' Point: New Directions for the Physics of Time*. New York: Oxford University Press, 1996.

Price, H. "Cosmology, Time's Arrow, and That Old Double Standard." In *Time's Arrows Today: Recent Physical and Philosophical Work on the Direction of Time*, edited by S. F. Savitt, 66–96. Cambridge: Cambridge University Press, 1997.

Price, H. "On the Origins of the Arrow of Time: Why There Is Still a Puzzle about the Low Entropy Past." In *Contemporary Debates in Philosophy of Science*, edited by C. Hitchcock, 240–55. Malden: Wiley-Blackwell, 2004.

Prigogine, I. *Thermodynamics of Irreversible Processes.* New York: John Wiley, 1955.

Prigogine, I. *From Being to Becoming: Time and Complexity in the Physical Sciences.* New York: W. H. Freeman, 1980.

Proust, M. *Swann's Way: In Search of Lost Time,* vol. 1 (*Du côté de chez Swann: À la recherche du temps perdu*). Translated by L. Davis. New York: Penguin Classics, 2004.

Putnam, H. "It Ain't Necessarily So." *Journal of Philosophy* 59, no. 22 (1962): 658–71.

Pynchon, T. *Slow Learner.* Boston: Back Bay Books, 1984.

Randall, L. *Warped Passages: Unraveling the Mysteries of the Universe's Hidden Dimensions.* New York: HarperCollins, 2005.

Regis, E. *What Is Life?: Investigating the Nature of Life in the Age of Synthetic Biology.* Oxford: Oxford University Press, 2009.

Reichenbach, H. *The Direction of Time.* Mineola: Dover, 1956.

Reichenbach, H. *The Philosophy of Space and Time.* Mineola: Dover, 1958.

Reid, M. J. "Is There a Supermassive Black Hole at the Center of the Milky Way?" (2008). http://arxiv.org/abs/0808.2624.

Reznik, B., and Aharonov, Y. "Time-Symmetric Formulation of Quantum Mechanics." *Physical Review A* 52 (1995): 2538–50.

Ridderbos, K., ed. *Time: The Darwin College Lectures.* Cambridge: Cambridge University Press, 2002.

Riess, A., et al., Supernova Search Team. "Observational Evidence from Supernovae for an Accelerating Universe and a Cosmological Constant." *Astronomical Journal* 116 (1998): 1009–38.

Rouse Ball, W. W. *A Short Account of the History of Mathematics,* 4th ed., reprinted 2003. Mineola, NY: Dover, 1908.

Rovelli, C. "Forget Time" (2008). http://arxiv.org/abs/0903.3832.

Rowling, J. K. *Harry Potter and the Half-Blood Prince.* New York: Scholastic, 2005.

Rukeyser, M. *Willard Gibbs.* Woodbridge: Ox Bow Press, 1942.

Sagan, C. *Contact.* New York: Simon and Schuster, 1985.

Savitt, S. F., ed. *Time's Arrows Today: Recent Physical and Philosophical Work on the Direction of Time.* Cambridge: Cambridge University Press, 1997.

Schacter, D. L., Addis, D. R., and Buckner, R. L. "Remembering the Past to Imagine the Future: The Prospective Brain." *Nature Reviews Neuroscience* 8 (2007): 657–61.

Schlosshauer, M. "Decoherence, the Measurement Problem, and Interpretations of Quantum Mechanics." *Reviews of Modern Physics* 76 (2004): 1267–1305.

Schneider, E. D., and Sagan, D. *Into the Cool: Energy Flow, Thermodynamics, and Life.* Chicago: University of Chicago Press, 2005.

Schrödinger, E. *What Is Life?* Cambridge: Cambridge University Press, 1944.

Seife, C. *Decoding the Universe: How the New Science of Information Is Explaining Everything in the Cosmos, from Our Brains to Black Holes.* New York: Viking, 2006.

Sethna, J. P. *Statistical Mechanics: Entropy, Order Parameters, and Complexity.* Oxford: Oxford University Press, 2006.

Shalizi, C. R. *Notebooks* (2009). http://www.cscs.umich.edu/~crshalizi/notebooks/.

Shannon, C. E. "A Mathematical Theory of Communication." *Bell System Technical Journal* 27 (1948): 379–423 and 623–56.

Singh, S. *Big Bang: The Origin of the Universe.* New York: Fourth Estate, 2004.

Sklar, L. *Physics and Chance: Philosophical Issues in the Foundations of Statistical Mechanics.* Cambridge: Cambridge University Press, 1993.

Smolin, L. *The Life of the Cosmos.* Oxford: Oxford University Press, 1993.

Snow, C.P. *The Two Cultures.* Cambridge: Cambridge University Press, 1998.

Sobel, D. *Longitude: The True Story of a Lone Genius Who Solved the Greatest Scientific Problem of His Time.* New York: Penguin, 1995.

Spergel, D. N., et al., WMAP Collaboration. "First Year Wilkinson Microwave Anisotropy Probe (WMAP) Observations: Determination of Cosmological Parameters." *Astrophysical Journal Supplement* 148 (2003): 175.

Steinhardt, P. J., and Turok, N. "Cosmic Evolution in a Cyclic Universe." *Physical Review D* 65 (2002): 126003.

Steinhardt, P. J., and Turok, N. *Endless Universe: Beyond the Big Bang.* New York: Doubleday, 2007.

Stoppard, T. *Arcadia,* in *Plays: Five.* London: Faber and Faber, 1999.

Strominger, A., and Vafa, C. "Microscopic Origin of the Bekenstein-Hawking Entropy." *Physics Letters B* 379 (1996): 99–104.

Styer, D. F. "Entropy and Evolution." *American Journal of Physics* 76 (2008): 1031–33.

Susskind, L. "The World as a Hologram." *Journal of Mathematical Physics* 36 (1995): 6377-96.
Susskind, L. *The Cosmic Landscape: String Theory and the Illusion of Intelligent Design.* New York: Little, Brown, 2006.
Susskind, L. *The Black Hole War: My Battle with Stephen Hawking to Make the World Safe for Quantum Mechanics.* New York: Little, Brown, 2008.
Susskind, L., and Lindesay, J. *An Introduction to Black Holes, Information, and the String Theory Revolution: The Holographic Universe.* Singapore: World Scientific, 2005.
Susskind, L., Thorlacius, L., and Uglum, J. "The Stretched Horizon and Black Hole Complementarity." *Physical Review D* 48 (1993): 3743-61.
Swinburne, R. *The Existence of God.* Oxford: Oxford University Press, 2004.
't Hooft, G. "Causality in (2+1)-Dimensional Gravity." *Classical and Quantum Gravity* 9 (1992): 1335-48.
't Hooft, G. "Dimensional Reduction in Quantum Gravity." In *Salamfestschrift: a Collection of Talks*, edited by A. Ali, J. Ellis, and S. Randjbar-Daemi. Singapore: World Scientific, 1993.
Tegmark, M. "The Interpretation of Quantum Mechanics: Many Worlds or Many Words?" *Fortschritte der Physik* 46 (1998): 855-62.
Thomson, W. "On the Age of the Sun's Heat." *Macmillan's* 5 (1862): 288-93.
Thorne, K. S. *Black Holes and Time Warps: Einstein's Outrageous Legacy.* New York: W. W. Norton, 1994.
Tipler, F. J. "Rotating Cylinders and the Possibility of Global Causality Violation." *Physical Review D* 9 (1974): 2203-6.
Tipler, F. J. "Singularities and Causality Violation." *Annals of Physics* 108 (1977): 1-36.
Tolman, R. C. "On the Problem of Entropy of the Universe as a Whole." *Physical Review* 37 (1931): 1639-60.
Toomey, D. *The New Time Travelers: A Journey to the Frontiers of Physics.* New York: W. W. Norton, 2007.
Toulmin, S. "The Early Universe: Historical and Philosophical Perspectives." In *The Early Universe*, Proceedings of the NATO Advanced Study Institute, held in Victoria, Canada, Aug. 17-30, 1986, edited by W. G. Unruh and G. W. Semenoff, 393. Dortrecht: D. Reidel, 1988.
Tribus, M., and McIrvine, E. "Energy and Information." *Scientific American* (August 1971): 179.
Ufflink, J. "Boltzmann's Work in Statistical Physics." *The Stanford Encyclopedia of Philosophy* (Winter 2008 edition), edited by Edward N. Zalta (2004). http://plato.stanford.edu/archives/win2008/entries/statphys-Boltzmann/.
Vilenkin, A. "The Birth of Inflationary Universes." *Physical Review D* 27 (1983): 2848-55.
Vilenkin, A. "Eternal Inflation and Chaotic Terminology" (2004). http://arxiv.org/abs/ gr-qc/0409055.
Vilenkin, A. *Many Worlds in One: The Search for Other Universes.* New York: Hill and Wang, 2006.
Viola, F. *From Eternity to Here: Rediscovering the Ageless Purpose of God.* Colorado Springs: David C. Cook, 2009.
Von Baeyer, H. C. *Warmth Disperses and Time Passes: The History of Heat.* New York: Modern Library, 1998.
Von Baeyer, H. C. *Information: The New Language of Science.* Cambridge, MA: Harvard University Press, 2003.
Vonnegut, K. *Slaughterhouse-Five.* New York: Dell, 1969.
Wald, R. W. "Asymptotic Behavior of Homogeneous Cosmological Models in the Presence of a Positive Cosmological Constant." *Physical Review D* 28 (1983): 2118-20.
Weinberg, S. *The First Three Minutes: A Modern View of the Origin of the Universe.* New York: Basic Books, 1977.
Weiner, J. *Time, Love, Memory: A Great Biologist and His Quest for the Origins of Behavior.* New York: Vintage, 1999.
Wells, H. G. *The Time Machine.* Reprinted in *The Complete Science Fiction Treasury of H. G. Wells* (1978). New York: Avendel, 1895.
West, G. B., Brown, J. H., and Enquist, B. J. "The Fourth Dimension of Life: Fractal Geometry and the Allometric Scaling of Organisms." *Science* 284 (1999): 1677-79.
Wheeler, J. A. "Time Today." In *Physical Origins of Time Asymmetry*, edited by J. J. Halliwell, J. Pérez-Mercader, and W. H. Zurek, 1-29. Cambridge: Cambridge University Press, 1994.
Wiener, N. *Cybernetics: or the Control and Communication in the Animal and the Machine.* Cambridge, MA: MIT Press, 1961.
Wikipedia contributors. "Time." *Wikipedia, The Free Encyclopedia.* http://en.wikipedia.org/wiki/Time (accessed January 6, 2009).
Wright, E. L. "Errors in the Steady State and Quasi-SS Models" (2008). http://www.astro.ucla.edu/~wright/stdystat.htm.
Wu, C. S., Ambler, E., Hayward, R. W., Hoppes, D. D., and Hudson, R. P. "Experimental Test of Parity Nonconservation in Beta Decay." *Physical Review* 105 (1957): 1413-15.

Zeh, H. D. *The Physical Basis of The Direction of Time*. Berlin: Springer-Verlag, 1989.
Zermelo, E. "Über einen Satz der Dynamik und die mechanische Warmtheorie." *Annalen der Physik* 57 (1896a): 485. Translated in Brush (2003) as "On a Theorem of Dynamics and the Mechanical Theory of Heat," 382.
Zermelo, E. "Über mechanische Erklärungen irreversibler Vorgänge." *Annalen der Physik* 59 (1896b): 793. Translated in Brush (2003) as "On the Mechanical Explanation of Irreversible Processes," 403.
Zurek, W. H. "Entropy Evaporated by a Black Hole." *Physical Review Letters* 49 (1982): 1683–86.
Zurek, W. H. *Complexity, Entropy, and the Physics of Information*. Boulder: Westview Press, 1990.

ACKNOWLEDGMENTS

Shepherding a book from conception to publication is a remarkably collaborative effort, and there are many people who deserve thanks for helping me along the way. While this book was in its very early stages, I had the good fortune to meet, fall in love with, and marry a person who just happened to be an extraordinarily talented science writer. All the thanks I can give to Jennifer Ouellette, who improved the book immensely and made the journey worthwhile.

I sent drafts of the manuscript to a large number of my friends, and they responded with remarkably good humor and irritatingly sensible suggestions for improvement. Enormous thanks to Scott Aaronson, Allyson Beatrice, Jennie Chen, Stephen Flood, David Grae, Lauren Gunderson, Robin Hanson, Matt Johnson, Chris Lackner, Tom Levenson, Karen Lorre, George Musser, Huw Price, Ted Pyne, Mari Ruti, Alex Singer, and Mark Trodden, for keeping me honest along the way. I suspect most of them will be writing books of their own in the near future, and I'll be happy to read all of them.

I've been talking about the arrow of time and other issues contained herein with my fellow scientists for years now, and it's impossible to disentangle who made what contribution to my thinking. In addition to the readers mentioned above, I'd like to thank Anthony Aguirre, David Albert, Andreas Albrecht, Tom Banks, Raphael Bousso, Eddie Farhi, Brian Greene, Jim Hartle, Kurt Hinterbichler, Tony Leggett, Andrei Linde, Laura Mersini, Ken Olum, Don Page, John Preskill, Iggy Sawicki, Cosma Shalizi, Mark Srednicki, Kip Thorne, Alex Vilenkin, and Robert Wald (plus others I've doubtless shamefully forgotten) for conversations over the years. I'd like to offer special thanks to Jennie Chen, who not only read the manuscript carefully but was a valued collaborator when I first started taking the arrow of time seriously.

More recently, I've been a neglectful collaborator myself, as I worked to finish this book while my colleagues forged ahead on our research projects. So thanks/ apologies to Lotty Ackerman, Matt Buckley, Claudia de Rham, Tim Dulaney, Adrienne Erickcek, Moira Gresham, Matt Johnson, Marc Kamionkowski, Sonny Mantry, Michael Ramsey-Musolf, Lisa Randall, Heywood Tam, Chien-Yao Tseng,

Ingunn Wehus, and Mark Wise, for putting up with me in recent days when my attention wasn't always fully on the task at hand.

Katinka Matson and John Brockman were instrumental in turning my original notions into a sensible idea for a book, and generally in making things happen. I first met my editor Stephen Morrow years before this book was conceived, and it was a pleasure to get the chance to work with him. Jason Torchinsky took my meager sketches and turned them into compelling illustrations. Somehow Michael Bérubé, through the mediation of Elliot Tarabour, managed to offer a review of the book before it was actually written. But for a work about the nature of time, what else should we expect?

I'm the kind of person who grows restless working at home or in the office for too long, so I frequently gather up my physics books and papers and bring them to a restaurant or coffee shop for a change of venue. Almost inevitably, a stranger will ask me what it is I'm reading, and—rather than being repulsed by all the forbidding math and science—follow up with more questions about cosmology, quantum mechanics, the universe. At a pub in London, a bartender scribbled down the ISBN number of Scott Dodelson's *Modern Cosmology*; at the Green Mill jazz club in Chicago, I got a free drink for explaining dark energy. I would like to thank every person who is not a scientist but maintains a sincere fascination with the inner workings of Nature, and is willing to ask questions and mull over the answers. Thinking about the nature of time might not help us build better TV sets or lose weight without exercising, but we all share the same universe, and the urge to understand it is part of what makes us human.

INDEX

1

The Fundamental Nature of Reality

I n the old Road Runner cartoons, Wile E. Coyote would frequently find himself running off the edge of a cliff. But he wouldn't, as our experience with gravity might lead us to expect, start falling to the ground below, at least not right away. Instead, he would hover motionless, in puzzlement; it was only when he realized there was no longer any ground beneath him that he would suddenly crash downward.

We are all Wile E. Coyote. Since human beings began thinking about things, we have contemplated our place in the universe, the reason why we are all here. Many possible answers have been put forward, and partisans of one view or another have occasionally disagreed with each other. But for a long time, there has been a shared view that there is some meaning, out there somewhere, waiting to be discovered and acknowledged. There is a point to all this; things happen for a reason. This conviction has served as the ground beneath our feet, as the foundation on which we've constructed all the principles by which we live our lives.

Gradually, our confidence in this view has begun to erode. As we understand the world better, the idea that it has a transcendent purpose seems increasingly untenable. The old picture has been replaced by a wondrous new one—one that is breathtaking and exhilarating in many ways, challenging and vexing in others. It is a view in which the world stubbornly refuses to give us any direct answers about the bigger questions of purpose and meaning.

The problem is that we haven't quite admitted to ourselves that this transition has taken place, nor fully accepted its far-reaching implications. The issues are well-known. Over the course of the last two centuries, Darwin has upended our view of life, Nietzsche's madman bemoaned the death of God, existentialists have searched for authenticity in the face of absurdity, and modern atheists have been granted a seat at society's table. And yet, many continue on as if nothing has changed; others revel in the new order, but placidly believe that adjusting our perspective is just a matter of replacing a few old homilies with a few new ones.

The truth is that the ground has disappeared beneath us, and we are just beginning to work up the courage to look down. Fortunately, not everything in the air immediately plummets to its death. Wile E. Coyote would have been fine if he had been equipped with one of those ACME-brand jet packs, so that he could fly around under his own volition. It's time to get to work building our conceptual jet packs.

What is the fundamental nature of reality? Philosophers call this the question of *ontology*—the study of the basic structure of the world, the ingredients and relationships of which the universe is ultimately composed. It can be contrasted with *epistemology*, which is how we obtain knowledge about the world. Ontology is the branch of philosophy concerned with the nature of reality; we also talk about "an" ontology, referring to a specific idea about what that nature actually is.

The number of approaches to ontology alive in the world today is somewhat overwhelming. There is the basic question of whether reality exists at all. A *realist* says, "Of course it does"; but there are also *idealists*, who think that capital-*M* Mind is all that truly exists, and the so-called real world is just a series of thoughts inside that Mind. Among realists, we have *monists*, who think that the world is a single thing, and *dualists*, who believe in two distinct realms (such as "matter" and "spirit"). Even people who agree that there is only one type of thing might disagree about whether there are fundamentally different kinds of properties (such as mental properties and physical properties) that those things can have. And even people who agree that there is only one kind of thing, and that the world is purely physical, might diverge when it comes to asking which aspects of that world are "real" versus "illusory." (Are colors real? Is consciousness? Is morality?)

Whether or not you believe in God—whether you are a *theist* or an *atheist*—is part of your ontology, but far from the whole story. "Religion" is a completely different kind of thing. It is associated with certain beliefs, often including belief in God, although the definition of "God" can differ substantially within religion's broad scope. Religion can also be a cultural force, a set of institutions, a way of life, a historical legacy, a collection of practices and principles. It's much more, and much messier, than a checklist of doctrines. A counterpart to religion would be *humanism*, a collection of beliefs and practices that is as varied and malleable as religion is.

The broader ontology typically associated with atheism is *naturalism*— there is only one world, the natural world, exhibiting patterns we call the "laws of nature," and which is discoverable by the methods of science and empirical investigation. There is no separate realm of the supernatural, spiritual, or divine; nor is there any cosmic teleology or transcendent purpose inherent in the nature of the universe or in human life. "Life" and "consciousness" do not denote essences distinct from matter; they are ways of talking about phenomena that emerge from the interplay of extraordinarily complex systems. Purpose and meaning in life arise through fundamentally human acts of creation, rather than being derived from anything outside ourselves. Naturalism is a philosophy of unity and patterns, describing all of reality as a seamless web.

Naturalism has a long and distinguished pedigree. We find traces of it in Buddhism, in the atomists of ancient Greece and Rome, and in Confucianism. Hundreds of years after the death of Confucius, a Chinese thinker named Wang Chong was a vocal naturalist, campaigning against the belief in ghosts and spirits that had become popular in his day. But it is really only in the last few centuries that the evidence in favor of naturalism has become hard to resist.

✳

All of these isms can feel a bit overwhelming. Fortunately we don't need to be rigorous or comprehensive about listing the possibilities. But we do need to think hard about ontology. It's at the heart of our Wile E. Coyote problem.

The last five hundred or so years of human intellectual progress have completely upended how we think about the world at a fundamental level. Our everyday experience suggests that there are large numbers of truly different *kinds of stuff* out there. People, spiders, rocks, oceans, tables, fire, air, stars—these all seem dramatically different from one another, deserving of independent entries in our list of basic ingredients of reality. Our "folk ontology" is pluralistic, full of myriad distinct categories. And that's not even counting notions that seem more abstract but are arguably equally "real," from numbers to our goals and dreams to our principles of right and wrong.

As our knowledge grows, we have moved by fits and starts in the direction of a simpler, more unified ontology. It's an ancient impulse. In the sixth century BCE, the Greek philosopher Thales of Miletus suggested that *water* is a primary principle from which all else is derived, while across the world, Hindu philosophers put forward *Brahman* as the single ultimate reality. The development of science has accelerated and codified the trend.

Galileo observed that Jupiter has moons, implying that it is a gravitating body just like the Earth. Isaac Newton showed that the force of gravity is universal, underlying both the motion of the planets and the way that apples fall from trees. John Dalton demonstrated how different chemical compounds could be thought of as combinations of basic building blocks called atoms. Charles Darwin established the unity of life from common ancestors. James Clerk Maxwell and other physicists brought together such disparate phenomena as lightning, radiation, and magnets under the single rubric of "electromagnetism." Close analysis of starlight revealed that stars are made of the same kinds of atoms as we find here on Earth, with Cecilia Payne-Gaposchkin eventually proving that they are mostly hydrogen and helium. Albert Einstein unified space and time, joining together matter and energy along the way. Particle physics has taught us that every atom in the periodic table of the elements is an arrangement of just three basic particles: protons, neutrons, and electrons. Every object you have ever seen or bumped into in your life is made of just those three particles.

We're left with a very different view of reality from where we started. At a fundamental level, there aren't separate "living things" and "nonliving things," "things here on Earth" and "things up in the sky," "matter" and "spirit." There is just the basic stuff of reality, appearing to us in many different forms.

How far will this process of unification and simplification go? It's impossible to say for sure. But we have a reasonable guess, based on our progress thus far: it will go all the way. We will ultimately understand the world as a single, unified reality, not caused or sustained or influenced by anything outside itself. That's a big deal.

※

Naturalism presents a hugely grandiose claim, and we have every right to be skeptical. When we look into the eyes of another person, it doesn't seem like what we're seeing is simply a collection of atoms, some sort of immensely complicated chemical reaction. We often feel connected to the universe in some way that transcends the merely physical, whether it's a sense of awe when we contemplate the sea or sky, a trancelike reverie during meditation or prayer, or the feeling of love when we're close to someone we care about. The difference between a living being and an inanimate object seems much more profound than the way certain molecules are arranged. Just looking around, the idea that everything we see and feel can somehow be explained by impersonal laws governing the motion of matter and energy seems preposterous.

It's a bit of a leap, in the face of all of our commonsense experience, to think that life can simply start up out of non-life, or that our experience of consciousness needs no more ingredients than atoms obeying the laws of physics. Of equal importance, appeals to transcendent purpose or a higher power seem to provide answers to questions to some of the pressing "Why?" questions we humans like to ask: Why this universe? Why am I here? Why anything at all? Naturalism, by contrast, simply says: those aren't the right questions to ask. It's a lot to swallow, and not a view that anyone should accept unquestioningly.

Naturalism isn't an obvious, default way to think about the world. The case in its favor has built up gradually over the years, a consequence of our relentless quest to improve our understanding of how things work at a deep level, but there is still work to be done. We don't know how the universe began, or if it's the only universe. We don't know the ultimate, complete laws of physics. We don't know how life began, or how consciousness arose. And we certainly haven't agreed on the best way to live in the world as good human beings.

The naturalist needs to make the case that, even without actually having these answers yet, their worldview is still by far the most likely framework in which we will eventually find them. That's what we're here to do.

❋

The pressing, human questions we have about our lives depend directly on our attitudes toward the universe at a deeper level. For many people, those attitudes are adopted rather informally from the surrounding culture, rather than arising out of rigorous personal reflection. Each new generation of people doesn't invent the rules of living from scratch; we inherit ideas and values that have evolved over vast stretches of time. At the moment, the dominant image of the world remains one in which human life is cosmically special and significant, something more than mere matter in motion. We need to do better at reconciling how we talk about life's meaning with what we know about the scientific image of our universe.

Among people who acknowledge the scientific basis of reality, there is often a conviction—usually left implicit—that all of that philosophical stuff like freedom, morality, and purpose should ultimately be pretty easy to figure out. We're collections of atoms, and we should be nice to one another. How hard can it really be?

It can be really hard. Being nice to one another is a good start, but it doesn't get us very far. What happens when different people have incompatible conceptions of niceness? Giving peace a chance sounds like a swell idea, but in the real world, there are different actors with different interests, and conflicts will inevitably arise. The absence of a supernatural guiding force doesn't mean we can't meaningfully talk about right and wrong, but it doesn't mean we instantly know one from the other, either.

Meaning in life can't be reduced to simplistic mottos. In some number of years I will be dead; some memory of my time here on Earth may linger, but I won't be around to savor it. With that in mind, what kind of life is worth living? How should we balance family and career, fortune and pleasure, action and contemplation? The universe is large, and I am a tiny part of it, constructed of the same particles and forces as everything else: by itself, that tells us precisely nothing about how to answer such questions. We're going to have to be both smart and courageous as we work to get this right.

Poetic Naturalism

O ne thing *Star Trek* never really got clear on was how transporter machines are supposed to work. Do they disassemble you one atom at a time, zip those atoms elsewhere, and then reassemble them? Or do they send only a blueprint of you, the information contained in your arrangement of atoms, and then reconstruct you from existing matter in the environment to which you are traveling? Most often the ship's crew talks as if your actual atoms travel through space, but then how do we explain "The Enemy Within"? That's the episode, you'll remember, in which a transporter malfunction causes two copies of Captain Kirk to be beamed aboard the *Enterprise*. It's hard to see how two copies of a person could be made out of one person-sized collection of atoms.

Fortunately for viewers of the show, the two copies of Kirk weren't precisely identical. One copy was the normal (good) Kirk, and the other was evil. Even better, the evil one quickly got scratched on the face by Yeoman Rand, so it wasn't hard to tell the two apart.

But what if they had been identical? We would then be faced with a puzzle about the nature of personal identity, popularized by philosopher Derek Parfit. Imagine a transporter machine that could disassemble a single individual and reconstruct multiple exact copies of them out of different atoms. Which one, if any, would be the "real" one? If there were just a single copy, most of us would have no trouble accepting them as the original person. (Using different atoms doesn't really matter; in actual human bodies, our atoms are lost and replaced all the time.) Or what if one copy were made of new atoms, while the original you remained intact—but the original suffered a tragic death a few seconds after the duplicate was made. Would the duplicate count as the same person?

All good philosophical fun and games of course, but without much relevance to the real world, at least not at our current level of technology. Or maybe not. There's an older thought experiment called the Ship of Theseus that raises some of the same issues. Theseus, the legendary founder of Athens, had an impressive ship in which he had fought numerous battles. To honor him, the citizens of Athens preserved his ship in their port. Occasionally a plank or part of the mast would decay beyond repair, and at some point that piece would have to be replaced to keep the ship in good order. Once again we have a question of identity: is it the same ship after we've replaced one of the planks? If you think it is, what about after we've replaced *all* of the planks, one by one? And (as Thomas Hobbes went on to ask), what if we then took all the old planks and built a ship out of them? Would that one then suddenly become the Ship of Theseus?

Narrowly speaking, these are all questions about identity. When is one thing "the same thing" as some other thing? But more broadly, they're questions about ontology, our basic view of what exists in the world. What kinds of things are there at all?

When we ask about the identity of the "real" Captain Kirk or Ship of Theseus, a whole bundle of unstated assumptions come along for the ride. We are assuming that there are things called "persons," and things called "ships," and that these things have some persistence over time. And everything goes swimmingly, until we come up against a puzzle, such as these duplication scenarios, that puts a strain on how we define these kinds of objects.

All this matters, not because we're on the verge of building a working transporter, but because our attempts to make sense of the big picture inevitably involve different kinds of overlapping ways of talking about the world. We have atoms, and we have biological cells, and we have human beings. Is the notion of "this particular human being" an important one to how we think about the world? Should categories like "persons" and "ships" be part of our fundamental ontology at all? We can't decide whether an

individual human life actually matters if we don't know what we mean by
"human being."

❋

As knowledge generally, and science in particular, have progressed over the
centuries, our corresponding ontologies have evolved from quite rich to
relatively sparse. To the ancients, it was reasonable to believe that there were
all kinds of fundamentally different things in the world; in modern
thought, we try to do more with less.

We would now say that Theseus's ship is made of atoms, all of which are
made of protons, neutrons, and electrons—exactly the same kinds of particles
that make up every other ship, or for that matter make up you and me. There
isn't some primordial "shipness" of which Theseus's is one particular example; there are simply arrangements of atoms, gradually changing over time.

That doesn't mean we can't talk about ships just because we understand
that they are collections of atoms. It would be horrendously inconvenient
if, anytime someone asked us a question about something happening in the
world, we limited our allowable responses to a listing of a huge set of atoms
and how they were arranged. If you listed about one atom per second, it
would take more than a trillion times the current age of the universe to
describe a ship like Theseus's. Not really practical.

It just means that the notion of a ship is a derived category in our ontology, not a fundamental one. It is a useful *way of talking* about certain subsets of the basic stuff of the universe. We invent the concept of a ship
because it is useful to us, not because it's already there at the deepest level
of reality. Is it the same ship after we've gradually replaced every plank? I
don't know. It's up to us to decide. The very notion of "ship" is something
we created for our own convenience.

That's okay. The deepest level of reality is very important; but all the
different ways we have of talking about that level are important too.

❋

What we're seeing is the difference between a rich ontology and a sparse
one. A rich ontology comes with a large number of different fundamental
categories, where by "fundamental" we mean "playing an essential role in
our deepest, most comprehensive picture of reality."

In a sparse ontology, there are a small number of fundamental categories (maybe only one) describing the world. But there will be very many ways of talking about the world. The notion of a "way of talking" isn't mere decoration—it's an absolutely crucial part of how we apprehend reality.

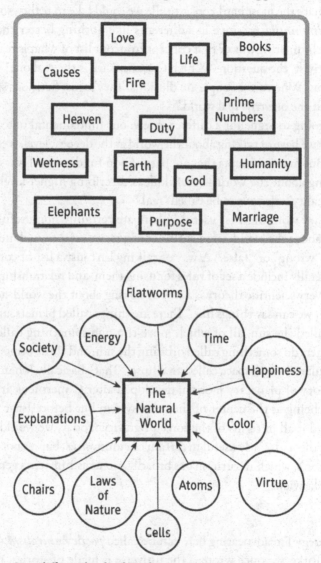

Two different kinds of ontologies, rich and sparse. Boxes are fundamental concepts, while circles are derived or emergent concepts—ways of talking about the world.

One benefit of a rich ontology is that it's easy to say what is "real"—every category describes something real. In a sparse ontology, that's not so clear. Should we count only the underlying stuff of the world as real, and all the different ways we have of dividing it up and talking about it as merely illusions? That's the most hard-core attitude we could take to reality, sometimes called *eliminativism*, since its adherents like nothing better than to go around eliminating this or that concept from our list of what is real. For an eliminativist, the question "Which Captain Kirk is the real one?" gets answered by "Who cares? People are illusions. They're just fictitious stories we tell about the one true real world."

I'm going to argue for a different view: our fundamental ontology, the best way we have of talking about the world at the deepest level, is extremely sparse. But many concepts that are part of non-fundamental ways we have of talking about the world—useful ideas describing higher-level, macroscopic reality—deserve to be called "real."

The key word there is "useful." There are certainly non-useful ways of talking about the world. In scientific contexts, we refer to such non-useful ways as "wrong" or "false." A way of talking isn't just a list of concepts; it will generally include a set of rules for using them, and relationships among them. Every scientific theory is a way of talking about the world, according to which we can say things like "There are things called planets, and something called the sun, all of which move through something called space, and planets do something called orbiting the sun, and those orbits describe a particular shape in space called an ellipse." That's basically Johannes Kepler's theory of planetary motion, developed after Copernicus argued for the sun being at the center of the solar system but before Isaac Newton explained it all in terms of the force of gravity. Today, we would say that Kepler's theory is fairly useful in certain circumstances, but it's not as useful as Newton's, which in turn isn't as broadly useful as Einstein's general theory of relativity.

※

The strategy I'm advocating here can be called *poetic naturalism*. The poet Muriel Rukeyser once wrote, "The universe is made of stories, not of atoms." The world is what exists and what happens, but we gain enormous insight by talking about it—telling its story—in different ways.

Naturalism comes down to three things:

1. There is only one world, the natural world.
2. The world evolves according to unbroken patterns, the laws of nature.
3. The only reliable way of learning about the world is by observing it.

Essentially, naturalism is the idea that the world revealed to us by scientific investigation is the one true world. The poetic aspect comes to the fore when we start talking about that world. It can also be summarized in three points:

1. There are many ways of talking about the world.
2. All good ways of talking must be consistent with one another and with the world.
3. Our purposes in the moment determine the best way of talking.

A poetic naturalist will agree that both Captain Kirk and the Ship of Theseus are simply ways of talking about certain collections of atoms stretching through space and time. The difference is that an eliminativist will say "and therefore they are just illusions," while the poetic naturalist says "but they are no less real for all of that."

Philosopher Wilfrid Sellars coined the term *manifest image* to refer to the folk ontology suggested by our everyday experience, and *scientific image* for the new, unified view of the world established by science. The manifest image and the scientific image use different concepts and vocabularies, but ultimately they should fit together as compatible ways of talking about the world. Poetic naturalism accepts the usefulness of each way of talking in its appropriate circumstances, and works to show how they can be reconciled with one another.

Within poetic naturalism we can distinguish among three different kinds of stories we can tell about the world. There is the deepest, most fundamental description we can imagine—the whole universe, exactly described in every microscopic detail. Modern science doesn't know what that description actually is right now, but we presume that there at least *is* such an underlying reality. Then there are "emergent" or "effective" descriptions, valid within some limited domain. That's where we talk about ships and people, macroscopic collections of stuff that we group into individual entities as part of this higher-level vocabulary. Finally, there are values: concepts of right and wrong, purpose and duty, or beauty and ugliness. Unlike higher-level scientific descriptions, these are not determined by the scientific goal of fitting the data. We have other goals: we want to be good people, get along with others, and find meaning in our lives. Figuring out the best way to talk about the world is an important part of working toward those goals.

Poetic naturalism is a philosophy of freedom and responsibility. The raw materials of life are given to us by the natural world, and we must work to understand them and accept the consequences. The move from description to prescription, from saying what happens to passing judgment on what should happen, is a creative one, a fundamentally human act. The world is just the world, unfolding according to the patterns of nature, free of any judgmental attributes. The world exists; beauty and goodness are things that we bring to it.

※

Poetic naturalism may seem like an appealing idea—or it may seem like an absurd bunch of hooey—but it certainly leaves us with a lot of questions. Most obviously, what is the unified natural world that underlies everything? We've been bandying about words like "atoms" and "particles," but we know from discussions of quantum mechanics that the truth is a bit more slippery than that. And we certainly don't claim to know the ultimate final Theory of Everything—so how much do we actually know? And what makes us think that it's enough to justify the dreams of naturalism?

There are equally many, if not more, questions about connecting that underlying physical world to our everyday reality. There are "Why?" questions: Why this particular universe, with these particular laws of nature? Why does the universe exist at all? There are also "Are you sure?" questions: Are we sure that a unified physical reality could naturally give rise to life as we know it? Are we sure it is sufficient to describe consciousness, perhaps the most perplexing aspect of our manifest world? And then there are the

"How?" questions: How do we decide what ways of talking are the best? How do we agree on judgmental questions about right and wrong? How do we find meaning and purpose in a world that is purely natural? Above all, how do we *know* any of this?

Our task is to put together a rich, nuanced picture that reconciles all the different aspects of our experience. To put ourselves in the right frame of mind, in the next few chapters we'll survey some of the ideas that helped set humanity on the road to naturalism.

SEAN CARROLL

"Carroll is a sure-footed guide through some of the most perplexing and fascinating insights of modern physics."

—Brian Greene, author of *The Elegant Universe*

P.O. 0003805514 20200124